THE LIVING EARTH

PLATE TECTONICS

UNRAVELING THE MYSTERIES OF THE EARTH

REVISED EDITION

JON ERICKSON

FOREWORD BY ERNEST H. MULLER, PH.D.

 Checkmark Books®

An imprint of Facts On File, Inc.

PLATE TECTONICS
Unraveling the Mysteries of the Earth, Revised Edition

For information contact:

Checkmark Books
An imprint of Facts On File, Inc.
11 Penn Plaza
New York, NY 10001

Library of Congress Cataloging-in-Publication Data

Erickson, Jon, 1948–
Plate tectonics : unraveling the mysteries of the earth / Jon Erickson ; foreword Ernest H. Muller—Rev. ed.
p. cm. — (The living earth)
Includes bibliographical references and index.
ISBN 0-8160-4327-2 (hardcover: alk. paper) ISBN 0-8160-4589-5 (pbk)
1. Plate tectonics. I. Title.

QE511.4.E75 2001
551.1′36—dc21 00-049039

Checkmark Books are available at special discounts when purchased in bulk quantities for businesses, associations, institutions, or sales promotions. Please contact our Special Sales Department at 212/967-8800 or 800/322-8755.

You can find Facts On File on the World Wide Web at http://www.factsonfile.com

Text design by Cathy Rincon
Cover design by Nora Wertz
Illustrations by Jeremy Eagle and Dale Dyer, © Facts On File

Printed in the United States of America

MP Hermitage 10 9 8 7 6 5 4 3 2 1
(pbk) 10 9 8 7 6 5 4 3 2 1

This book is printed on acid-free paper.

CONTENTS

TABLES

ACKNOWLEDGMENTS

The author thanks the National Aeronautics and Space Administration (NASA), the National Oceanic and Atmospheric Administration (NOAA), the U.S. Air Force, the U.S. Department of Agriculture–Forest Service, the U.S. Department of Energy, the U. S. Geological Survey (USGS), and the U.S. Navy for providing photographs for this book.

The author also thanks Mr. Frank K. Darmstadt, Senior Editor, and Ms. Cynthia Yazbek, Associate Editor, and the rest of the staff at Facts On File for their invaluable assistance in the publication of this book.

FOREWORD

From time immemorial, people's perceptions of Earth have been framed largely by personal experience and by conversation among neighbors, later by broader communication and across linguistic barriers. With modern advances in science and information technology, knowledge of Earth seems sometimes on the verge of bursting ahead of popular comprehension. Whereas Earth has been commonly conceived of as stable and unresponsive, science recognizes it as dynamic and responsive in degree far beyond general awareness. As an example, recent discovery of 10 new planetary bodies outside our solar system once again marks an advance of science ranging ahead of common experience.

In language scientifically correct, yet designed for the typical inquiring mind, Jon Erickson seeks in *Plate Tectonics* to broaden popular awareness of the expanded temporal and spatial dimensions of geologic science. In sequence and informal manner, the opening chapter credits scientists who, in recent centuries, have recognized the relationships and formulated the principles upon which geology is based. The reader is introduced early to conjectural primitive Archean events that account for differentiation of Earth's crust, mantle, and lithosphere.

Movement of crustal plate over semimolten mantle rock at a depth below Earth's surface comprises the essence of plate tectonics. This accounts directly or indirectly for essentially all near-surface movement and deformation as represented in the geologic record. On a major scale, these changes involve displacement of continents, changing of ocean basins, and building of mountains—subjects treated in successive chapters. On a more local scale, plate tectonics accounts for volcanism and earthquakes. Even more subtle and indirect consequences of plate tectonics are variations in ocean currents and atmospheric climates that have been made apparent in the distribution of marine fossils and archeological remains. Although geology literally refers to earth science, its scope is stretched in the last chapter as plate tectonics theory is invoked for assessing the nature of neighboring planets.

—Ernest H. Muller, Ph. D.

INTRODUCTION

All aspects of the Earth's history and structure, including its majestic mountains, giant rift valleys, and deep ocean basins, were fashioned by mobile crustal plates. The Earth's outer shell is made up of a dozen or so plates composed of the upper mantle, or lithosphere, and the overlying continental and oceanic crust. The plates ride on the semimolten rocks of the upper mantle and carry the continents along with them. The movement of the plates accounts for all geologic activity taking place on the planet's surface. In this manner, plate tectonics is continously changing and rearranging the face of the Earth.

The interaction of lithospheric plates plays a fundamental role in global tectonics and accounts for many geologic processes that shape the surface of the planet. The theory of plate tectonics clearly explains how these changes take place. When two plates collide, they create mountain ranges on the continents and volcanic islands on the ocean floor. When an oceanic plate slides beneath a continental plate, it produces sinuous mountain chains. The breakup of a plate creates new continents and oceans. The rifting and patching of the continents have been ongoing processes practically since the very beginning.

The text begins with the discoveries leading to the theory of plate tectonics. It follows the evolution of the planet from the earliest construction of the crust to the present, defines the force that drives the continents around the surface of the Earth, and examines the mobile crustal plates, their interactions, and their effects on the planet. It reveals how the best evidence for plate tectonics was found lying on the bottom of the ocean and how new oceanic crust is created at midocean ridges.

The discussion continues with an exploration of the ocean floor for evidence of plate subduction, the counterpart of seafloor spreading. After explaining the basics of plate tectonics, it applies that knowledge to the subject of mountain building, the most visible manifestations of the powerful tectonic forces that shape the Earth. After seeing how plate tectonics has played a fundamental role in shaping the Earth, the text explains how it has made our world a living planet and examines the critical cycles that are important to life on Earth.

This revised and updated edition is a much expanded examination of plate tectonics. The book makes a difficult-to-understand and sometimes controversial theory easily comprehensible to the lay reader. Science enthusiasts will particularly enjoy this fascinating subject and gain a better understanding of how the forces of nature have shaped the Earth. Students of geology and earth science will find this a valuable reference book to further their studies. Readers will enjoy this clear and easily readable text that is well illustrated with fascinating photographs, detailed illustrations, and helpful tables. A comprehensive glossary is provided to define difficult terms, and a bibliography lists references for further reading. Perhaps no other theory has changed the fundamental thinking about our living Earth as has plate tectonics.

1

CONTINENTAL DRIFT

DISCOVERY OF PLATE TECTONICS

When early cartographers drew the first world maps, they noticed something very peculiar. The continents of Africa and South America seemed to fit together as though they were pieces of a giant jigsaw puzzle. Many other clues, such as matching mountain ranges and identical life-forms on continents widely separated by ocean, led to speculation that once all the lands were united into a single large continent that subsequently broke apart. This was a preposterous notion to most scholars, who then devised all sorts of theories to explain these strange phenomena. This chapter examines the discoveries leading to the theory of plate tectonics.

THE BIRTH OF GEOLOGY

The ancient Greeks sought logical or scientific explanations for the violent and destructive movements that often plagued civilization. The sixth-century B.C. Greek philosopher Thales held the view that the world floated on water, which accounted for the new springs that often spurt water and mud during and after an earthquake. The fifth-century B.C. Greek philosopher Anaxagoras proposed the idea that earthquakes resulted when sections of the earth cracked and caved in. After observing earthquakes and volcanic eruptions, Anaxagoras concluded they were the aftermath of fires that raged inside volcanoes, forcing their fragile crust to tremble, crack, and finally collapse.

One of the first to inquire about the origin of volcanoes was the fourth-century B.C. Greek philosopher Plato. He believed that volcanoes were produced when hot winds were imprisoned under great pressure inside immense subterranean caverns. When the winds blew across a great burning river, Plato called the Pyriphlegathon, they caught fire. The fiery winds then rose to the surface and shot out of volcanic mountains.

The Greek geographer Strabo claimed that volcanoes were giant safety valves that periodically released the pressure building up inside the Earth. Strabo was the first to recognize that Mount Vesuvius (Fig. 1), near Naples, Italy, was actually a volcano and not just an ordinary mountain. The citizens of Pompeii discovered this to their peril when the mountain blew up on August 24, A.D. 79, killing 16,000 people.

The Roman statesman and philosopher Seneca put forward a more contemporary idea about volcanism. He believed that volcanoes were vents through which the melted matter of subterranean reservoirs erupted onto the Earth's surface. The Roman author Pliny the Younger became the first volcanologist and described in explicit detail the A.D. 79 eruption of Mount Vesuvius. Unfortunately, his uncle, Pliny the Elder was studying the volcano up close and was killed by the eruption.

In the late Renaissance, a rebirth of inquiry about natural phenomena took hold of Europe after nearly 1,000 years of silence during the Middle Ages. Around the beginning of the 18th century, the French chemist Nicolas Lemery delved into why volcanoes erupted. He observed how a mixture of iron filings, sulfur, and water spontaneously combusted and gave off steam and hot projected matter. In Lemery's view, sulfur fermented in the depths of the Earth to produce earthquakes and conflagrations. He concluded that volcanoes were the product of fermentation and combustion of certain matter when they contacted air and water.

The French naturalist Georges de Buffon championed this idea. He concluded that the center of volcanic activity was not deep down in the bowels of the Earth but instead near the surface, where it was exposed to wind and rain. This theory became known as neptunism, named for Neptune, the Roman god of the sea. The German geologist Abraham Werner became one of its prime exponents. He maintained that once the mineral pyrite was exposed to water, it heated up and ignited coal. The burning coal melted nearby rocks, which then erupted onto the Earth's surface.

Figure 1 *The eruption of Mount Vesuvius in 1944.*

(Photo courtesy USGS)

The Scottish geologist James Hutton, known today as the father of modern geology, argued against neptunism. He advanced an opposing theory called plutonism, named for Pluto, the Greek god of the underworld. Hutton's discovery of unconformities, where ancient sedimentary strata were upturned, eroded, and blanketed by younger deposits (Fig. 2), suggested the history of the Earth is exceedingly long and complex. Hutton believed contemporary rocks at the surface were formed by the waste of older rocks laid down in the sea, consolidated under great pressure, and upheaved by the expanding power of the Earth's subterranean heat. His theory was based on the premise that the depths of the Earth are in a constant state of turmoil and that molten matter rises to the surface through cracks or fissures, resulting in an erupting volcano. However, the theory was not put to the test until the first serious study of the Mount Vesuvius volcano was conducted in the middle 19th century, providing conclusive evidence supporting the theory of plutonism.

Recognition of the slow processes by which geologic forces operate led James Hutton to propose the theory of uniformitarianism in 1785. Simply stated, it means that the present is the key to the past. In other words, the forces that shaped the Earth are uniform and operate in the same manner and at the same rate today as they did long ago. Therefore, the Earth had to be very ancient for these processes to take place. Hutton envisioned the prime mover behind these slow changes to be the Earth's own internal heat. Early geologists recognized that rocks were molten in the Earth's interior, and this observation was manifested by volcanic eruptions (Fig. 3).

As further proof, temperatures in deep mines increased with greater depth, indicating that rocks grew progressively hotter toward the center of the Earth. Hutton called this phenomenon the Earth's great heat engine. He believed that the heat was left over from an earlier time when the planet was in a molten state. The British geologist Sir Charles Lyell, born in 1797, the

Figure 2 *Angular unconformity, a gap in geologic time, in a small mesa near San Lorezo, Socorro County, New Mexico.*

(Photo by R. H. Chapman, courtesy USGS)

__Figure 3__ The November 1968 eruption of Cerro Negro in west-central Nicaragua, which resembles a chain of volcanoes similar to the Cascade Range of the Pacific Northwest.

(Photo courtesy USGS)

same year that Hutton died, continued with Hutton's work and gained worldwide acceptance for the theory of uniformitarianism. Lyell gathered many observations about rocks and landforms of western Europe. He showed they were the products of the same processes in existence today provided they were given enough time.

In 1830, Lyell took these ideas one step further by proposing that rock formations and other geologic features took shape, eroded, and reformed at a constant rate throughout time according to the theory of uniformitarianism.

Many geologists, however, felt this theory was not fully adequate to explain all the geologic forces at work. Events in the past appeared not to be slowly evolving but occurring rather suddenly. This opposing view called catastrophism had as its most ardent supporter the French geologist Georges Cuvier, who believed the Earth's history was a series of catastrophes. Adherents pointed to gaps in the geologic record and the extinctions of large numbers of species.

Other theories underpinning the foundation of modern geology also developed over time. The 17th-century Danish physician and geologist Nicholas Steno recognized that in a sequence of layered rocks, undeformed by folding or faulting, each layer was formed after the one below it and before the one above it. This observation led Steno to propose the law of superposition. This law might seem obvious to us today, but during his time it was hailed as an important scientific discovery. Steno also put forward the principle of original horizontality. It states that sedimentary rocks were initially laid down horizontally in the ocean, and subsequent folding and faulting uplifted them out of the sea and inclined them at steep angles.

If angled rocks are overlain by horizontal strata, the gap in the chronological record is known as an angular unconformity. Also, if a body of igneous rocks derived from molten magma cuts across the boundaries of other rock units, it is younger than those it intercepts. This is the principle of crosscutting relationship. It states that granitic intrusions, from the injection of magma into preexisting rocks, are younger than the rocks they invade. A sequence of rocks placed into their proper order is called a stratigraphic cross section.

The development of a geologic time scale applicable around the world required the matching of rock units from one locality with those of similar age at another site in a process called rock correlation. Geologists can trace a bed or a series of beds from one outcrop to another by recognizing certain distinguishable features in the rocks. By correlating rocks from one place to another over a wide area, a comprehensive view of the geologic history of a region is obtained.

Matters are complicated, however, if faulting occurs in the area. One block of a rock sequence might be down-dropped in relation to the other (Fig. 4) or might have been thrust on top of another. Rocks that occur in repetitive sequences of sandstone, shale, and limestone complicate correlation even further. A stratum folded over on itself contains rock units that are completely reversed, making matters even more confusing. Although these methods might be sufficient for tracing rock formations over relatively short distances, they are inadequate for matching rocks over great distances, such as between continents.

To correlate between widely separated areas, geologists had to rely on the fossil content of the rocks, indicating they were the same age. The branch

Figure 4 A down-dropped block forms a grabben (top), and an upthrust block forms a horst (bottom).

of geology devoted to the study of ancient life based on fossils is called paleontology. A fossil is the remains or traces of an organism preserved from the geologic past. A plant or an animal must be buried rapidly without oxygen or bacteria to prevent decay. If given enough time, the remains of an organism are modified, often becoming petrified and literally turned into stone.

The existence of fossils has been known since the ancient Greeks. The sixth-century B.C. Greek philosopher Xenophanes first speculated that fossil seashells in the mountains meant the latter originated in the sea. The fifth-century B.C. Greek historian Hirodotus reasoned that the inland hills once existed underwater because he found fossils of seashells embedded in rocks high up in the mountains. The fourth century B.C. Greek philosopher Aristotle clearly recognized that certain fossils such as fish bones were the remains of organisms. However, he generally believed that fossils were placed into the rocks by supernatural causes.

Figure 5 *Carbonized plant remains of a middle Pennsylvanian fern of the Kanawha series, West Virginia.*

(Photo courtesy USGS)

Around A.D. 1500, the Italian artist and scientist Leonardo da Vinci claimed that fossils were the remains of once-living organisms and not inorganic substances as was previously thought. He also believed the presence of fossil seashells in the mountains was evidence that the distribution of lands and seas had changed through time. Geologists thought the Earth underwent periods of catastrophic death of all life, after which life began anew. This also explained the abundance of fossils at certain stages in the geologic record. By the 1700s, most geologists began to accept fossils as the remains of extinct organisms because they closely resembled living species (Fig. 5).

The significance of fossils as a geologic tool was discovered by the English geologist William Smith in the late 18th century. He recognized that sedimentary rocks in widely separated areas could be identified by their distinctive fossil content. These observations led Smith to one of the most important and basic principles of historical geology. Fossilized organisms succeed one another in a definite and determinable order. Therefore, various geologic periods could be recognized by their distinctive fossils. Smith made the most significant contribution to the understanding of fossils when he proposed the law of faunal succession, which stated that rocks could be placed into their proper time sequence by studying their fossil content. This revealed an important clue to the understanding of geologic history.

These findings became the basis for the establishment of the geologic time scale (Table 1) and the beginning of modern geology. The major geologic periods were delineated by 19th-century geologists mostly in Great Britain and western Europe. The periods were named for localities with the best rock exposures. For example, the Jurassic period took its name from the Jura Mountains in Switzerland. To develop a geologic time scale applicable over the entire world, rocks of one locality are correlated or matched with rocks of similar age in another location.

MATCHING COASTLINES

Early geographers were often puzzled over the way the bulge on the east coast of South America fits almost precisely into the cleft on the west coast of Africa. In 1620, the British philosopher Sir Francis Bacon noticed similarities between the New World and the Old World. Isthmuses and capes looked much the same. Both continents were broad and extended toward the north but narrow and pointed toward the south. He also noticed that the Atlantic coastlines of South America and Africa seemed to match. Georges de Buffon suggested that Europe and North America had once been joined because of similarities between their plants and animals.

The French moralist François Placet suggested that the New and Old Worlds were once joined and then were separated by the biblical Flood. Geologists in the 17th and 18th centuries argued that the Flood was so devastating that it broke up old continents into entirely new ones. Ironically, modern geology emerged from a failed attempt to interpret rocks and landforms as products of the Great Flood. By expanding on this idea, the 19th-century German naturalist-explorer Alexander von Humboldt suggested that a massive tidal wave surged across the globe and carved out the Atlantic Ocean like a giant river valley, leaving the continents divided with opposing shorelines. Humboldt studied the landforms and plant and animal life in South America. He noted that mountain ranges on the eastern coast resembled those on the western coast of Africa. He also

TABLE 1 THE GEOLOGIC TIME SCALE

Era	Period	Epoch	Age (Millions of Years)	First Life Forms	Geology
		Holocene	0.01		
	Quaternary				
		Pleistocene	3	Humans	Ice age
Cenozoic		Pliocene	11	Mastodons	Cascades
		Neogene			
		Miocene	26	Saber-toothed tigers	Alps
	Tertiary	Oligocene	37		
		Paleogene			
		Eocene	54	Whales	
		Paleocene	65	Horses, Alligators	Rockies
	Cretaceous		135		
				Birds	Sierra Nevada
Mesozoic	Jurassic		210	Mammals	Atlantic
				Dinosaurs	
	Triassic		250		
	Permian		280	Reptiles	Appalachians
	Pennsylvanian		310		Ice age
				Trees	
	Carboniferous				
Paleozoic	Mississippian		345	Amphibians	Pangaea
				Insects	
	Devonian		400	Sharks	
	Silurian		435	Land plants	Laursia
	Ordovician		500	Fish	
	Cambrian		570	Sea plants	Gondwana
				Shelled animals	
			700	Invertebrates	
Proterozoic			2500	Metazoans	
			3500	Earliest life	
Archean			4000		Oldest rocks
			4600		Meteorites

noticed a striking resemblance in the geologic strata of the two continents as well.

Even nonscientists put forth theories for continental drift to the consternation of geologists. In the mid-19th century, the American writer Antonio Snider suggested that as the Earth cooled and crystallized from a molten state, most of the continental material gathered on one side, making the planet lopsided and unbalanced. This created such internal stresses that a large continent cracked wide open and hot lava bled through the fissures. Meanwhile, the rains of the Great Flood came, and the raging waters pushed apart the segments of the broken continent to their present positions. Snider cited as evidence for a single large landmass coal beds in Africa and South America that were the same age and contained similar fossils, dating to the Carboniferous period.

In 1871, George Darwin, son of the British naturalist Charles Darwin, speculated on the origin of the moon. He thought the moon was torn out of the Earth (Fig. 6). While the planet was in its early molten state, it was spinning wildly on its axis, causing a bulge at the equator to be flung into space. Another idea suggested the moon was torn out of the Earth by the tidal pull of the sun or a passing star. In 1882, the British scientist Osmond Fisher suggested that when the Moon was plucked out of the Earth, it left a great scar that formed the Pacific Basin, which quickly filled with molten magma from the Earth's interior. As the upper fluid layers flowed into this cavity, the cooling solid crust floating on top broke up, and part of it was pulled toward the cavity like a raft riding on a river of molten rock.

In 1885, the Austrian geologist Edward Suess demonstrated how the continents of the Southern Hemisphere were united into a composite landmass he called Gondwanaland, now called Gondwana (Fig. 7), named for an ancient region of east-central India. The large continent comprised Africa, South America, Australia, Antarctica, and India. Suess named the northern landmass Laurasia, for the Laurentian province of Canada and the continent of Eurasia. Laurasia comprised North America, Europe, and Asia. During Suess's day, geologists thought that as the Earth was cooling, it shriveled like a baked apple, accounting for the formation of mountain belts.

Biologists and geologists alike were bewildered by the similarity between fossils and both living plants and animals in the Old and the New Worlds. One theory contended that the continents did not actually move away from each other but that the ocean basins between them developed from sinking land, which once linked the continents. Since most geologists of that time thought the Earth was contracting as the interior cooled and shrank, they argued that blocks of crust fell inward to fill the void spaces and in the process created the ocean basins. The theory allowed life-forms a convenient means of traveling from one continent to another. However, it placed severe limits on the age of the Earth, making it much younger than it truly was.

The belief that the Earth was shrinking to maintain its internal temperature was soon discarded with the discovery of radioactive decay near the turn of the 20th century. In 1896, the French physicist Henri Becquerel discovered that uranium spontaneously emitted radiation called gamma rays. In 1897, the British physicist J. J. Thomsom identified the electron as the negatively charged particle orbiting the atom. In 1898, the British physicist Ernest Rutherford discovered alpha and beta particles emanating from radioactive elements. That same year, Pierre and Marie Curie isolated many radioactive elements including radium. The decay of radioactive elements in the Earth's interior therefore supplied the energy required to maintain the Earth's temperature. Many geologists then did a complete about-face and suggested that instead of shrinking, the Earth was expanding to rid itself of the excess heat.

Figure 6 *Apollo spacecraft photograph of an earthrise over the lunar surface.*

(Photo courtesy USGS Earthquake Information Bulletin 52)

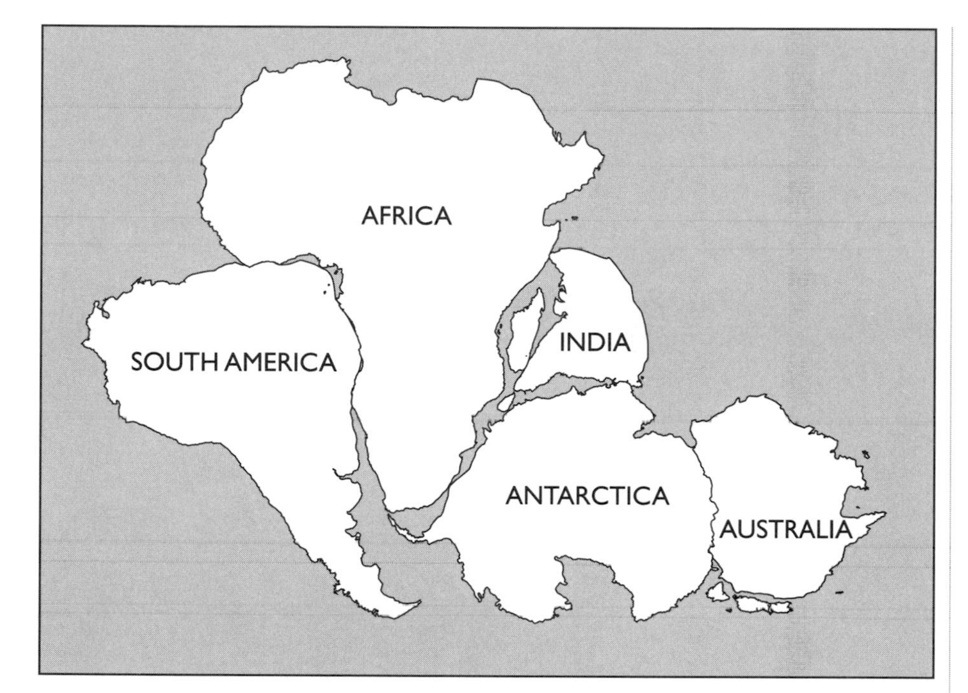

Figure 7 *During the Paleozoic, the southern continents combined into Gondwana.*

EVIDENCE FOR DRIFT

The major problem with early theories concerning the separation of the continents was that the phenomenon was thought to have commenced early in the Earth's history. Therefore, geologists were forced to devise complex theories to account for the similarity among plant and animal fossils that have long been separated by oceans. For so large a variety of species to have evolved along parallel lines in such diverse environments over such a lengthy period seemed highly unlikely.

In the frigid cliffs of the Transantarctic Mountains of Antarctica, scientists discovered a fossilized jawbone and canine tooth belonging to the mammal-like reptile *Lystrosaurus* (Fig. 8). This unusual-looking animal, with large down-pointing tusks, lived around 160 million years ago. The only other known lystrosaurus fossils were found in China, India, and southern Africa. The idea that this freshwater reptile somehow crossed the salty oceans that separated the southern continents seemed very implausible. Instead, its discovery on the frozen wastes of Antarctica was hailed as definite proof for the existence of Gondwana.

In 1985, a small fossilized opossum tooth dating about 37 million years old was discovered in central Siberia. Opossums originated in North America during the late Cretaceous, about 85 million years ago. The animal could have taken a northern route through Asia, leading eventually to Australia, or taken the direct southern route through South America (Fig. 9). The latter hypothesis is

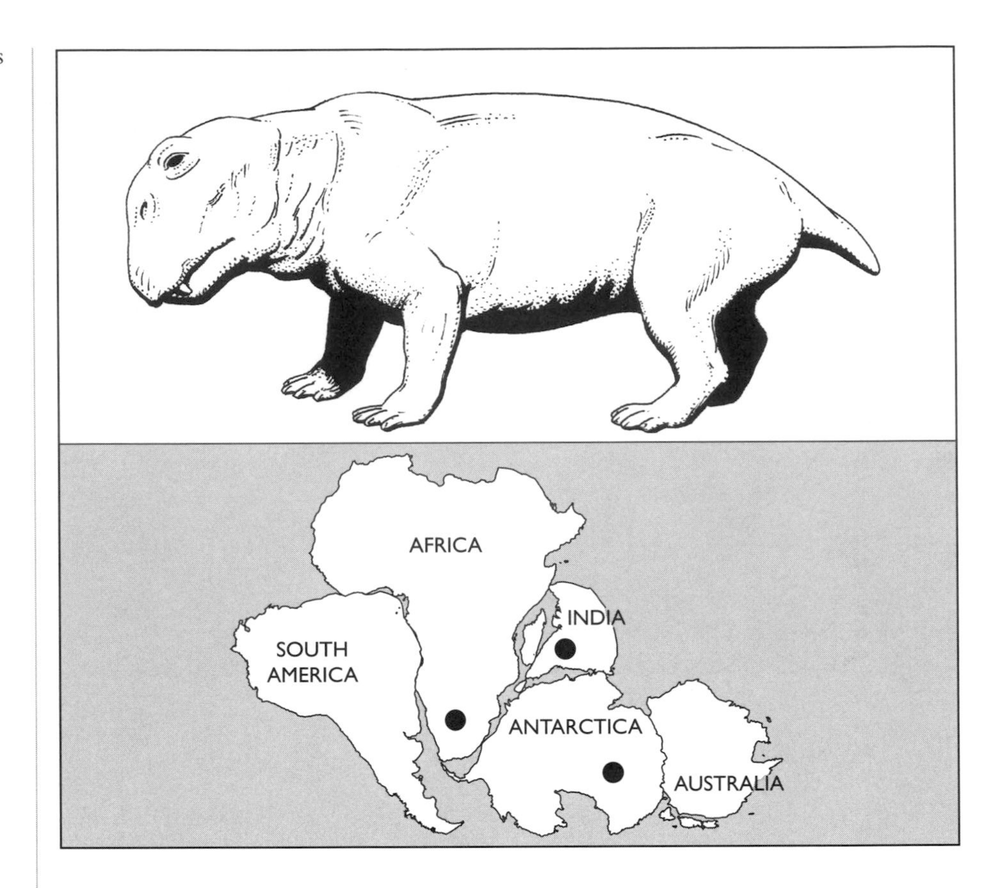

Figure 8 Lystrosaurus *and fossil sites in Gondwana.*

supported by a fossil of a South American marsupial found in Antarctica, which would have acted as a land bridge between the southern tip of South America and Australia. This lends further support to the existence of Gondwana.

Additional evidence of continental drift includes the fossils of a 270-million-year-old reptile called *Mesosaurus* (Figs. 10 and 11) found in Brazil and South Africa. This two-foot-long reptile lived in shallow, freshwater lakes and therefore could not have possibly swum across the thousands of miles that separated the two continents. Fossils of the late Paleozoic fern *Glossopteris* (Fig. 12) were found in Australia and India but were conspicuously missing on the northern continents. The Northern Hemispheric continents were themselves distinct, blanketed by luxuriant forests of tropical vegetation. This observation suggested that two large landmasses existed, with one in the Southern Hemisphere and the other in the Northern Hemisphere. The continents were separated by a wide sea, and their breakup appeared to occur relatively late in Earth history.

Not only did plant and animal life on widely dispersed continents seem to have common ancestors, but the older rocks were remarkably similar as well. Furthermore, matching rocks of several mountain ranges were present.

The Cape Mountains in South Africa connected with the Sierra Mountains south of Buenos Aires, Argentina. Matches were also found between mountains in Canada, Scotland, and Norway. Not only were the rock strata the same type and age, but they were laid down in the exact same order.

The continents of Africa, South America, Australia, India, and Antarctica showed evidence of contemporaneous glaciation in the late Paleozoic, around 270 million years ago. This is indicated by deposits of glacial till and grooves in the ancient rocks excavated by boulders embedded in slowly moving masses of ice. The lines of ice flow were away from the equator and toward the poles. This would be highly unlikely if the continents were situated as they are today since glacial centers do not exist on the equator. Thus, the continents

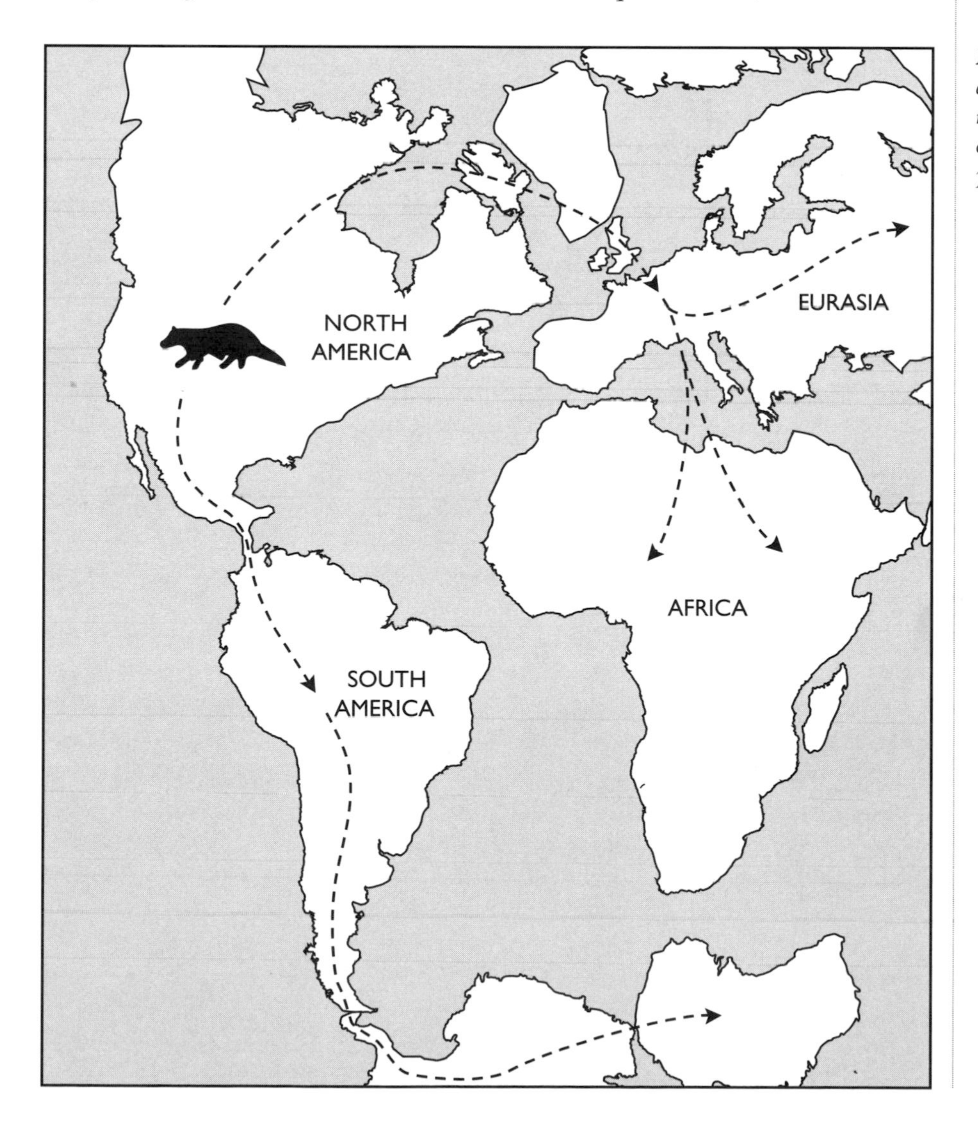

Figure 9 *A map indicating the dispersion of marsupials to other parts of the world 80 million years ago.*

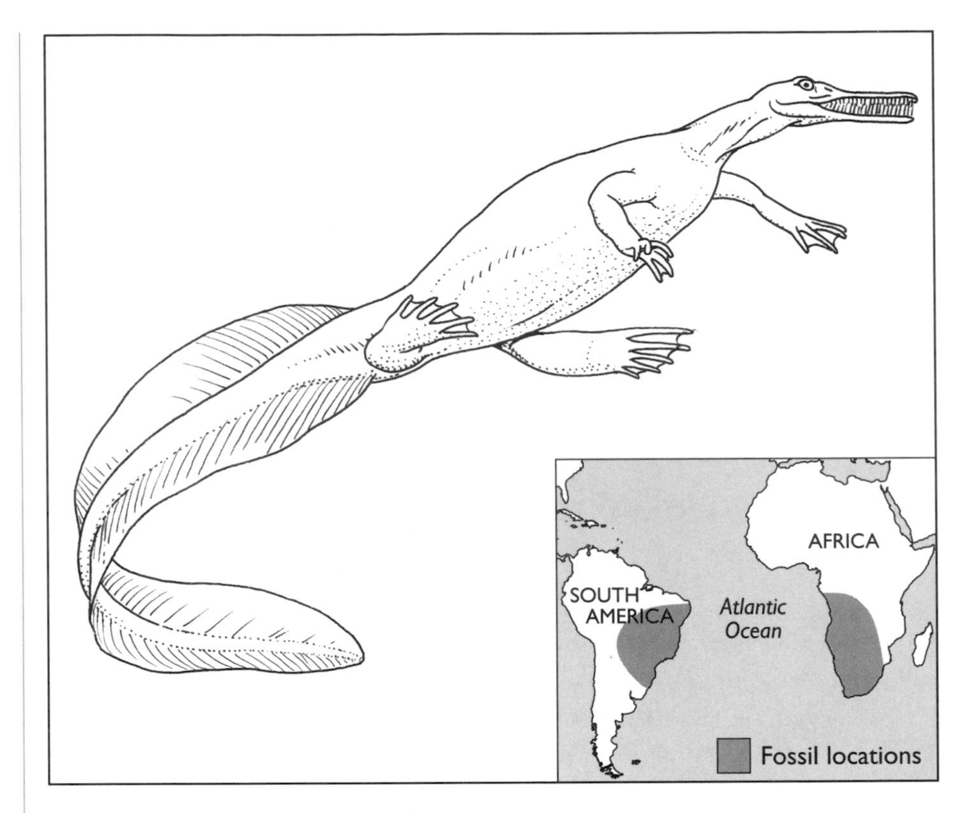

Figures 10, 11
Mesosaur fossils found in Africa and South America are strong evidence for continental drift

must have been joined so that the ice moved across a single landmass, radiating outward from a glacial center over the South Pole.

Strange, out-of-place boulders called glacial erratics, composed of rock types not found elsewhere on one continent, matched rocks on the opposing continent. The glacial deposits were overlain by thick sequences of terrestrial deposits, which in turn were overlain by massive outpourings of basalt lava flows. Overlying these deposits were coal beds containing similar fossilized plant material.

Glacial deposits in the equatorial areas suggested that in the past, these regions were much colder. Coral reefs and coal deposits found in the north polar regions indicated a previously tropical climate. Moreover, in the arctic regions, salt deposits indicated an ancient desert climate. Either the climate in the past changed dramatically or the continents changed position with respect to the equator.

CONTINENTAL MOVEMENTS

Early 20th-century geologists still held to the belief that narrow land bridges spanned the distances between continents. Geologists used the similarity of fossils in South America and Africa to support the existence of a land bridge. The idea was that the continents were always fixed and that land bridges rose from the ocean floor to enable species to migrate from one continent to another. Later, the land bridges sank out of sight beneath the surface of the sea.

However, a search for evidence of land bridges by sampling the ocean floor failed to turn up even a trace of sunken land. New theories were devised that also cast doubt on the existence of land bridges. The American geologist Frank Taylor was opposed to the idea that continents sank into the ocean floor simply because they were lighter than the underlying basalts. Geologists had known for some time that lighter continental crust could not possibly sink into heavier oceanic crust.

In 1908, Taylor suggested an alternative explanation for the formation of mountains based on continental movements. He thought that two great landmasses located at the poles slowly crept toward the equator and their collision shoved up great blocks of crust into mountains. He also described an undersea mountain range between South America and Africa, known today as the Mid-Atlantic Ridge. He believed it was a line of rifting between the two con-

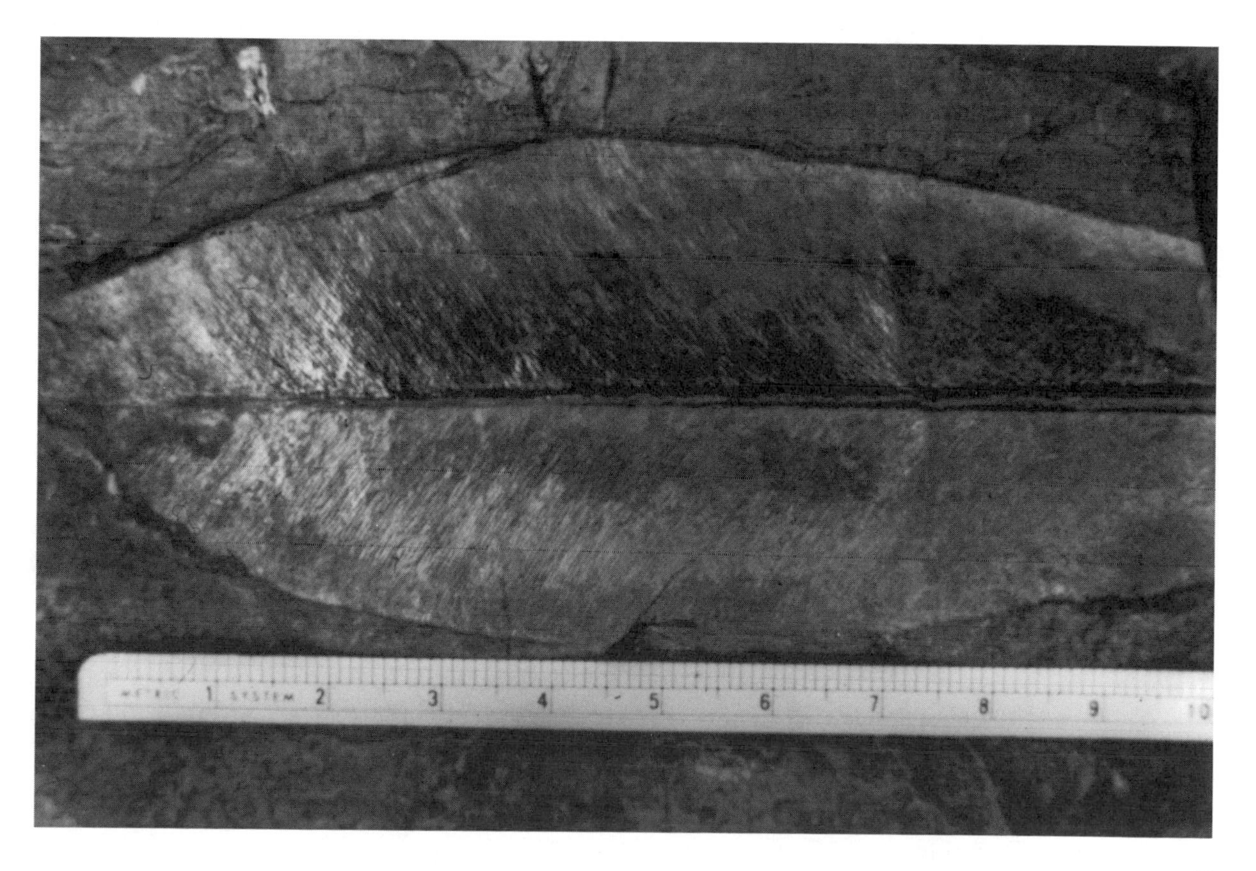

Figure 12 *Fossil* Glossopteris *leaf, whose existence on the southern continents is strong evidence for Gondwana.*

(Photo by D. L. Schmidt, courtesy USGS)

tinents. The ridge remained stationary, while the two continents slowly crept away from it in opposite directions.

The German meteorologist and Arctic explorer Alfred Wegener was intrigued by the high degree of correspondence between the shapes of continental coastlines on either side of the Atlantic Ocean and by the similarity of fossils in South America and Africa, originally used to support the theory of land bridges. He argued that a land bridge was not possible because the continents stand higher than the seafloor for the simple reason that they are composed of light granitic rocks that float on the denser basaltic rocks of the upper mantle. Wegener thought that lighter rocks could not possibly sink into heavier ones because of their greater buoyancy. The most likely explanation was that the continents were once connected and subsequently drifted apart.

Wegener published his continental drift theory in 1915. He believed that 200 million years ago, all the lands comprised a single large continent he named Pangaea (Fig. 13), from the Greek meaning "all lands." The rest of the world was covered by an ocean he called Panthalassa, Greek meaning "universal sea." The breakup of Pangaea and the drifting of the continents created the Atlantic and Indian Oceans. Wegener supported his hypothesis with an impressive collection of evidence, including the geometric fit of continental margins, matching mountain chains on opposite continents, corresponding rock successions, similar ancient climatic conditions, and identical species on continents now widely separated by ocean. The continents were likened to torn pieces of a newspaper—not only did the edges match when fit together but the printed words fit as well.

Before Wegener introduced his theory, the formation of mountain ranges was poorly understood. Geologists just assumed that mountains formed when the molten crust solidified and shriveled up. After making more extensive studies of mountain ranges, however, they were forced to conclude that the folding of rock layers was much too intense (Fig. 14). It required a considerably more rapid cooling and contraction than was possible. Moreover, under these conditions, mountains would have been scattered evenly throughout the world instead of concentrating in a few long mountain belts. One of the first comprehensive models of how mountains evolve was the geographic cycle published in 1899. It proposed a hypothetical life cycle for mountain ranges, from a violent birth caused by a brief but powerful spurt of uplift followed by a gradual death from slow but persistent erosion.

Alfred Wegener provided an alternative explanation for the development of mountain ranges based on continental motions. As the continents traveled across the ocean floor following their breakup, they encountered increasing resistance that forced the leading edges to crumble, fold back, and thrust upward. Wegener pointed to the long, sinuous Rocky and Andes Mountains in North and South America as classic examples of this type of folded mountain belt.

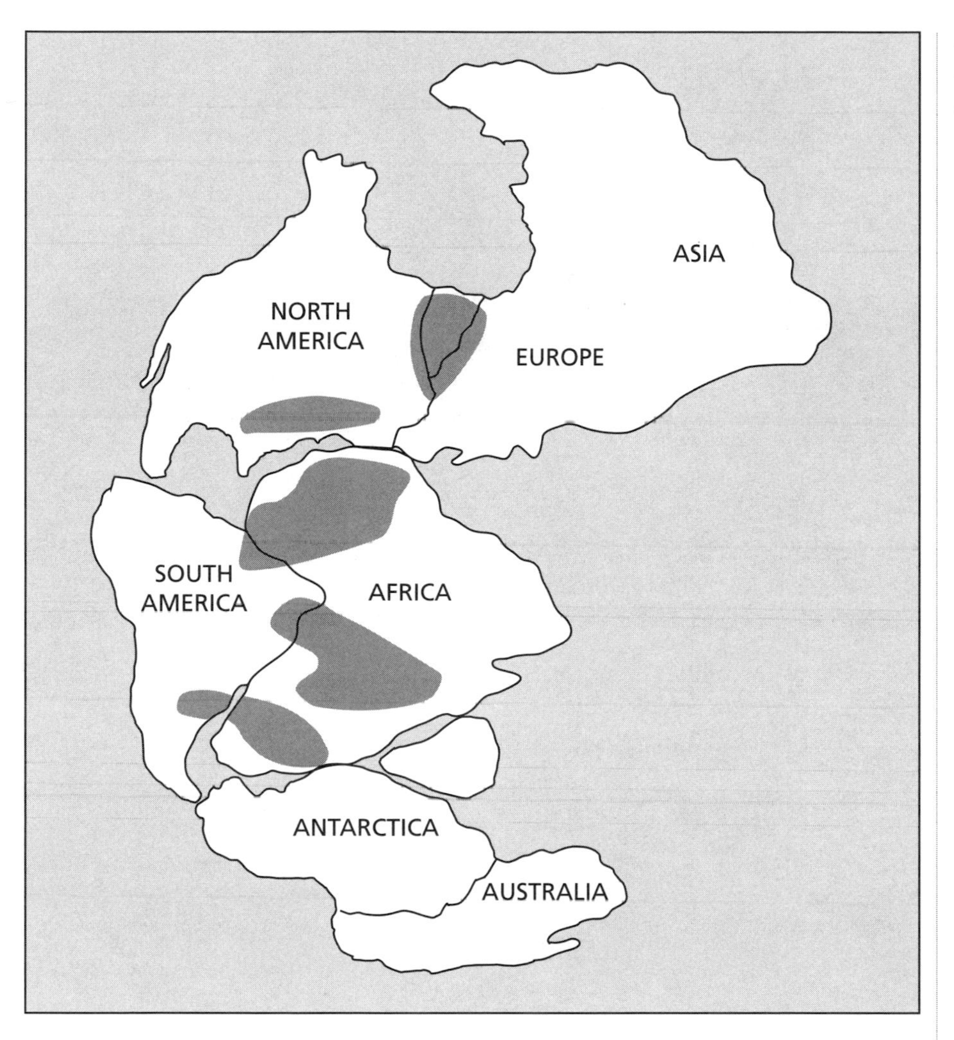

Figure 13 *Pangaea showing matching geologic provinces.*

Wegener's continental drift theory drew furious fire from most contemporary geologists. They questioned whether the soft, light rocks of the continents could penetrate the hard, dense rocks of the ocean floor. They opposed the idea that the breakup of the continents occurred so late in geologic history. Geologists also insisted that erosion was responsible for the matching coastlines of South America and Africa and that the close fit of the continents was merely a coincidence.

Wegener also failed to devise a satisfactory mechanism for the movement of the continents. He thought that the rotation of the Earth provided the necessary force. His contention was that as the Earth spins on its axis, centrifugal force makes the outer layers fly outward, pushing them away from the poles, producing a bulge at the equator. The continents would simply slide off

Figure 14 *Folded Cambrian limestone on the south side of Scapegoat Mountain, Lewis and Clark County, Montana.*

(Photo by M. R. Mudge, courtesy USGS)

the slope of the bulge under the force of gravity. To account for the westerly drift of the continents, Wegener proposed that the gravitational pull of the Sun and Moon act on the land similar to the way they cause the ebb and flow of ocean tides.

Other scientists countered Wegener with opposing scientific data. Geophysicists calculated that these external forces were much too weak to account for the movements of continents in many directions for such a short period. A tremendous amount of energy would be required to budge the continents. Calculations of the Earth's heat flow suggested the continents were formed from the mantle below. Therefore, if continental drift occurred, uneven patterns of heat flow would exist, and no such anomaly was found. This placed severe restrictions on continental drift, for the continents were much thicker and therefore anchored securely in place.

Studies of deep earthquakes at the ocean-continent boundaries, especially around the Pacific, demonstrated the deep structure of the continents with their roots well embedded in the upper mantle. The Yugoslavian seismologist Andrija Mohorovicic first discovered the division between the mantle and the crust in 1909, known as the Mohorovicic discontinuity or simply Moho (Fig. 15). By studying seismic waves generated by earthquakes, seismologists could determine certain properties of the Earth's interior. They proved that the iron core is composed of an inner solid portion and an outer liquid layer. The core is surrounded by a semisolid mantle, covered by a thin crust, giving the Earth a structure similar to that of an egg.

Because Wegener worked so hard to prove his theory, he tended to exaggerate and saw evidence for continental drift where none existed. By using inaccurate land surveys in Greenland, he calculated its drift at a phenomenal rate of more than 100 feet a year, which would have it circle the Earth every million years. When Wegener died during an expedition to Greenland in 1930, the continental drift theory largely died with him.

The rejection of continental drift is a clear example of how old scientific dogmas become entrenched. Once a theory is accepted as fact by the scientific community, in this case the issue of land bridges, it becomes a solemn doctrine. More effort is put into supporting the theory than into disproving it. Geologists were caught in the middle of a scientific revolution and a crisis of contradiction in which hard evidence was simply ignored, fearing that years of painstaking research would have to be discarded. As with Charles Darwin's theory of evolution, if the continental drift theory was ever spoken of, it was held up to ridicule and contempt and considered a classic scientific blunder. Wegener was also considered an outsider and not part of the geologic fraternity. This might have worked to his advantage, however, for he was less reluctant to dismiss outmoded geologic thinking.

By the late 1940s, Wegener's work was completely discredited by the scientific community, mostly in the Northern Hemisphere. Scientists in the Southern Hemisphere were too overwhelmed with evidence to the contrary because of many similarities between the South American and African continents. Geologists who argued in favor of continental drift risked ridicule by their colleagues and loss of fame and recognition. Yet in the midst of such antagonism, the revival of continental drift came from a totally unexpected quarter.

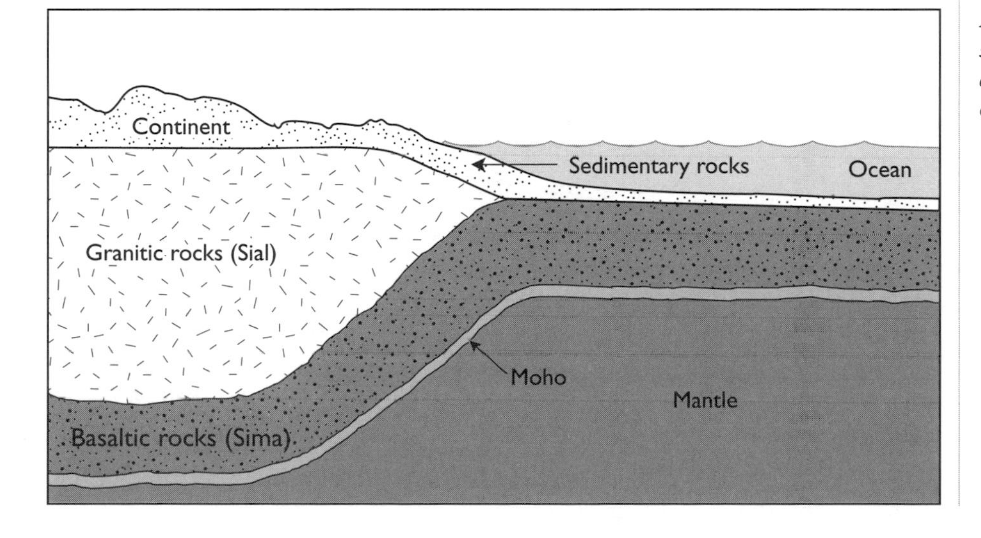

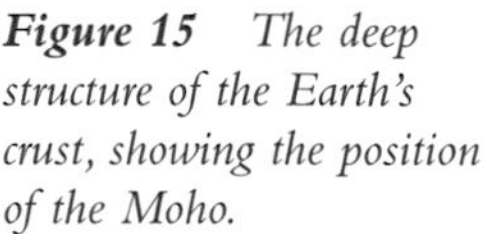
Figure 15 *The deep structure of the Earth's crust, showing the position of the Moho.*

PALEOMAGNETISM

Further evidence for continental drift was found in studies of the Earth's geomagnetic field, which is believed to originate from electric currents in the core. These currents could be generated by several means, such as the chemical differences between the inner and outer core. Once they are initiated, the currents are amplified by the dynamo effect created by the core as it rotates, which acts similar to an electrical generator. Since the core is composed of iron and nickel, both good electrical conductors, the electric currents passing through it sets up a weak magnetic field within the core. The motion of the core reinforces the current, which in turn generates a larger magnetic field.

Because the Earth rotates, the magnetic field tends to be aligned in one direction. Apparently, the Earth's magnetic field remains stable for long periods. Then for unknown reasons, the electric currents fail and the magnetic field collapses. Eventually, the field is regenerated with opposite polarity with north where south used to be and vice versa. For over a century, scientists have known that when iron-rich basalt lava cools, the magnetic fields of its iron molecules are aligned with the Earth's magnetic field like miniature bar magnets. In essence, they are sort of fossilized compasses, pointing in whatever direction north happens to be at the time of deposition.

When sensitive magnetic recording instruments called magnetometers were first taken into the field in the 1950s, scientists in England were very much perplexed by their findings. Rocks formed 200 million years ago showed a magnetic inclination (the downward pointing of the needle of a vertically held compass) of 30 degrees north, whereas England's present inclination is 65 degrees north. The inclination is almost 0 degrees at the equator and 90 degrees at the poles. The only conclusion that could be drawn from the data was that England must have once been farther south.

To test this astounding discovery further, the scientists took their instruments to India's Deccan Plateau. Rocks dating 150 million years old showed a magnetic inclination of 64 degrees south. Rocks that were 50 million years old indicated an inclination of only 26 degrees south. What was even more astonishing was that rocks dating 25 million years old completely reversed their magnetic inclination and read 17 degrees north. Apparently, India was once in the Southern Hemisphere and had crossed over the equator into the Northern Hemisphere to its present location as part of Asia (Fig. 16).

Skeptics pointed out that the same phenomenon could occur simply by shifting the Earth's magnetic poles. Such polar wandering would have altered the direction of the Earth's magnetic field, and records of these changes would be permanently locked up in the rocks. Therefore, rocks formed at different times in England and India could have been imprinted with different inclinations without moving an inch. Thus, evidence for continental drift was used

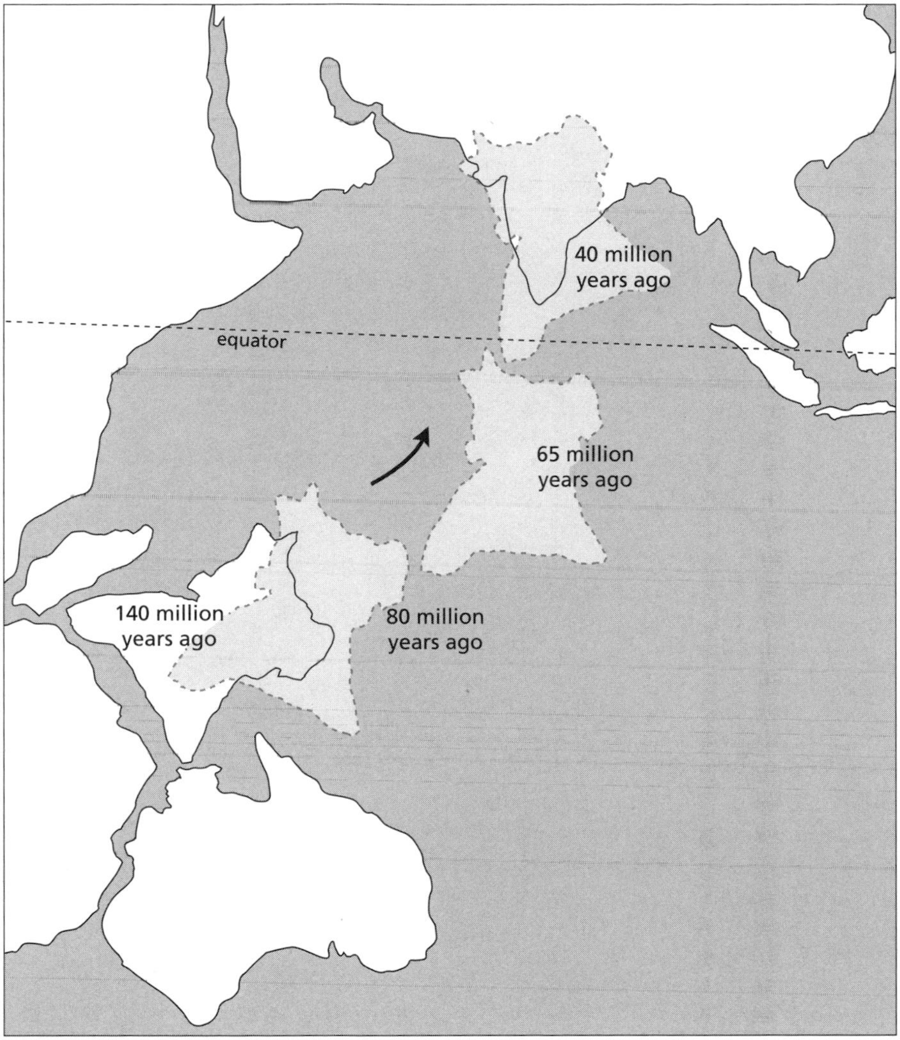

Figure 16 *The drift of India, which collided with Asia about 40 million years ago.*

instead to show that the North Pole had actually wandered some 13,000 miles over the last billion years, from western North America, across the northern Pacific Ocean and northern Asia, finally coming to rest at its present location in the Arctic Ocean.

When similar magnetometer measurements were taken in North America, however, the results came as a complete surprise. Although the polar paths derived from data on Eurasia and North America both were much the same shape and had a common point of origin at the North Pole, the curves gradually veered away from each other (Fig. 17). Only by hypothetically joining the continents together did the two curves overlap. Thus, in their attempts

Figure 17 *Polar wandering curves for North America (top curve) and Europe (bottom curve) veer apart due to continental drift.*

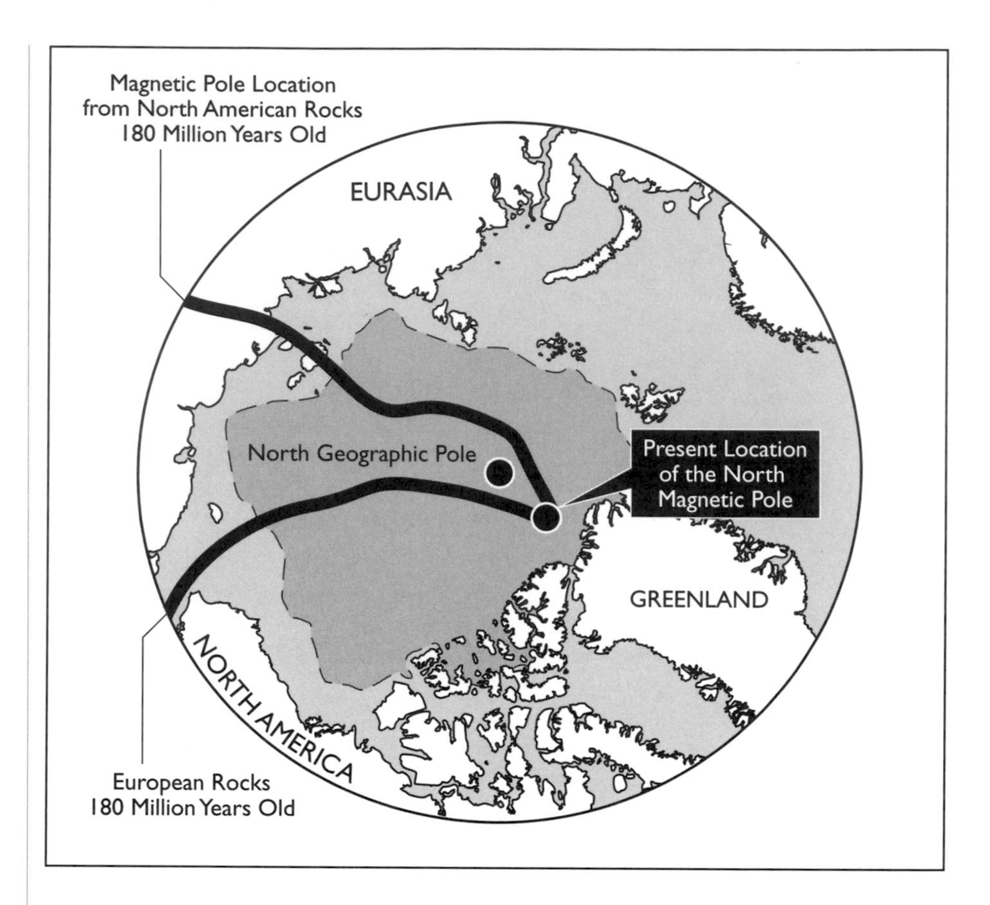

to disprove continental drift, scientists inadvertently provided the strongest evidence in its favor.

LITHOSPHERIC PLATES

Only after overwhelming geologic and geophysical evidence was collected from the ocean floor supporting the theory of continental drift did geologists finally abandon the archaic thinking of earlier centuries. By the late 1960s, most geologists in the Northern Hemisphere, who had fought hard against the theory, joined their southern colleagues, who for some time were convinced of the reality of continental drift.

The generally accepted model, described in detail in the next chapter, is that the present continents were sutured together into the supercontinent Pangaea during the Permian and early Triassic. In the late Triassic and early Jurassic, Pangaea began to rift apart along a fracture zone now represented by the Mid-Atlantic Ridge. The breakup finally became complete sometime during the early Cretaceous. Since then, the continents bordering on the Atlantic have been drifting away from each other (Fig. 18).

In spite of the mounting evidence supporting continental drift, skeptics continued to doubt the breakup and drift of the continents. They questioned whether the currents in the Earth's interior were powerful enough to propel the continents around. Even some supporters of continental drift thought this energy source might not be sufficient and suggested that an additional mechanism, such as gravity, was needed to move the continents. Furthermore, if Earth processes were essentially uniform throughout geologic time, why did this event happen so late in the Earth's history? Were earlier episodes of continental collision and breakup evident in the past?

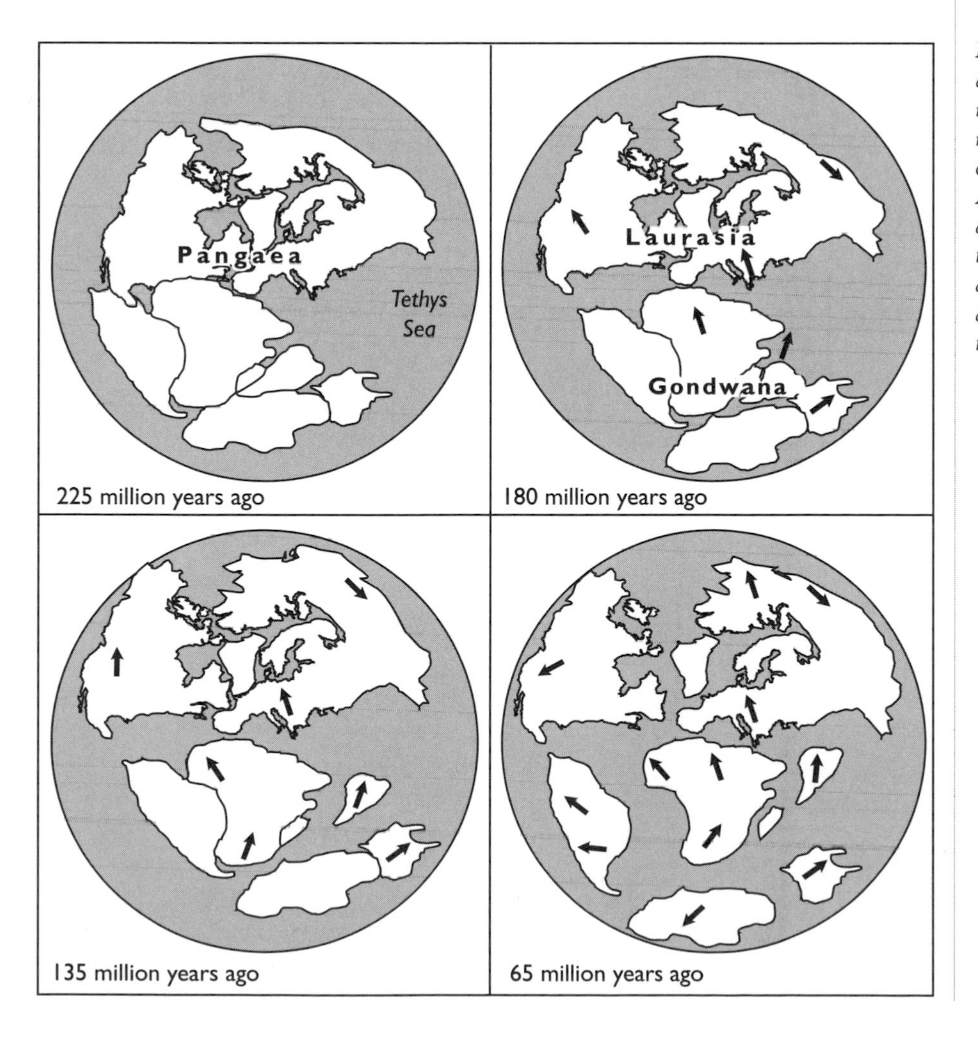

Figure 18 *The breakup and dispersal of the continents: (1) Pangaea 200 million years ago, (2) the opening of the North Atlantic 160 million years ago, (3) the continents at the height of the dinosaur age 80 million years ago, and (4) the present locations of the continents.*

Despite these objections, the overwhelming majority of earth scientists accepted continental drift as a scientific certainty. So convincing was the evidence it led to an entirely new way of looking at the Earth called the theory of plate tectonics. Tectonics, from the Greek *tekton* meaning "builder," is any process by which the Earth's surface is shaped. The plate tectonics theory was first publicized by the British geophysicists Dan McKenzie and Robert Parker in 1967. It incorporated the processes of seafloor spreading and continental drift into a single comprehensive model. Therefore, all aspects of the Earth's history and structure could be unified by the revolutionary concept of movable plates. The boundaries of the plates were marked by well-defined earthquake zones, and analysis of earthquakes around the Pacific Ocean revealed a consistent direction of crustal movement.

The theory was also developed independently by the American geophysicist Jason Morgan at Princeton University. He regarded the Earth's outer shell as neither rigid nor fixed but made up of several movable plates about 60 miles thick. Each plate was composed of the lithosphere, the solid portion of the upper mantle, and the overlying continental or oceanic crust. The plate boundaries were the midocean ridges, where new oceanic crust is created as the plates are pulled apart; the transform faults, where the plates slide past each other, often wrenching the ocean floor in the process; and the deep-sea trenches, where the plates are subducted or absorbed into the mantle and destroyed. The plates ride on the asthenosphere, the semimolten layer of the upper mantle. The plates carry the continents along with them somewhat like rafts of rock riding on a sea of molten magma.

After discussing the evolution of plate tectonic theory, the next chapter will show how plate tectonics shaped the Earth over its long history.

2

HISTORICAL TECTONICS

AN OVERVIEW OF EARTH HISTORY

Plate tectonics has been shaping the Earth almost from the very beginning. Continents were set adrift practically since they first formed some 4 billion years ago. This is manifested by ancient granites discovered in the Northwest Territories of Canada. They suggest that the formation of the crust was well underway by this time because the presence of granites indicates crustal generation. The continental crust was only about one-tenth its present size and contained slivers of granite that drifted freely over the Earth's watery surface.

As time progressed, these slices of crust began to slow their erratic wanderings and combined into larger landmasses. Continuous bumps and grinds built up the crust. By the time Earth was half its current age, the landmass occupied up to one-quarter the total surface area. Also during this time, plate tectonics began to operate extensively, and the Earth as we know it began to take shape. This chapter follows the evolution of the planet from the earliest construction of the crust to the present.

ARCHEAN TECTONICS

The first 2 billion years of Earth history, known as the Archean eon, was a time when the Earth's interior was hotter, the crust was thinner and therefore more unstable, and the crustal plates were highly mobile. The Earth was in a great upheaval and subjected to extensive volcanism and a massive meteorite bombardment, which had a major effect on the development of life early in the planet's history.

A permanent crust began to form about 4 billion years ago. Ancient metamorphosed granite in the Great Slave region of Canada's Northwest Territories, called Acasta Gneiss, indicates that substantial continental crust had formed by this time, comprising some 20 percent of the present landmass. Its presence leaves little doubt that at least small patches of continental crust existed during the first billion years. The metamorphosed marine sediments of the Isua Formation in a remote mountainous region in southwest Greenland suggest the existence of a saltwater ocean by at least 3.8 billion years ago.

Only three sites located in Canada, Australia, and Africa contain rocks exposed on the surface during the early Precambrian that have not changed significantly throughout geologic time. Most other rocks have either eroded, were metamorphosed, or melted entirely, leaving only a few untouched. Few rocks on Earth date beyond 3.7 billion years, however, suggesting that little continental crust was generated until after that time or was recycled before then.

The crust was probably far too hot for plate tectonics to operate effectively due to more vertical bubbling than horizontal sliding. Basalt, a dark dense volcanic rock, formed most of the early crust, both on the continents and under the oceans. The crust was thin and highly unstable. When tectonic activity melted and remelted the basaltic crust, granites formed out of the recycled rock, producing scattered blocks embedded in the basalt called rockbergs.

The formative Earth was subjected to massive volcanism and meteorite bombardment that repeatedly destroyed the crust. Therefore, the first 700 million years of Precambrian time, called the Azoic eon, is missing from the geologic record. Heavy turbulence in the mantle with a heat flow three times greater than today resulted in violent agitation on the surface. This resulted in a sea of molten and semimolten rock broken up by giant fissures, from which fountains of lava spewed skyward.

About 4 billion years ago, during the height of the great meteorite bombardment, a massive asteroid landed in what is presently central Ontario, Canada. The impact created a crater up to 900 miles wide and might have triggered the formation of continental crust by melting vast amounts of basalt, converting it into granitic rock. Slices of granitic crust combined into stable bodies of basement rock, upon which all other rocks were deposited. The basement rocks formed the nuclei of the continents and are presently exposed in broad, low-lying, domelike structures called shields (Fig. 19). The shields are extensive, uplifted areas surrounded by sediment-covered bedrock called continental platforms. These platforms are broad, shallow depressions of basement complex comprising crystalline rock filled with nearly flat-lying sedimentary rocks.

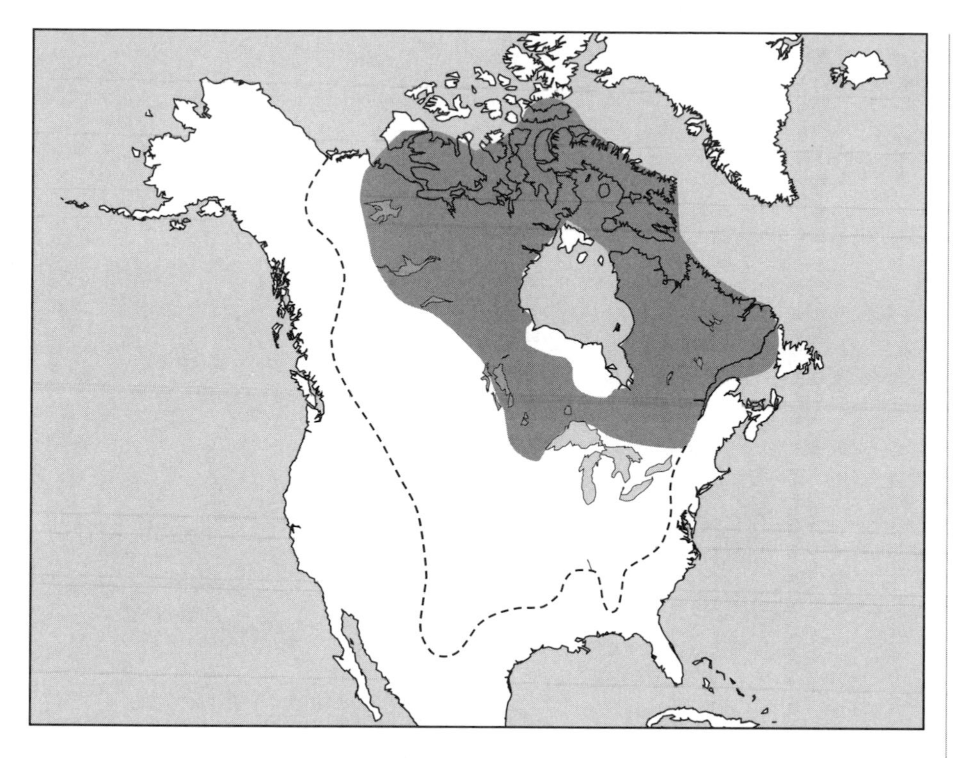

Figure 19 *The Canadian Shield (darkened area) and platforms (enclosed in dashed line).*

Archean greenstone belts are dispersed among and around the shields. They are composed of a jumble of metamorphosed (recrystallized) lava flows and sediments possibly from chains of volcanic islands called island arcs caught between colliding continents. Greenstone belts are zones where protocontinents rifted apart and opened basins filled with lavas. The basins were later shut again, and the volcanic rocks were compressed and folded into narrow belts. Their green color is derived from abundant chlorite, a greenish micalike mineral. The greenstone belts cover an area of several hundred square miles and are surrounded by immense expanses of gneiss, the metamorphic equivalent of granite and the predominant Archean rock type.

Caught in the greenstone belts were ophiolites, from the Greek *ophis,* meaning "serpent," due to their similar color. They are slices of ocean floor shoved up onto the continents by drifting plates and are as much as 3.6 billion years old. Ophiolites provide the best evidence for ancient plate motions. They are vertical cross sections of oceanic crust peeled off during plate collisions and plastered onto continents. This resulted in a linear formation of greenish volcanic rocks along with light-colored masses of granite and gneiss, common igneous and metamorphic rocks. Pillow lavas, which are tubular bodies of basalt extruded undersea, are also found in the greenstone belts. This signifies the volcanic eruptions took place on the ocean floor.

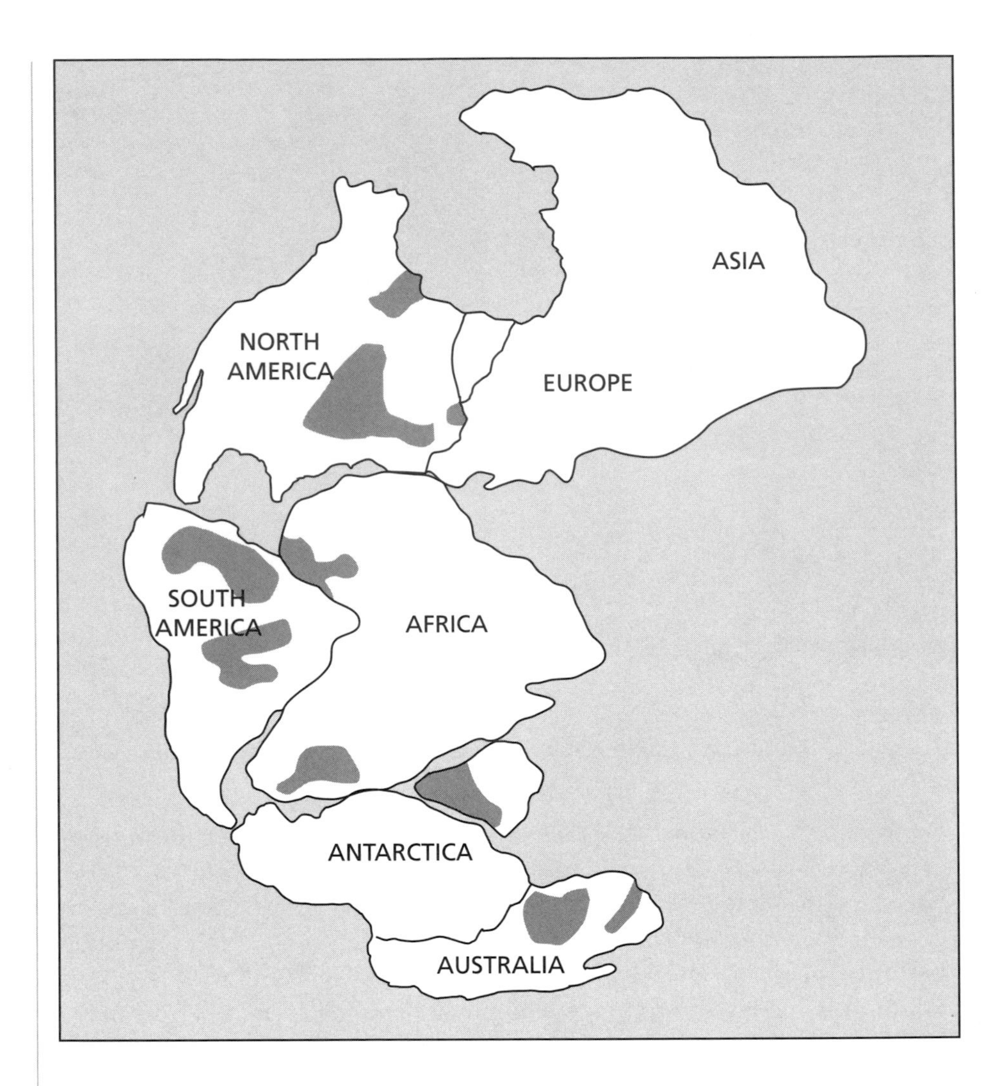

Figure 20 *Archean greenstone belts comprise the ancient cores of the continents.*

Greenstone belts are found in all parts of the world and occupy the ancient cores of the continents (Fig. 20). They are of particular interest to geologists not only as evidence for Archean plate tectonics but also because they contain most of the world's gold. India's Kolar greenstone belt holds the richest gold deposits. It is some 3 miles wide and 50 miles long. It formed when two plates clashed about 2.5 billion years ago. The two pieces of continental crust crunched together from the east and west, squeezing up a band of seafloor between them. This slice of oceanic crust makes up the belt and differs in composition and density from the surrounding continental crust. In Africa, the best gold deposits are in rocks as old as 3.4 billion years, and most South African gold mines are found in greenstone belts. In North America, the best gold mines are in the greenstone belts of the Great Slave region of northwest Canada, where well over 1,000 deposits are known.

In the Barberton Greenstone Belt of South Africa lies a thick, widespread bed of silicate spherules. These are small, glassy beads believed to have originated from the melt of an immense meteorite impact between 3.5 and 3.2 billion years ago. This was a time when large impacts were quite numerous and played a prominent role in shaping the Earth's surface. Because greenstone belts are essentially Archean in age, their disappearance from the geologic record around 2.5 billion years ago marked the end of the eon.

PROTEROZOIC TECTONICS

The Proterozoic eon, from 2.5 billion to about 570 million years ago, witnessed a dramatic change in the Earth as it matured from a tumultuous adolescence to a more tranquil adulthood. At the beginning of the eon, nearly three-quarters of the present landmass was in existence. Continents were more stable and melded into a single large supercontinent. The collisions forced up mountain ranges. The seams joining the landmasses remain as cores of ancient mountains called orogens, from the Greek *oros* meaning "mountain." The interval was possibly the most energetic period of tectonic activity, which rapidly built new continental crust. Extensive volcanic activity, magmatic intrusions, and rifting and patching of the crust built up the continental interiors, while erosion and sedimentation extended the continental margins outward. The global climate of the Proterozoic was significantly cooler. The Earth experienced its first major glaciation between 2.2 and 2.4 billion years ago.

Most of the material presently locked up in sedimentary rocks was at or near the surface by the beginning of the Proterozoic. Thick deposits of Proterozoic sediments were derived from Archean granites. Ample sources of Archean rocks were available for erosion and redeposition into Proterozoic rock types. The weathering of Archean rocks during the Proterozoic produced solutions of calcium carbonate, magnesium carbonate, calcium sulfate, and sodium chloride. These, in turn, precipitated into limestone, dolomite, gypsum, and halite. Sediments derived directly from primary sources are called wackes, often described as dirty sandstone. Most Proterozoic wackes composed of sandstones and siltstones originated from Archean greenstones. Another common rock type was a fine-grained metamorphosed rock called quartzite, derived from the erosion of siliceous grainy rocks such as granite and arkose, a coarse-grained sandstone with abundant feldspar.

Conglomerates, which are consolidated equivalents of gravels, were also abundant during the Proterozoic. Nearly 20,000 feet of Proterozoic sediments lie in the Uinta Range of Utah (Fig. 21). The Montana Proterozoic belt system contains sediments more than 11 miles thick. The Proterozoic is also known for its terrestrial redbeds composed of sandstones and shales cemented by iron oxide, which colored the rocks red. Their appearance, around 1 billion years ago, signifies that the atmosphere contained substantial amounts of oxygen by this time, which oxidized the iron similar to the rusting of metal.

The continents were composed of odds and ends of Archean cratons, which were the first pieces of land to appear. These were comprised of blocks of granitic crust welded together to form the cores of the continents. Cratons consist of ancient igneous and metamorphic rocks, whose composition is remarkably similar to their modern equivalents. The existence of cratons early in Earth history suggests a fully operating rock cycle was already in place.

Many cratons throughout the world were assembled at about the same time. The original cratons formed within the first 1.5 billion years and totaled about 10 percent of the present landmass. These ancient, stable rock masses in the continental interiors contain the oldest rocks on Earth (Fig. 22), dating back to 4 billion years. They are composed of highly altered granite and metamorphosed marine sediments and lava flows. They originated from intrusions of magma into the primitive oceanic crust. This allowed the magma to cool slowly and separate into a light component, which rose toward the surface, and a heavy component, which settled to the bottom of the magma chamber.

Some magma also seeped through the crust, where it poured out as lava onto the ocean floor. Successive intrusions and extrusions of magma built up new

Figure 21 *Uinta Mountains, Summit County, Utah.*

(Photo by W. R. Hansen, courtesy USGS)

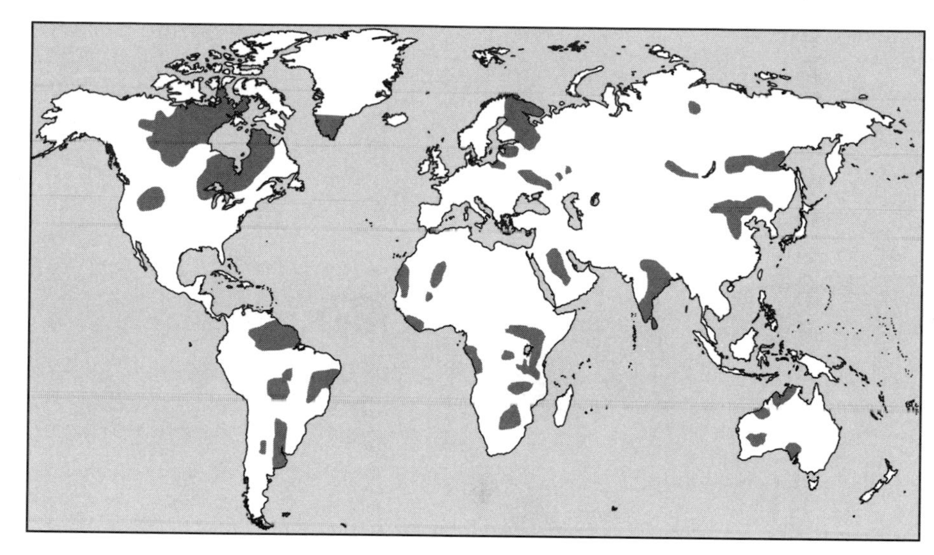

Figure 22 *The world's oldest rocks.*

crust until it finally broke the surface of a global sea. Unlike the volcanic islands, which existed for only a short time, these thin slivers of land became a permanent part of the landscape since cratons were lighter and more buoyant than the oceanic crust. If a craton were forced into the mantle under the pull of gravity, it would simply bob up again like a cork on the end of a fishing line.

The cratons numbered in the dozens. They ranged in size from larger than India to smaller than Madagascar, which are themselves continental fragments. The cratons moved about freely on the molten rocks of the upper mantle called the asthenosphere, from the Greek *asthenes,* meaning "weak." They were independent minicontinents that periodically collided with and rebounded off each other. All cratons eventually coalesced into a single large landmass several thousand miles wide. The point at which the cratons collided with one another caused a slight crumpling at the leading edges of the cratons, forming small parallel mountain ranges perhaps only a few hundred feet high. The sutures joining the landmasses are still visible today as cores of ancient mountains more than 2 billion years old.

Volcanoes were highly active on the cratons. Lava and ash continuously built them upward and outward. New crustal material was also added to the interior of the cratons by magmatic intrusions composed of molten crustal rocks recycled through the upper mantle. This effectively cooled the mantle, causing the cratons to slow their erratic wanderings. The average rate of continental growth was perhaps as much as one cubic mile a year. The constant rifting and patching of the interior along with sediments deposited along the continental margins eventually built the landmass outward until its area was nearly equal to the total area of all present-day continents.

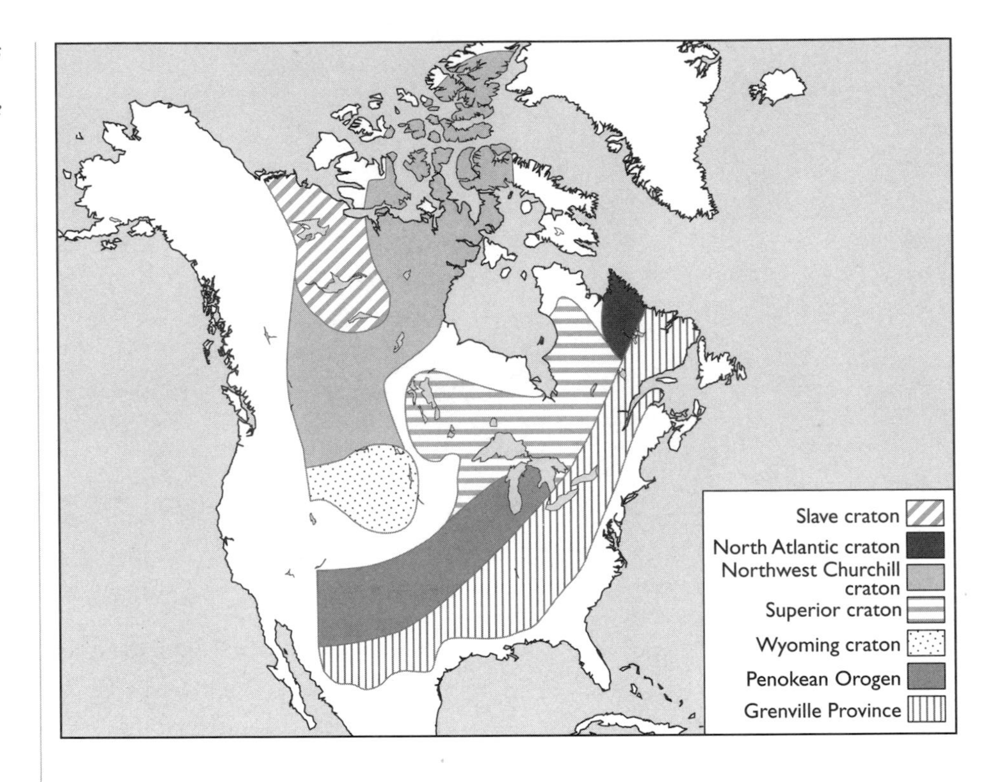

Figure 23 *The cratons that comprise the North American continent came together some 2 billion years ago.*

The North American continent comprises seven cratons, forming central Canada and the northern-central United States (Fig. 23). These cratons assembled around 2 billion years ago, making North America the oldest continent. Successive continental collisions added new crust to the growing proto-North American continent known as Laurentia. Most of the continent, comprising the interior of North America, Greenland, and northern Europe, evolved in a relatively brief period of only 150 million years. A major part of the continental crust underlying the United States from Arizona to the Great Lakes to Alabama formed in one great surge of crustal generation around 1.8 billion years ago that has no equal. This was possibly the most energetic period of tectonic activity and crustal generation in Earth history. More than 80 percent of all continental mass was created at that time. The best exposure of these Precambrian metamorphic rocks is the Vishnu Schist on the floor of the Grand Canyon (Fig. 24). The assembled North American continent was stable enough to resist another billion years of jostling and rifting. It continued to grow by plastering bits and pieces of continents and island arcs to its edges.

After the rapid continent building, the interior of Laurentia experienced extensive igneous activity that lasted from 1.6 to 1.3 billion years ago. A broad belt of red granites and rhyolites, which are igneous rocks formed by molten magma solidifying below ground as well as on the surface, extended several thousand miles across the interior of the continent from southern California to Labrador. The Laurentian granites and rhyolites are unique due to their sheer volume. This suggests the continent stretched and thinned almost to the breaking point. These rocks are presently exposed in Missouri, Oklahoma, and a few other localities. However, they are buried under sediments up to a mile thick in the center of the continent. In addition, vast quantities of molten basalt poured from a huge tear in the crust running from southeast Nebraska into the Lake Superior region about 1.1 billion years ago. Arcs of volcanic rock also weave through central and eastern Canada down into the Dakotas.

Half a billion years ago, North America was a lost continent, drifting on its own. The rest of the Earth's landmass, however, combined into a supercontinent. The African and South American continents did not aggregate until about 700 million years ago. Over the past half-billion years, about a dozen individual continental plates welded together to form Eurasia. It is the youngest and largest modern continent. It is still being pieced together with chunks of crust arriving from the South, riding on highly mobile tectonic plates.

Late in the Proterozoic, the Earth experienced at least four major ice ages between 850 and 600 million years ago. Around 670 million years ago, thick ice sheets spread over much of the landmass during perhaps the greatest period of glaciation the Earth has ever known. This period is called the Varanger ice age. At this time, all continents were assembled into a supercontinent named Rodinia (Fig. 25), Russian for "motherland." It might have wandered over one

Figure 24 *Precambrian Vishnu Schist, Grand Canyon National Park, Arizona.*

(Photo by R. M. Turner, courtesy USGS)

Figure 25 *The supercontinent Rodinia 700 million years ago.*

of the poles and collected a thick sheet of ice. When the ice age ended, life took off in all directions, resulting in many unique and bizarre creatures.

The presence of large amounts of volcanic rock near the eastern edge of North America implies the continent was once the core of a larger supercontinent. The central portion of the supercontinent was far removed from the cooling effects of subducting plates, where the Earth's crust sinks into the mantle. As a result, the interior of the supercontinent heated and erupted with volcanism. The warm, weakened crust consequently broke into possibly four or five major continents between 630 and 560 million years ago. The breakup produced extensive continental margins. This increased the habitat area and prompted the greatest explosion of new species the world has ever known.

PALEOZOIC TECTONICS

The Paleozoic era, from about 570 million to about 250 million years ago, is generally divided into two time units of nearly equal duration. The lower

Paleozoic consists of the Cambrian, Ordovician, and Silurian periods. The upper Paleozoic comprises the Devonian, Carboniferous, and Permian periods. The first half of the Paleozoic was tectonically quiet, with little mountain building and volcanic activity.

During the late Precambrian and early Cambrian, Rodinia rifted apart. In the process, it opened a proto-Atlantic Ocean called the Iapetus Sea. The rifting formed extensive inland seas. This submerged most of Laurentia some 540 million years ago, as evidenced by the presence of Cambrian seashores in places such as Wisconsin. It also flooded the ancient European continent called Baltica. The Iapetus was similar in location and size as the North Atlantic and was dotted with volcanic islands. It resembled the present-day southwestern Pacific Ocean.

During the Cambrian, continental motions assembled the present continents of Africa, South America, Australia, Antarctica, and India into Gondwana, much of which was in the south polar region from the Cambrian to the Silurian. A major mountain building episode from the Cambrian to the middle Ordovician deformed areas among all continents comprising Gondwana, indicating their collision during this interval. Extensive igneous activity and metamorphism accompanied the mountain building at its climax.

In the Ordovician, North and South America apparently abutted one another (Fig. 26). A limestone formation in Argentina contains a distinctive trilobite species typical of North America but not of South America. The fos-

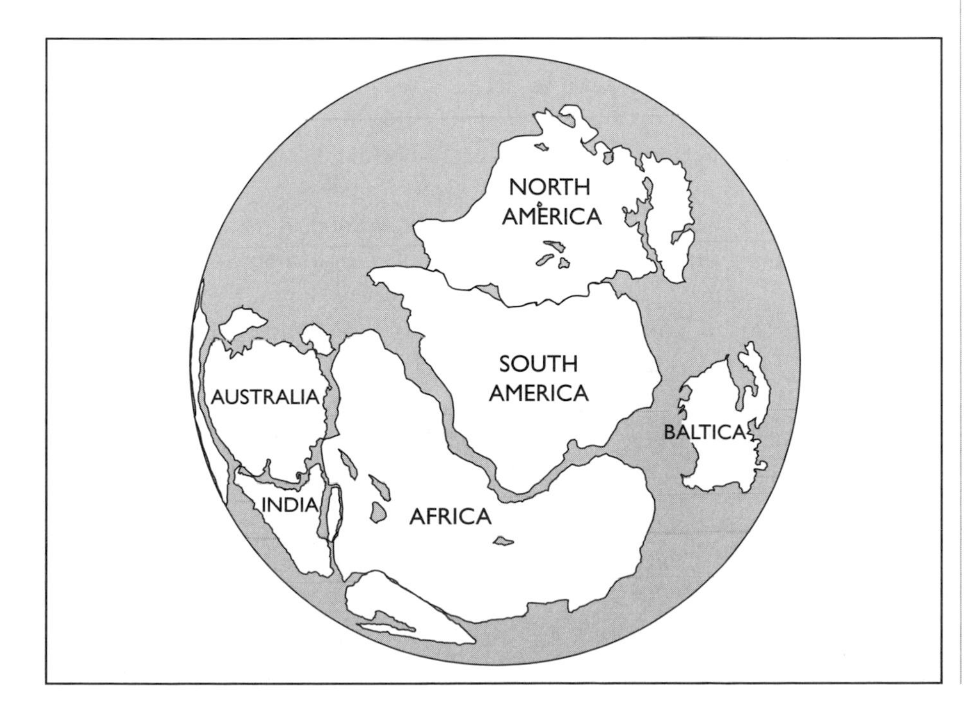

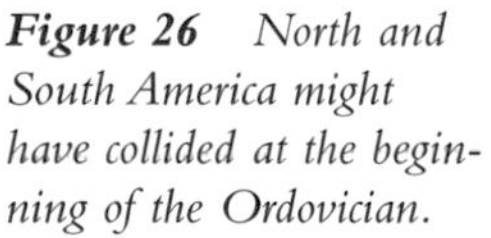
Figure 26 *North and South America might have collided at the beginning of the Ordovician.*

sil evidence suggests that the two continents collided about 500 million years ago, creating an ancestral Appalachian range along eastern North America and western South America long before the present Andes formed. Later, the continents rifted apart, transferring a slice of land containing trilobite fauna from North America to South America.

During the Silurian, all northern continents collided to form Laurasia. It included what is now North America, Greenland, Europe, and Asia. Laurasia and Gondwana were separated by a large body of water called the Tethys Sea, named for the mother of the seas in Greek mythology. The Tethys held thick deposits of sediments washed off the continents. The continents were lowered by erosion. Shallow seas flooded inland, covering more than half the present land area. The weight of the sediments formed a deep depression in the ocean crust called a geosyncline. The sedimentary rocks were later uplifted into great mountain belts surrounding the Mediterranean Sea when Africa slammed into Europe.

Continental movements are thought to be responsible for a period of glaciation during the late Ordovician around 440 million years ago. The study of magnetic orientations in rocks from many parts of the world indicates the positions of the continents relative to the magnetic poles at various times in geologic history. The paleomagnetic studies in Africa were very curious, however. The northern part of the continent was placed directly over the South Pole during the Ordovician, which led to widespread glaciation.

Evidence for this period of glaciation came from an unexpected source. In the middle of the Sahara Desert, geologists exploring for oil stumbled upon a series of giant grooves that appeared to be cut into the underlying strata by glacial ice. The scars were created by rocks embedded at the base of glaciers as they scraped the bedrock. Further evidence that the Sahara had once been covered by thick sheets of ice included erratic boulders carried long distances by the glaciers and sinuous sand deposits from glacial outwash streams.

The second half of the Paleozoic followed a Silurian ice age. During this time, Gondwana wandered into the south polar region around 400 million years ago and acquired a thick sheet of ice. Glacial centers expanded in all directions. Ice sheets covered large portions of east-central South America, South Africa, India, Australia, and Antarctica (Fig. 27). During the early part of the glaciation, the maximum glacial effects occurred in South America and South Africa. Later, the chief glacial centers switched to Australia and Antarctica, providing strong evidence that the southern continents wandered locked together over the South Pole.

In Australia, Silurian-age marine sediments were found interbedded with glacial deposits and tillites, composed of glacially deposited boulders and clay. These were separated by seams of coal, indicating that periods of glaciation were punctuated by warm interglacial spells, when extensive forests grew. The Karroo Series in South Africa is composed of a sequence of late Paleozoic

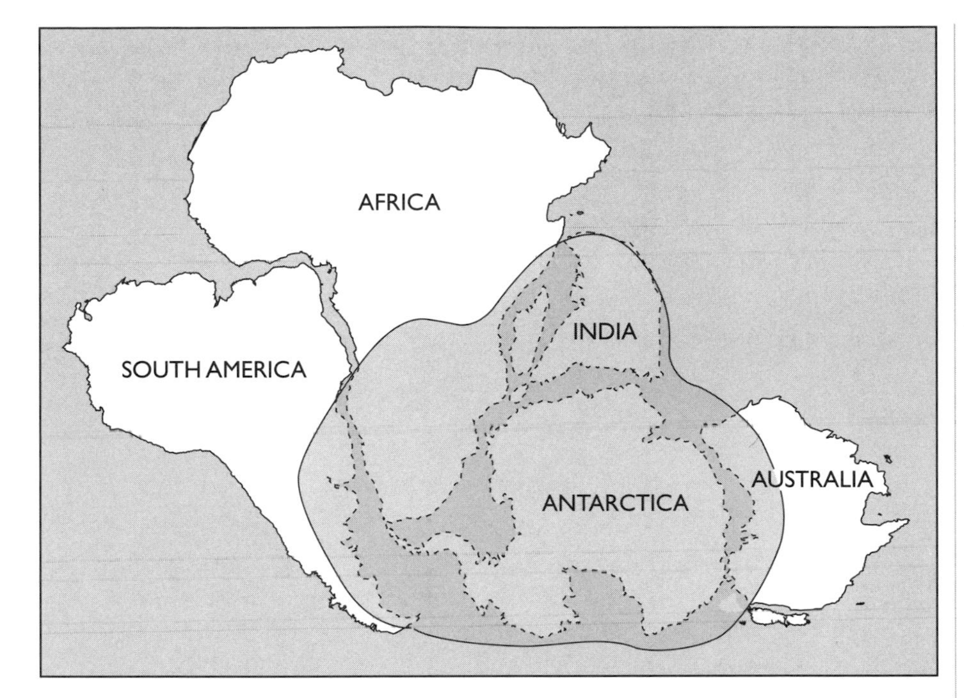

Figure 27 The extent of late Paleozoic glaciation in Gondwana.

lava flows, tillites, and coal beds, reaching a total thickness of 20,000 feet. Between layers of coal were fossil leaves of the extinct fern *Glossopteris.* Because this plant is found only on the southern continents, it is among the best evidence for Gondwana.

During the Carboniferous, ocean levels lowered and the continents rose. The inland seas departed and were replaced by great swamplands, where vast coal deposits accumulated. Extensive forests and swamps grew successively on top of one another. They continued to add to thick deposits of peat, which were buried under layers of sediment and compressed into lignite and bituminous and anthracite coal. The widespread distribution of evaporite deposits in the Northern Hemisphere, coal deposits in the Canadian Arctic, and extensive carbonate reefs suggest a warm climate and desert conditions over large areas. Warm temperatures of the past are generally recognized by abundant marine limestones, dolomite, and calcareous shales. A coal belt, extending from northeastern Alaska across the Canadian archipelago to northernmost Russia, suggests that vast swamps were prevalent in these regions.

Beginning in the late Devonian and continuing into the Carboniferous, Gondwana and Laurasia converged into the supercontinent Pangaea (Fig. 28). It extended practically from pole to pole and comprised some 40 percent of the Earth's total surface area. The climate on Pangaea was one of extremes. The northern and southern regions were as cold as the arctic, and the interior was

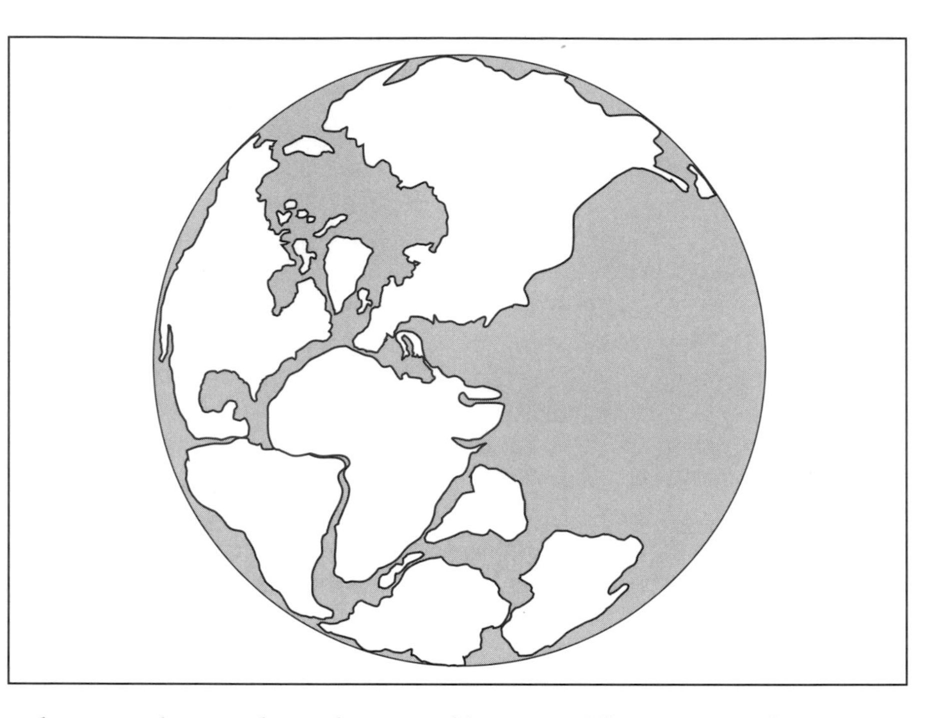

Figure 28 *The supercontinent Pangaea extended almost from pole to pole.*

as hot as a desert, where almost nothing grew. The massing of continents together created an overall climate that was hotter, drier, and more seasonal than at any other time in geologic history.

A single great ocean called Panthalassa stretched uninterrupted across the rest of the planet. Over the ensuing time, smaller parcels of land contin-

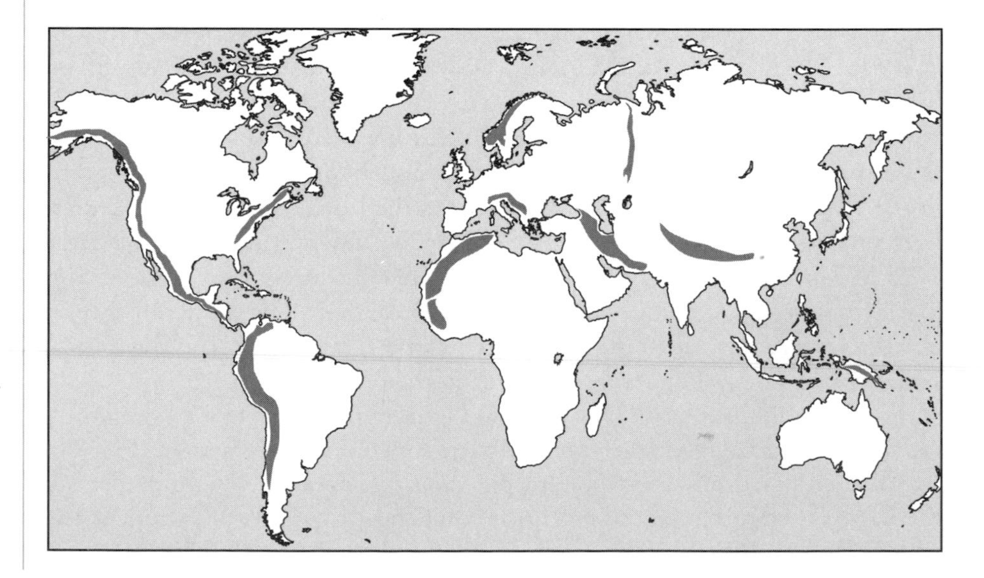

Figure 29 *Major mountain ranges resulting from continental collisions.*

ued to collide with the supercontinent until it reached peak size by the end of the Triassic. Volcanic eruptions were prevalent due to frequent continental collisions. These collisions crumpled the crust and pushed up huge masses of rocks into several mountain belts throughout many parts of the world (Fig. 29). The sediments in the Tethys Sea separating Gondwana and Laurasia were squeezed and uplifted into various mountain belts, including the Ouachitas and Appalachians of North America and the ancestral Hercynian Mountains of southern Europe. As the continents rose higher and the ocean basins dropped lower, the land became dryer and the climate grew colder, especially in the southernmost lands, which were covered with glacial ice.

MESOZOIC TECTONICS

At the beginning of the Mesozoic era, from about 250 million to 65 million years ago, all the continents were consolidated into Pangaea. About midway through the Mesozoic, the supercontinent began to break up. At the end of the era, the separated continents were well along toward their present locations. The breakup of Pangaea created three major bodies of water, including the Atlantic, Arctic, and Indian Oceans. The climate was exceptionally mild for an unusually long period, possibly due to increased volcanic activity and its resultant greenhouse effect. The once towering mountain ranges of North America and Europe were toppled by erosion. Reef building was intense in the Tethys Sea, and thick deposits of limestone and dolomite were laid down by lime-secreting organisms. These beds were later uplifted to form the dolomite and limestone Alps.

In the early Triassic, the great glaciers of the previous ice age melted, and the seas began to warm. The energetic climate facilitated the erosion of the high mountain ranges of North America and Europe. Seas retreated from the continents as they continued to rise. Widespread deserts covered the land. Abundant terrestrial redbeds and thick beds of gypsum and salt were deposited into the abandoned basins. Redbeds covered much of North America, including the Colorado Plateau and a region from Nova Scotia to South Carolina. A preponderance of red rocks composed of sandstones and shales are exposed in the mountains and canyons in the western United States (Fig. 30). Redbeds were also common in Europe, especially northwestern England.

Such wide occurrences of red sediments were probably due to massive accumulations of iron supplied by one the most intense periods of igneous activity the world has ever known. These large volcanic eruptions created a series of overlapping lava flows. They gave many exposures a terracelike appearance known as traps, from the Dutch word for "stairs." Huge lava flows and granitic intrusions invaded Siberia. Extensive lava flows covered South

Figure 30 *Chugwater redbeds on the east side of Red Fork of the Powder River, Johnson County, Wyoming.*

(Photo by N. H. Darton, courtesy USGS)

America, Africa, and Antarctica as well. Southern Brazil was paved over with some 750,000 square miles of basalt, constituting one of the largest lava fields in the world.

Great floods of basalt, in places 2,000 feet or more thick, flowed over large parts of Brazil and Argentina. This occurred when the South American plate overrode the Pacific plate, causing it to subduct and melt to feed magma chambers underlying active volcanoes. Huge basalt flows also spanned from Alaska to California. In addition, massive granitic intrusions produced the huge Sierra Nevada batholith in California. These tremendous outpourings of basalt reflected one of the greatest crustal movements in the history of the planet. The continents probably traveled much faster than they do today because of more vigorous plate motions, resulting in greater volcanic activity.

Near the close of the Triassic and continuing into the Jurassic, Pangaea rifted apart into the present continents. North and South America separated. India, nestled between Africa and Antarctica, separated from the two continents and moved northward toward southern Asia. A great rift separated the North American continent from Eurasia. The rift was later breached and flooded with seawater, forming the infant North Atlantic Ocean. North America drifted westward as the North Atlantic continued to widen at the expense of the Pacific. South America separated from Africa like a zipper opening from south to north. Antarctica, still attached to Australia, swung away from Africa toward the southeast, forming the proto–Indian Ocean.

The Pacific plate was no larger than the United States during the initial breakup of Pangaea in the early Jurassic. About 190 million years ago, the Pacific plate might have begun as a tiny microplate, which is a small block of oceanic crust that often lies at the junction between two or three major plates. The rest of the ocean floor consisted of other unknown plates that have long since disappeared as the Pacific plate continued to grow. Because of the effects of seafloor spreading in the Atlantic and plate subduction in the Pacific, the crust of the Pacific Basin is no older than middle Jurassic in age.

Much of western North America was assembled from oceanic island arcs and other crustal debris skimmed off the Pacific plate as the North American plate continued heading westward. Northern California is a jumble of crust assembled only a few hundred million years ago. A nearly complete slice of ocean crust 2.7 billion years old, the type shoved up onto the continents by drifting plates, sits in the middle of Wyoming. Similarly, at Cape Smith on Hudson Bay lies a 2-billion-year-old piece of oceanic crust squeezed onto the land—an indication that continents collided and closed an ancient ocean.

During the Jurassic and continuing into the Cretaceous, the Western Interior Cretaceous Sea flowed into the west-central portions of North America (Fig. 31). Accumulations of marine sediments eroded from the Cordilleran highlands to the west (sometimes referred to as the ancestral Rockies). They were deposited onto the terrestrial redbeds of the Colorado Plateau, forming the Jurassic Morrison Formation, which is well-known for fossil dinosaur bones (Fig. 32). The continents were flatter, mountain ranges were lower, and sea levels were higher. The Appalachians, upraised by a continental collision between North America and Northwest Africa, were eroded down to stumps by the Cretaceous.

In the late Cretaceous and early Tertiary, land areas were inundated by the ocean. It flooded continental margins and formed great inland seas, some of which split continents in two. Great deposits of limestone and chalk were laid down in the interior seas of Europe and Asia, which is how the period got its name, from the Latin *creta* meaning "chalk." Seas divided North America in the Rocky Mountain and high plains regions. The interior seas that flooded North America were filled with thick deposits of sediment that are presently exposed as impressive cliffs in the American West (Fig. 33).

Eastern Mexico, southern Texas, and Louisiana were also flooded. South America was cut in two in the region that later became the Amazon Basin. Eurasia was split by the joining of the Tethys Sea and the newly formed Arctic

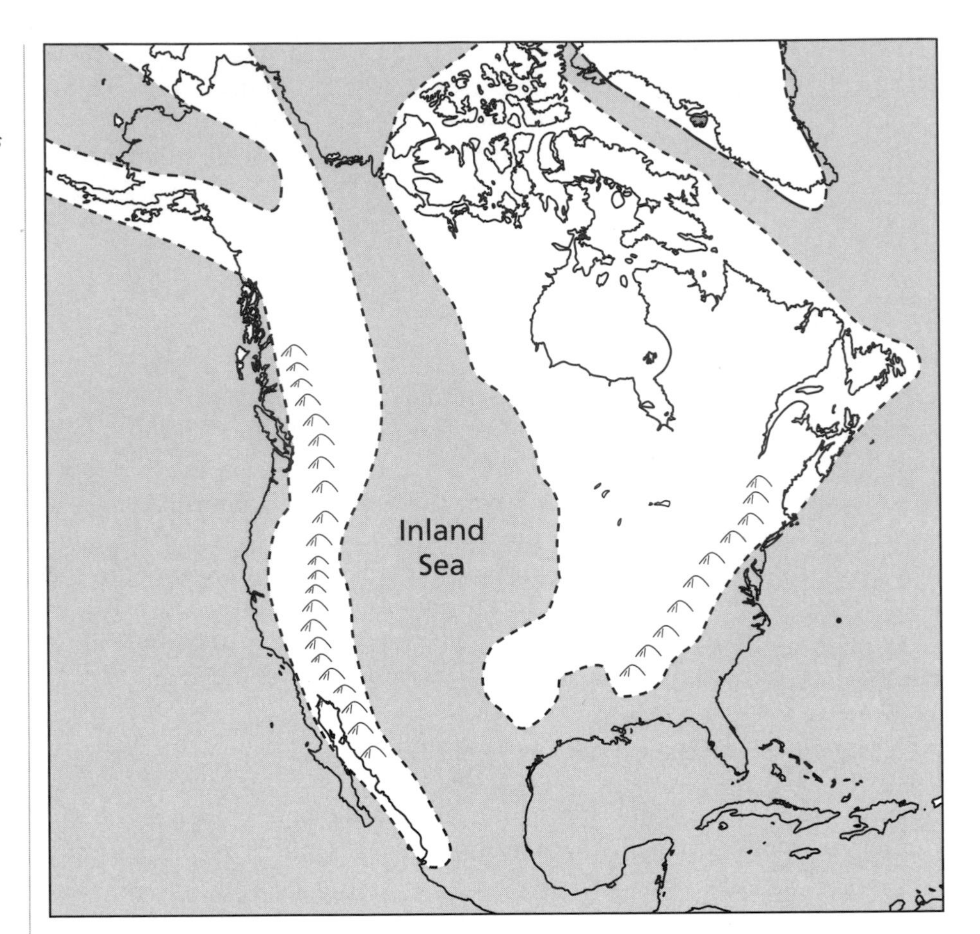

Figure 31 *The Cretaceous interior sea of North America where thick deposits of sediments were laid down.*

Ocean. The Tethys Sea provided a wide gulf between the northern and southern landmasses and continued to fill with thick layers of sediment. Seas also invaded Africa and Australia. The oceans of the Cretaceous were also interconnected in the equatorial regions by the Tethys and Central American seaways, providing a unique circumglobal ocean current system that made the climate equable.

Reef building was intense in the Tethys Sea. Thick deposits of limestone and dolomite were laid down by lime-secreting organisms in the interior seas of Europe and Asia. These deposits were later uplifted during one of geologic history's greatest mountain building episodes. The rim of the Pacific Basin became a hotbed of geologic activity. Practically all mountain ranges facing the Pacific Ocean and island arcs along its perimeter developed during this period.

When the Cretaceous ended, North America and Europe were no longer in contact except for a land bridge that spanned across Greenland, the world's largest island. Greenland separated from North America and

Norway during the early part of the Tertiary. The Bering Strait between Alaska and Asia narrowed, creating a nearly landlocked Arctic Ocean. The South Atlantic continued to widen at a rate of over an inch per year. South America and Africa were now separated by more than 1,500 miles of ocean. Africa moved northward, leaving Antarctica still joined to Australia far behind. As the African continent approached southern Europe, it began to pinch off the Tethys Sea.

Meanwhile, India continued to narrow the gap between southern Asia. The crust rifted open on the west side of India. Massive amounts of molten rock poured onto the landmass, blanketing much of west-central India, known as the Deccan Traps. Over a period of several million years, about 100 individual basalt flows produced more than 350,000 cubic miles of lava, totaling up to 8,000 feet thick. Continental rifting at the same time began separating Greenland from Norway and North America. The rifting poured out great

Figure 32 *A dinosaur boneyard in an outcrop of Morrison Formation at the Howe Ranch quarry near Cloverly, Wyoming.* (Photo by G. E. Lewis, courtesy USGS)

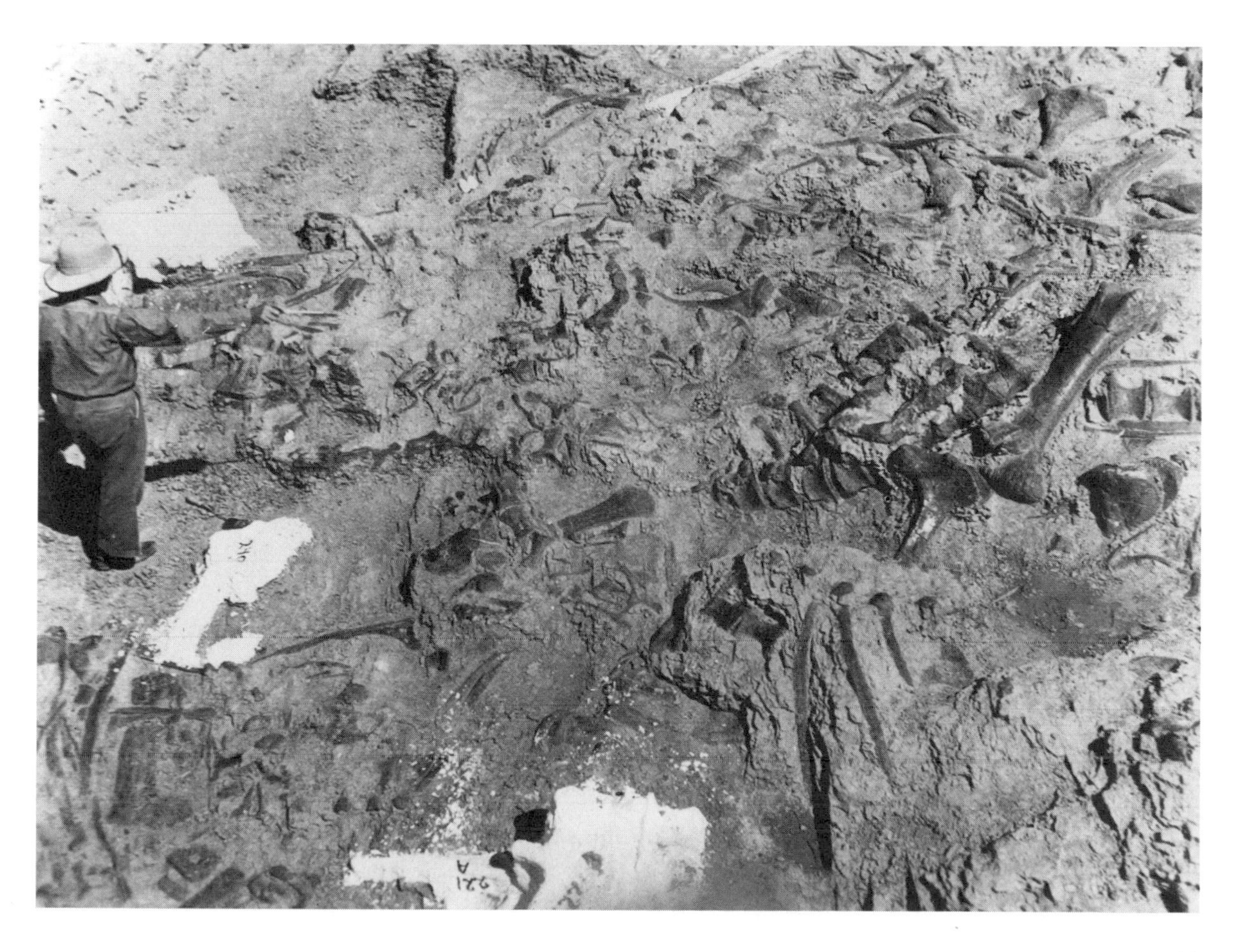

Figure 33 *Sandstone cliffs of the Windgate Formation in the Colorado Monument, Mesa County, Colorado.* (Photo by J. R. Stacy, courtesy USGS)

flood basalts across eastern Greenland, northwestern Britain, northern Ireland, and the Faeroe Islands between Britain and Iceland.

CENOZOIC TECTONICS

The Cenozoic period, from about 65 million years ago to the present, is best known for its intense mountain building. Highly active tectonic forces established much of the terrain features found on Earth today. Volcanic activity was extensive throughout the world, which might explain in part why the Earth grew so warm during the Eocene. A band of volcanoes stretching from Colorado to Nevada produced a series of very violent eruptions between 30 million and 26 million years ago. The tall volcanoes of the Cascade Range from northern California to Canada erupted in great profusion one after another.

TABLE 2 THE DRIFTING OF THE CONTINENTS

	Age (Millions of Years)	Gondwana	Laurasia
Quaternary	3		Opening of Gulf of California
Pliocene	11	Begin spreading near Galápagos Islands	Change of spreading directions in eastern Pacific
		Opening of the Gulf of Aden	
Miocene	26		Birth of Iceland
		Opening of Red Sea	
Oligocene	37		
		Collision of India with Eurasia	Beginning spreading in Arctic Basin
Eocene	54		Separation of Greenland from Norway
		Separation of Australia from Antarctica	
Paleocene	65		
			Opening of the Labrador Sea
		Separation of New Zealand from Antarctica	
			Opening of the Bay of Biscay
		Separation of Africa from Madagascar and South America	
			Major rifting of North America from Eurasia
Cretaceous	135		
		Separation of Africa from India, Australia, New Zealand, and Antarctica	
Jurassic	180		
			Beginning separation of North America from Africa
Triassic	250		

Figure 34 *Columbia River basalt, looking downstream from Palouse Falls, Franklin Whitman counties Washington.*

(Photo by F. O. Jones, courtesy USGS)

Great outpourings of basalt covered Washington, Oregon, and Idaho, creating the Columbia River Plateau (Fig. 34). Massive floods of lava poured onto an area of about 200,000 square miles and in places reached 10,000 feet thick. In a brief moment, volcanic eruptions spewed forth batches of basalt as large as 1,200 cubic miles, forming lava lakes up to 450 miles across. Extensive volcanism also covered the Colorado Plateau and the Sierra Madre region in northwest Mexico.

A collision between the African plate with the Eurasian plate about 50 million years ago squeezed out the Tethys. This resulted in a long chain of mountains and two major inland seas, the ancestral Mediterranean and a composite of the Black, Caspian, and Aral Seas called the Paratethys. This composite sea covered much of eastern Europe. Thick sediments accumulating over millions of years on the bottom of the Tethys buckled into long belts of mountain ranges on the northern and southern flanks. The Alps of northern Italy formed when the Italian prong of the African plate thrust into the European plate.

The collision of India with southern Asia, around 45 million years ago, uplifted the tall Himalaya Mountains and the broad three-mile-high Tibetan Plateau, whose equal has not existed on this planet for more than a billion years. The mountainous spine that runs along the western edge of South America forming the Andes Mountains continued to rise throughout much of the Cenozoic due to the subduction of the Nazca plate beneath the South American plate. The melting of the subducting plate fed magma chambers with molten rock, causing numerous volcanoes to erupt in one fiery outburst after another.

The Rocky Mountains, extending from Mexico to Canada, were pushed up during the Laramide orogeny (mountain building episode) from the late Cretaceous to the Oligocene. A large part of western North America was uplifted, and the entire Rocky Mountain Region was raised about a mile above sea level. Great blocks of granite soared high above the surrounding terrain. To the west in the Basin and Range Province, the crust was pulled apart and down-dropped in some places below sea level.

Around 30 million years ago, the North American continent began to approach the East Pacific Rise spreading center, the counterpart of the Mid-Atlantic Ridge. The first portion of the continent to override the axis of seafloor spreading was the coast of southern California and northwest Mexico. As the rift system and the subduction zone converged, the intervening oceanic plate was consumed in a deep trench. The sediments in the trench were caught in the big squeeze and heaved up to form the Coast Ranges of California. At about the same time, Baja California was ripped from the mainland to form the Gulf of California, providing a new outlet to the sea for the Colorado River as a prelude before cutting the mile-deep Grand Canyon. Similarly, Arabia split off from Africa to form the Red Sea.

In the northwestern United States and British Columbia, the northern part of the East Pacific Rise was consumed in a subduction zone located beneath the continent. As the 50-mile-thick crustal plate was forced down into the mantle, the heat melted parts of the descending plate and the adjacent lithospheric plate, forming pockets of magma. The magma melted through to the surface and formed the volcanoes of the Cascade Range. These

Figure 35 *The May 18, 1980 eruption of Mount St. Helens, Skamania County, Washington.*
(Photo courtesy USGS)

TABLE 3 HISTORY OF THE DEEP CIRCULATION OF THE OCEAN

Age (Millions of Years Ago)	Event
3	An Ice Age overwhelms the Northern Hemisphere.
3–5	Arctic glaciation begins.
15	The Drake Passage is open; the circum-Antarctic current is formed. Major sea ice forms around Antarctica, which is glaciated, making it the first major glaciation of the modern Ice Age. The Antarctic bottom water forms. The snow limit rises.
25	The Drake Passage between South America and Antarctica begins to open.
25–35	A stable situation exists with possible partial circulation around Antarctica. The equatorial circulation is interrupted between the Mediterranean Sea and the Far East.
35–40	The equatorial seaway begins to close. There is a sharp cooling of the surface and of the deep water in the south. The Antarctic glaciers reach the sea with glacial debris in the sea. The seaway between Australia and Antarctica opens. Cooler bottom water flows north and flushes the ocean. The snow limit drops sharply.
> 50	The ocean could flow freely around the world at the equator. Rather uniform climate and warm ocean occur even near the poles. Deep water in the ocean is much warmer than it is today. Only alpine glaciers exist on Antarctica.

erupted one after another in a great profusion, with Mount St. Helens among the most active of them all (Fig. 35).

South America was temporarily connected to Antarctica by a narrow, curved land bridge. Antarctica and Australia then broke away from South America and moved eastward, eventually separating themselves. Antarctica moved toward the South Pole, while Australia continued in a northeasterly direction.

Near the end of the Eocene, global temperatures dropped significantly. Antarctica wandered over the South Pole and acquired a thick blanket of ice that buried most of its terrain features (Fig. 36). Glaciers also grew for the first time in the highest ramparts of the Rocky Mountains. At times, Alaska connected with east Siberia at the Bering Strait, closing off the Arctic Basin from warm ocean currents. This resulted in the accumulation of pack ice. A permanent ice cover did not develop over the North Pole,

Figure 36 *A volcano practically buried in ice is evidence that tectonics were once active on Antarctica.*

(Photo by W. B. Hamilton, courtesy USGS)

however, until about 8 million years ago. At that time, Greenland acquired its first major ice sheet.

Between 3 and 4 million years ago, the Panama Isthmus separating North and South America was uplifted due to colliding oceanic plates. A land bridge formed between the continents. This halted the flow of cold-water currents from the Atlantic into the Pacific. The closing off of the Arctic Ocean from warm Pacific currents, might have initiated the Pleistocene glacial epoch. Never has permanent ice existed at both poles in Earth history. This also suggests that the planet has been steadily cooling down since the Cretaceous, when the dinosaurs held dominion over the world.

After seeing how plate tectonics has operated throughout Earth history, the next chapter will discuss how convection currents in the mantle move the lithospheric plates around.

3

CONVECTION CURRENTS

THE DRIVING FORCE

This chapter defines the force that drives the continents around the surface of the Earth. The collision and separation of continents are thought to be controlled by convection currents in the mantle. Convection is the motion within a fluid medium resulting from a difference in temperature from top to bottom (Fig. 37). Fluid rocks in the mantle receive heat from the core, ascend to the surface, dissipate heat to the lithosphere, cool, and descend to the core again to pick up more heat. The cycling of heat within the mantle is the main driving force behind all tectonic activity and, for that matter, all other activities taking place on Earth.

Rapid mantle convection leads to the breakup of supercontinents. This compresses the ocean basins, causing a rise in sea level and a transgression of the seas onto the land. An increase in volcanism floods the continental crust with vast amounts of lava during the early stages of many rifts. The rise in volcanism also increases the carbon dioxide content of the atmosphere, resulting in a strong greenhouse effect that promotes warm conditions worldwide.

The second phase of the continent cycle is a time of low mantle convection. This results in the assembly of continents into a supercontinent. This

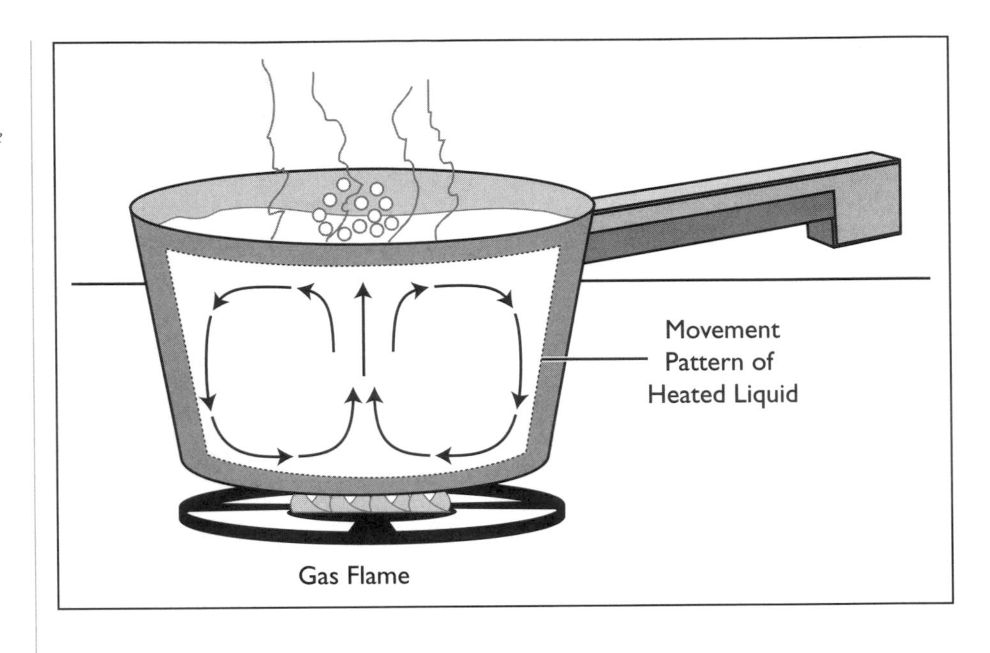

Figure 37 *The movement of water in a pot heated by a gas burner demonstrates the principle of convection.*

process widens the ocean basins, causing a drop in sea level and a regression of the seas from the land. Furthermore, a decrease in volcanism and a reduction of atmospheric carbon dioxide results in the development of an icehouse effect that leads to colder global temperatures, possibly spawning glaciation.

HEAT FLOW

The formation of molten rock and the rise of magma to the surface results from the exchange of heat within the Earth's interior. The Earth is continuously losing heat energy from the interior to the surface through its outer shell, or lithosphere, at a steady rate. About 70 percent of the heat flow results from seafloor spreading (described in Chapter 5). Most of the rest is due to volcanism in the Earth's subduction zones (described in Chapter 6). Volcanic eruptions, however, represent only highly localized and spectacular releases of this energy (Fig. 38).

The total heat loss does little toward heating the surface of the planet. On the surface, the vast majority of the heat comes from the sun and is thousands of times greater than that from the Earth's interior. Heat flow cannot be measured directly. It depends on the temperature gradient and thermal conductivity of the rocks. As a rule, rocks are good thermal insulators, and continental crust is a better insulator than oceanic crust. If this were not so, the interior heat would have long since escaped into space, the Earth's core and mantle would have cooled and solidified, and the continents would have stopped dead in their tracks.

Over the past several years, scientists have been taking the Earth's temperature. Thousands of heat flow measurements have been logged from around the world on land as well as at sea. On the ocean floor, a long, hollow cylinder is plunged into the soft sediments. Inside the buried cylinder, the temper-

Figure 38 *A large eruption cloud from the July 22, 1980 eruption of Mount St. Helens.*

(Photo courtesy USGS)

ature of the sediments is measured at intervals along its vertical length with fixed electrical thermometers. On land, exploratory boreholes are drilled into the crust, and thermometers are placed at various levels. Temperature measurements are also taken at different levels in mines.

Although heat flow patterns on the continents differ from those on the ocean floor, the average heat flow through both is similar. Heat flow decreases with an increase in age. The continental shields and the oldest ocean basins are colder than midocean ridges and young mountain ranges. The average global heat flow is roughly equivalent to the thermal output of a 300-watt lightbulb spread over an area of about the size of a football field. As a comparison, the same amount of energy from the sun would cover roughly 10 square feet.

Heat flow patterns also can supplement seismic data from earthquake waves to determine the thickness of the lithosphere in various parts of the world. Seismic waves resemble sound waves bouncing around inside the Earth that give a sort of X-ray picture of the planet's interior (Fig. 39). Once a geothermal gradient has been established for a region, scientists extrapolate the data to predict at what depth partial melting takes place in the mantle. Both

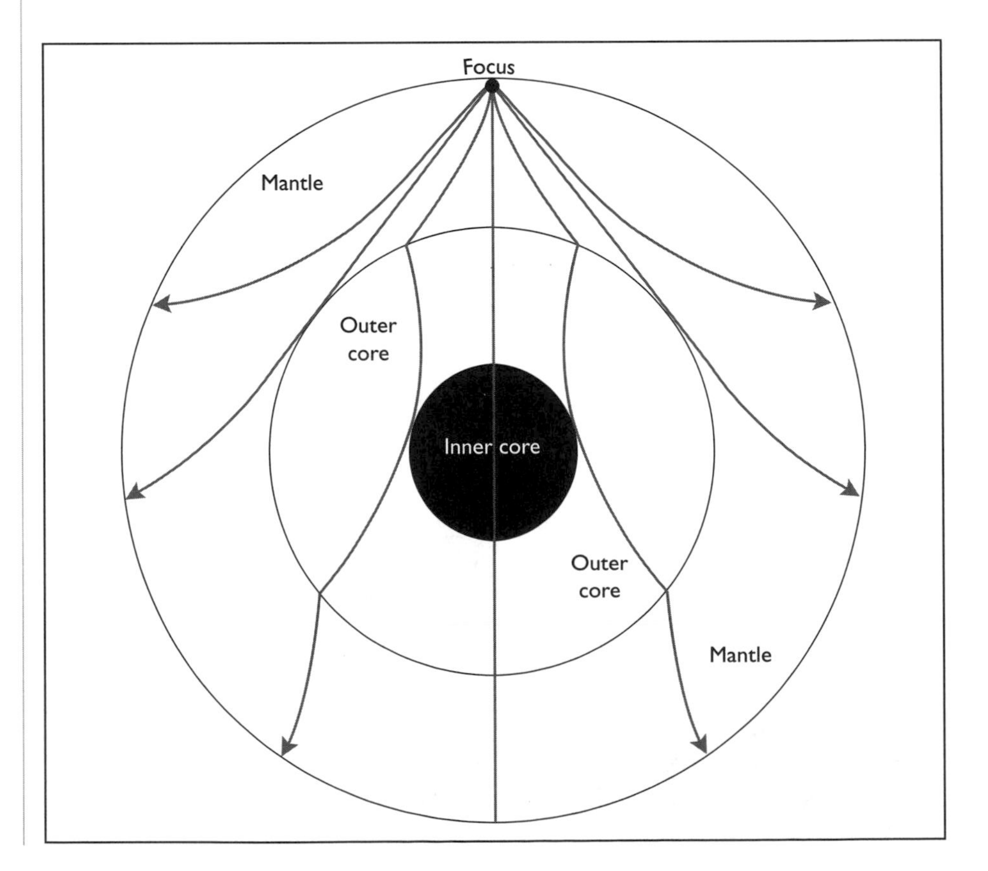

Figure 39 *Earthquake waves traveling through the interior of the Earth determine its structure.*

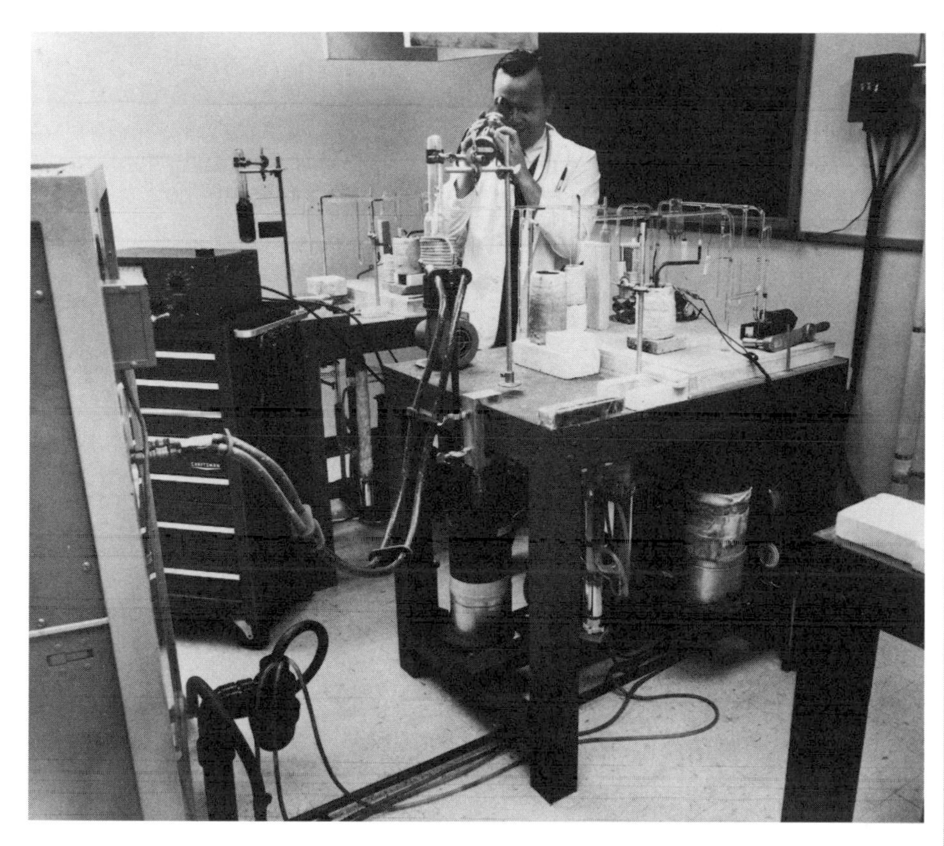

Figure 40 *A scientist dating a sample by the radiocarbon method.*

(Photo courtesy USGS)

seismic data and heat flow patterns show that the oceanic plates thicken as they age. They range from a few miles thick soon after formation at midocean spreading ridges to more than 50 miles thick in the oldest ocean basins, where the heat flow is the lowest.

The plates become thicker as they move away from a midocean spreading ridge. This occurs as material from the asthenosphere, a semimolten layer between the mantle and lithosphere, adheres to the underside of the plates and is transformed into new oceanic lithosphere. The continental plates vary in thickness from 25 miles in the young geologic provinces, where the heat flow is high, to 100 miles or more under the continental shields, where the heat flow is much lower. The shields are so thick they can actually scrape the bottom of the asthenosphere. The drag acts as an anchor to slow the motion of the plate.

Most of the Earth's thermal energy is generated by radioactive isotopes, mainly uranium, thorium, and potassium. The ratio of these elements to their stable daughter products also offers an accurate method for dating rocks (Fig. 40 and Table 4). The concentration of radioactive elements is greater in the

TABLE 4 FREQUENTLY USED RADIOISOTOPES FOR GEOLOGIC DATING

Radioactive Parent	Half-life (Years)	Daughter Product	Rocks and Minerals Commonly Dated
Uranium 238	4.5 billion	Lead 208	Zircon, uraninite, pitchblende
Uranium 235	713 million	Lead 207	Zircon, uraninite, pitchblende
Potassium 40	1.3 billion	Argon 40	Muscovite, biotite, hornblende, glauconite, sanidine, volcanic rock
Rubidium 87	47 billion	Strontium 87	Muscovite, biotite, lepidolite, microcline, glauconite, metamorphic rock
carbon 14	5,730 million	Nitrogen 14	All plant and animal material

crust than in the mantle. This suggests that they segregated early in the Earth's history. Presently, 40 percent of the heat flow at the surface is generated within the crust, which accounts for its high geothermal gradients.

When the Earth began to melt and segregate into its various layers, more than twice as much heat was generated by radioactivity than is produced today. Since the formation of the Earth, half the radioactive isotopes have decayed into stable elements. The increased heat production was probably due to a rise in heat flow on the surface. This also implies that the original lithospheric plates were thinner, smaller, and more numerous than they are today. The asthenosphere underlying the plates was probably much more active in the past. Therefore, the plates might have moved around more vigorously, causing many calamities on the Earth's surface.

THE CORE

The core is a little over half the diameter of the Earth and constitutes about 17 percent its volume and about 34 percent its mass. The inner core, composed of iron-nickel silicates, solidified into a sphere about 1,500 miles in diameter. The iron crystals in the inner core are aligned to give the iron a grainy appearance somewhat like that of wood. The outer core is about 1,400 miles thick and is composed mostly of molten iron that flows as easily as water. The pressure at the top of the outer core is 1.5 million atmospheres (atmospheric pressure at sea level). The pressure increases to 3.5 million atmospheres at the top of the inner core with only a slight increase in pressure toward the center. The temperature of the core ranges from about 4,500 degrees Celsius on the surface to about 7,000 degrees at the boundary between the inner and outer core, with little change in temperature toward the center.

This two-part structure of the core is responsible for generating a strong magnetic field due to differences in rotation rate, temperature, density, and chemistry between the inner and outer cores. The core and mantle are not tightly coupled, so rotating formations of fluid in the core drift westward as the Earth rotates. Furthermore, the solid inner core is rotating eastward faster than the rest of the planet. Since the outer core is a good conductor of electricity, electric currents set up a magnetic field, which is reinforced by the rotation due to the geodynamo effect similar to a giant electrical generator.

With the advent of modern sophisticated computer technology, scientists can take pictures of the Earth's core by using seismic tomography. This is a geologic version of the CAT scan used in medicine in which an image of a slice of a human organ can be constructed from X rays passed through it. Like CAT scans, seismic tomography scans are produced by computers. In this case, though, they combine information from earthquake seismic waves that travel deep within the Earth's interior. By analyzing where the waves change speed as they travel thousands of miles into the planet, scientists can determine where the mantle meets the core.

The surface of the fluid core is not smooth but has a broken topography (Fig. 41) consisting of rises taller than Mount Everest and depressions deeper than the Grand Canyon. This relief is caused by the rising and sinking of the overlying mantle by convection currents. The peaks are created by rising hot currents in the mantle, which draw the core upward along with them. The valleys are formed when cold material in the mantle sinks and presses against the core, creating abysses. Peaks in the core rise beneath eastern Australia, the North Atlantic, Central America, South-Central Asia, and the Northeast Pacific where a mountain rises six miles beneath the Gulf of Alaska. Valleys underlie Europe, Mexico, the southwest Pacific, and the East Indies, which lie above a canyon six miles deep.

Discarded materials might be building up a slag heap on top of the core, creating continents at the core-mantle boundary that might interfere with the geomagnetic field as well as heat flowing from the core to the mantle. This debris might also account for a relief on the surface of the core of several hundred feet. Moreover, these continent-like masses appear to have roots that extend into the core, the way continents on the Earth's surface have roots that extend deep into the mantle. The shielding of heat from the core could also have a major influence on volcanism, especially concerning its role in global tectonics.

THE MANTLE

Scientists know more about the far reaches of outer space than they do about the vast interior of the Earth. Even the deepest drill holes penetrate at most

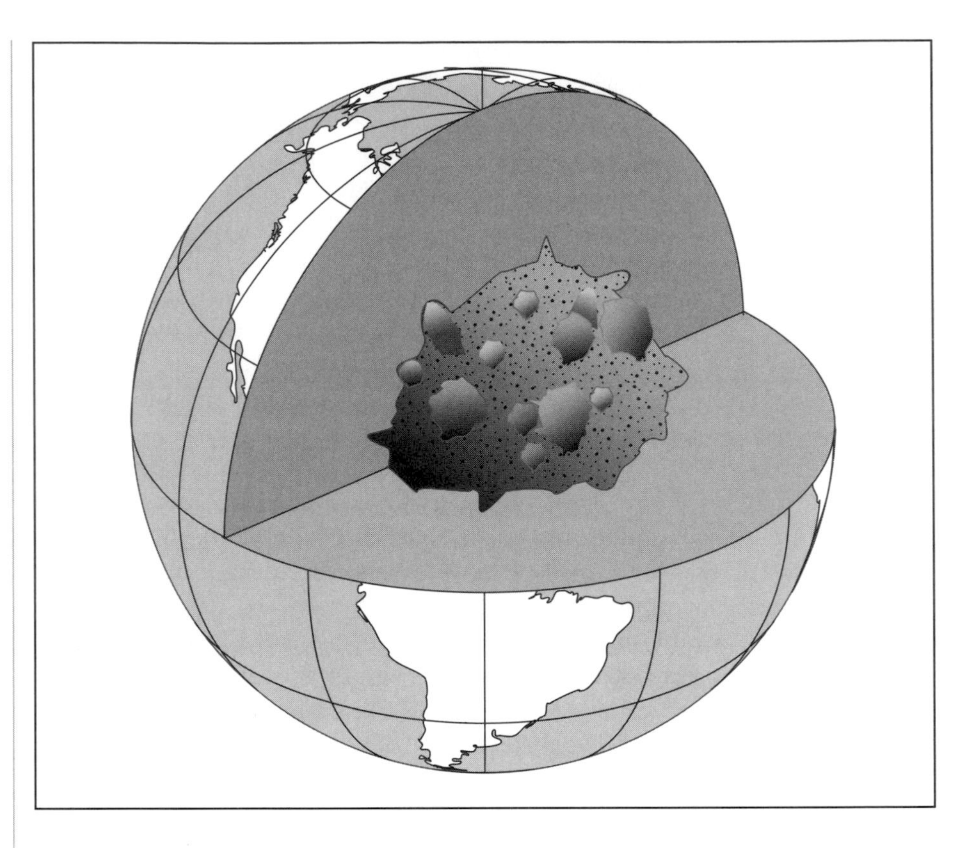

Figure 41 *The core has bumps and grooves created by convection currents in the overlying mantle. The topography shown is highly exaggerated.*

only a few miles into the Earth's thin outer membrane, and none has ever reached the 1,865-mile-thick mantle. The Earth is mostly mantle (Fig. 42), which comprises nearly half its radius, 83 percent its volume, and 67 percent its mass. The lower mantle begins at the top of the core and extends to about 410 miles beneath the Earth's surface. It is composed of primitive rock that has not changed significantly since the early history of the Earth and therefore represents what much of the planet was like in its infancy.

By comparison, the rocks of the upper mantle have been continuously changing composition and crystal structure with time. The two layers have remained separated for billions of years and are the geologic equivalents of oil on water. Over the 4.6-billion-year history of the Earth, the upper mantle has lost much of its volatiles and other important elements, which have congregated in the crust, ocean, and atmosphere. Although the mantle is solid rock, the intense heat causes it to flow slowly. However, exactly how the mantle moves remains one of our planet's great mysteries.

The lower mantle is possibly 30 times more viscous than the upper mantle, which would significantly slow mixing between the two layers or prevent it entirely. This higher viscosity would make the lower mantle much more sluggish compared with the upper mantle. Mantle rocks are composed of iron-magnesium silicates in a partially molten or plastic state, which allows them to flow at a rate of perhaps several inches per year. Large-scale convection currents transport heat away from the core and distribute it along the surface of the mantle. This transfer of heat appears to be responsible for the operation of plate tectonics.

All activity taking place on the surface of the Earth is therefore an outward expression of the great heat engine that drives the mantle. Continents separate, move about, and collide. Mountains rise skyward, and ocean trenches sink downward. Volcanoes erupt, and faults quake (Fig. 43). The mantle played a major role in shaping the Earth and giving it a unique character. Large amounts of gases and water vapor sweated out of the mantle through volcanic eruptions in what geologists term the "big burp," providing the Earth with an atmosphere and ocean. The mantle also supplied the rocks for the crust and the carbon for life.

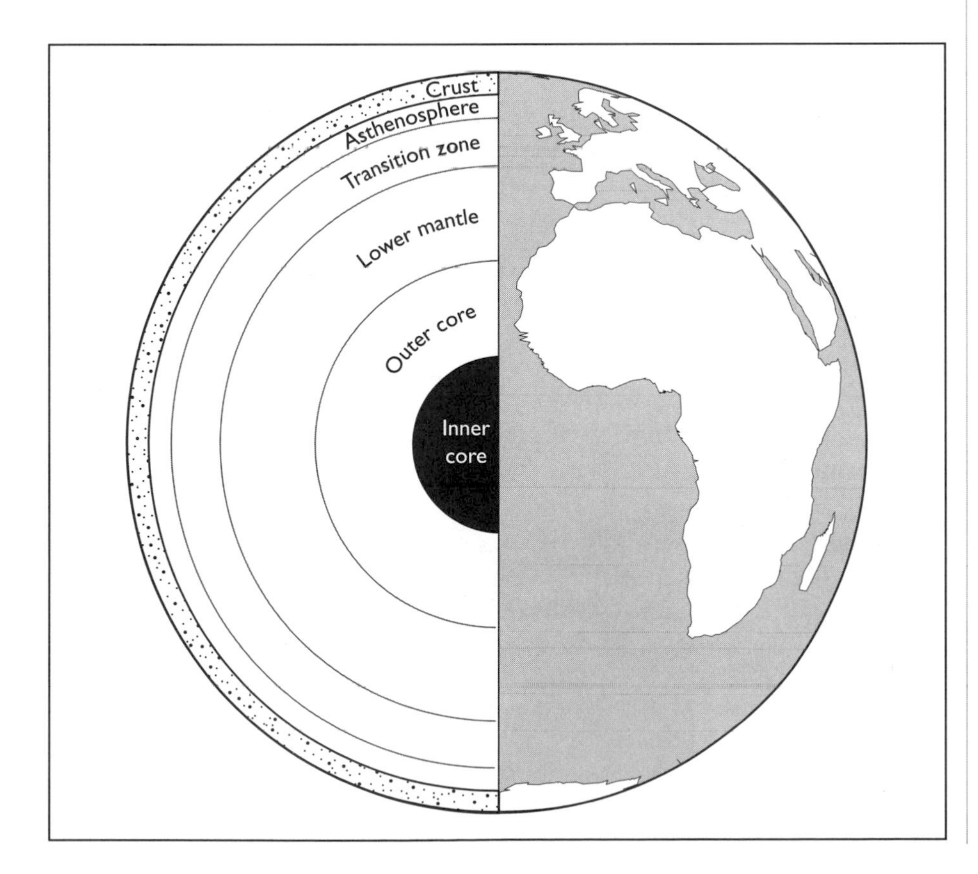

Figure 42 *The structure of the Earth, showing the core, mantle, and crust.*

Figure 43 *Secondary cracks in pavement from the 1906 San Francisco earthquake in California.* (Photo by G. K. Gilbert, courtesy USGS)

The mantle is a thick shell of red-hot rocks, separating the intensely hot metallic core from the cooler crust. It starts at an average depth of about 25 miles below the surface and continues to a depth of about 1,800 miles. The upper mantle has a layered structure composed of a rigid lithosphere, a soft region or asthenosphere on which the lithospheric plates ride, and a lower region that is in a plastic state and might be in convective flow. However, the continents are not buoyed up by the asthenosphere, which offers only passive resistance to sagging. The mantle is composed of iron-magnesium–rich silicate minerals. These minerals correspond to the rock peridotite, whose name comes from the transparent green gemstone peridot.

When starting at the surface, the temperature within the Earth increases rapidly with depth mostly due to the relative abundance of radioactive elements in the Earth's crust. At a depth of about 70 miles, the minerals that make up the major constituents of the upper mantle begin to melt. The temperature is about 1,200 degrees at this semimolten asthenosphere. The temperature in the mantle increases gradually to about 2,000 degrees at a depth of 300 miles. It then increases more rapidly to the top of the core. Between 70 and 80 percent of the mantle's heat is generated internally by radiogenic sources, mostly in the lower mantle, that produce energy by radioactive decay. The rest comes from the core.

The increase in temperature and pressure with depth results in a change in the mineral structure of the mantle rocks, called a phase transition. Phase transitions in the upper mantle occur at depths of about 45, 200, and 410 miles. These depths correspond to the boundaries of the lithosphere, asthenosphere, and upper mantle. The phase boundaries also correspond to the depths where earthquake wave velocities change, indicating that the upper mantle might have a layered structure. In a descending lithospheric plate, the first two phases occur at relatively shallow depths because of the plate's lower temperature. As the plate continues to descend, it changes to denser mineral forms, which help to heat the plate and speed its assimilation into the mantle.

The density of the mantle increases about 60 percent from top to bottom, and its rocks are compressible. By comparison, the crust is little more than an obscure layer of light rocks, making up only about 1 percent of the Earth's radius. The crust also contains a thin film of ocean and atmosphere, which are themselves distillates of the mantle. The driving force that moves the continents around arises from within the mantle, making it one powerful heat engine.

The only window scientists actually have on the mantle is through kimberlite pipes named for the South African town of Kimberley, where they were first found. Kimberlite pipes are cores of ancient, extinct volcanolike structures that extend deep into the upper mantle as much as 150 miles below the surface and have been exposed by erosion. They bring diamonds to the surface from deep below by explosive eruptions. The pipes are mined extensively for these gems throughout Africa and other parts of the world. Diamonds are produced when a pure form of carbon is subjected to extreme temperature and pressure, conditions found only in the mantle. Most economic kimberlite pipes are cylindrical or slightly conical and range from 1,000 to 5,000 feet across.

During the evolution of the Earth, beginning about 4.6 billion years ago, the mantle has had a profound influence on the conditions on the planet's surface. The mantle produced magma that rose to the surface and solidified. These basalt lava flows piled up layer by layer to form the early crust. The surface of the Earth was shaped by the action of the mantle moving very slowly below the crust. Without this movement, erosion would wear down mountains to the level of the sea within a space of a mere 100 million years. The surface of the Earth would then be a vast, featureless plain, unbroken by mountains and valleys. The planet would have no volcanoes, earthquakes, or plate tectonics. The Earth would indeed be a very uninteresting place in which to live—if life could exist under these conditions.

MANTLE CONVECTION

The mantle has now cooled to a semisolid or plastic state, except for a relatively thin layer of partially melted rock between 70 and 150 miles below the surface. This layer is called the low-velocity zone due to an abrupt decrease in the speed of seismic waves from earthquakes traveling through it. This zone is also equivalent to the asthenosphere in the plate tectonics model. The presence of water and carbon dioxide in the upper mantle acts as a catalyst that aids in the partial melting of rocks at lower pressures and makes the asthenosphere flow easily.

Heat transferred from the mantle to the asthenosphere causes convective currents to rise and travel laterally when reaching the underside of the lithosphere. Upon giving up their heat energy to the lithosphere, the currents cool and descend back into the mantle, similar to the way air currents operate in the atmosphere. If any cracks or areas of weakness occur in the lithosphere, the convective currents spread the fissures wider apart to form rift systems. Here the largest proportion of the Earth's interior heat is lost to the surface as magma flows out of the rift zones to form new oceanic crust.

Convection currents transport heat by the motion of mantle material, which in turn drives the plates. The mantle convection currents are believed to originate more than 410 miles below the surface. The deepest known earthquakes are detected at this level. Since almost all large earthquakes are triggered by plate motions, the energy they release must come from the forces that drive the plates. At the plate boundaries where one plate dives under another, the sinking slab meets great resistance to its motion at a depth of about 410 miles, the boundary between the upper and lower mantle, where the slabs tend to pile up.

However, sinking ocean crust has been known to breach this barrier and sink as much as 1,000 miles or more below the surface. Seismic images of mantle downwelling beneath the west coast of the Americas show a slab of subducting Pacific Ocean floor diving down to the very bottom of the mantle. Another slab of ancient ocean floor is sinking under the southern margin of Eurasia and is thought to be the floor of the Tethys, an ancient sea that separated India and Africa from Laurasia. Ocean slabs are sinking into the mantle beneath Japan, eastern Siberia, and the Aleutian Islands as well.

If a slab should sink as far as the bottom of the lower mantle, it might provide the source material for mantle plumes called hot spots. If all oceanic plates were to sink to this level, a volume of rock equal to that of the entire upper mantle would be thrust into the lower mantle every billion years. In order for the two mantle layers to maintain their distinct compositions, one floating on the other, some form of return flow back to the upper mantle would be needed. Hot spot plumes seem to fulfill this function.

The mantle rocks are churning over very slowly in large-scale convection loops (Fig. 44). They travel only a couple of inches a year, about the same as plate movements, providing no slippage occurs at the contact between plate and mantle. The convection currents might take hundreds of millions of years to complete a single loop. Some of these loops can be extremely large in the horizontal dimension and correspond to the dimensions of the associated plate. In the case of the Pacific plate, the loop would have to reach some 6,000 miles across.

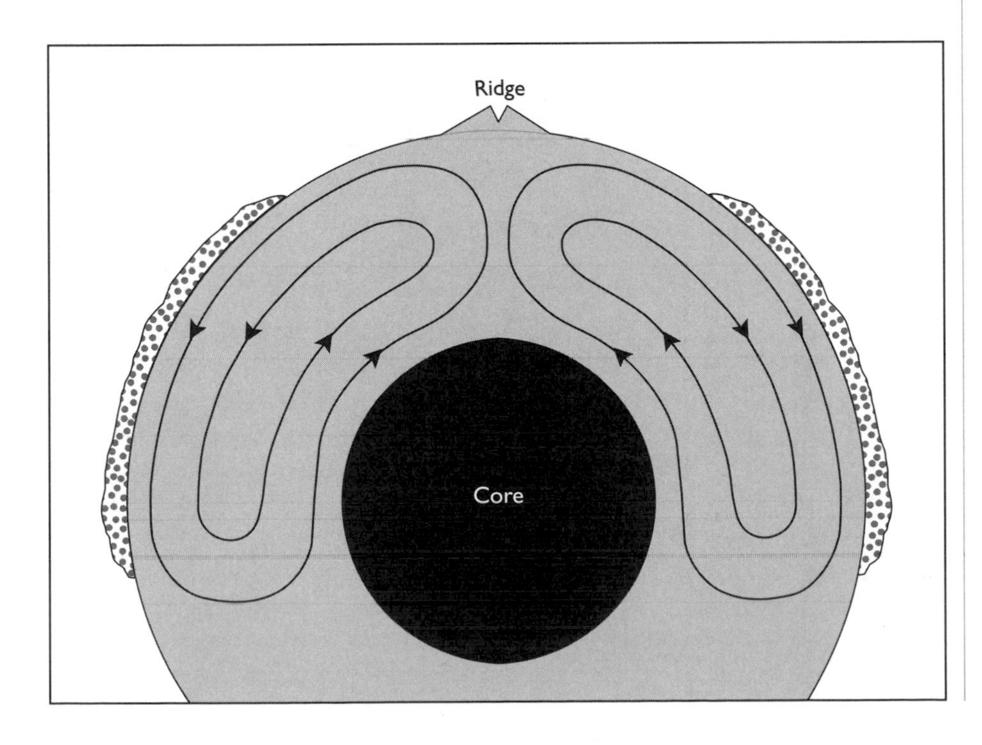

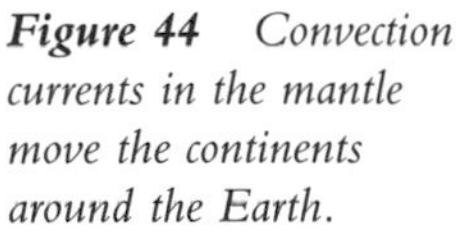
Figure 44 *Convection currents in the mantle move the continents around the Earth.*

Besides these large-scale features, small-scale convection cells might exist. Their horizontal dimensions would be comparable to a depth of about 410 miles, corresponding to the thickness of the upper mantle. These convection cells would act like rollers beneath a conveyor belt to propel the plates forward. Hot material rises from within the mantle and circulates horizontally near the Earth's surface. The top 30 miles or so cools to form the rigid plates, which carry the crust around. The plates complete the mantle convection by plunging back into the Earth's interior. Thus, they are merely surface expressions of mantle convection.

The convection cells might also be responsible for rising jets of lava that create chains of volcanoes, such as the Hawaiian Islands (Fig. 45). A strong mantle current possibly runs beneath the islands and disrupts the plume of ascending hot rock. Instead of rising vertically, the plume is

Figure 45 *Photograph of the Hawaiian Island chain looking south, taken from the space shuttle. The main island Hawaii is in the upper portion of the photograph.*
(Photo courtesy NASA)

sheared into discrete blobs of molten rock that climb like balloons in the wind. Each small plume created a line of volcanoes pointing in the direction of the movement of the underlying mantle. This might explain why the Hawaiian volcanoes do not line up exactly and why they erupt dissimilar lavas.

Another model suggests that the driving mechanism involves some 20 thermal plumes in the mantle, each several hundred miles wide. All upward movement of mantle material would then originate at the top of the core. As the material cools near the surface, it descends back to the core, where it is reheated. However, the mantle might not be viscous or fluid enough to provide convection. The heat might simply travel directly by conduction from the hotter lower level to the colder upper level.

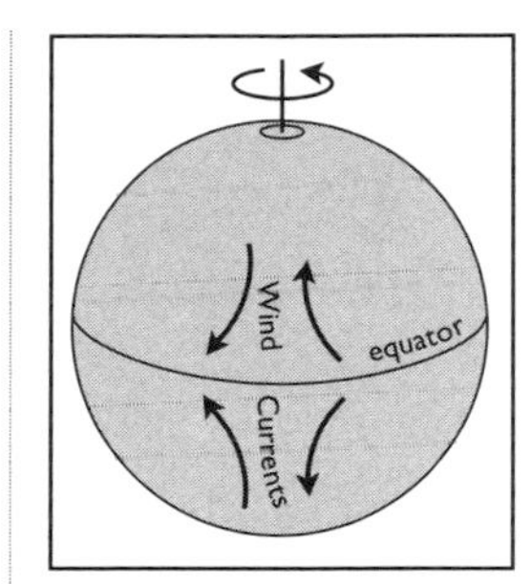

Figure 46 *The Coriolis effect causes equatorward currents to flow to the west and poleward currents to flow to the east due to the planet's rotation.*

The Earth's rotation is expected to influence convection in the mantle. This is similar to its influence on air and ocean currents by the Coriolis effect, which bends poleward-flowing currents to the east and equatorward-flowing currents to the west (Fig. 46). Yet the rotation does not seem to affect the mantle. Even if convective flow occurred, it might not exist in neat circular cells but instead create eddy currents. If so, the flow would thus become turbulent and extremely complex. Furthermore, the mantle is not heated only from below but, like the crust, it is also heated from within by radioactive decay. This further complicates the development of convection cells and causes distortion. The interior of the cells would no longer be passive but provide a significant portion of the heat as well.

Laboratory experiments and computer modeling have shown that under certain conditions, small- and large-scale convections can occur in a highly viscous fluid heated from below as well as from within. For stable convection cells, however, the width generally has to be equal to the depth. To develop continental-size cells under these conditions, the depth would have to extend to the top of the core. If the flow extends throughout the mantle, it must be almost entirely driven by internal heating, which could greatly distort the cell and make it unstable.

Since slabs of sinking material meet great resistance at a depth of about 410 miles, convection would have to be limited to the region above this depth. Yet large-scale convections do account for the geophysical observations, especially the motion of the surface plates. Large-scale motions might be superimposed onto small-scale convective features similar to the circulation of the atmosphere and oceans. To test this hypothesis, geophysical evidence must prove that small-scale flow does indeed exist.

Because the plates are so rigid, they tend to mask most of the effects that might be associated with small-scale convection. Two important effects that can be observed, however, are the variation in gravity and the greater-than-expected heat flows on the ocean floor. Variances in gravity are usually due to differences in density of the mantle and associated deformations on the surface, such as variations in the depth of the ocean. This results from the cooling and shrinking of a plate as it moves away from a midocean spreading ridge, where the growth of the plate begins with upwelling of hot material from the mantle. Usually, a large magma chamber lies under a ridge where the lithosphere is created at a fast rate, and a narrow magma chamber lies under a slow spreading center.

Furthermore, because all plates cool in much the same manner, the depth of the ocean should be a function of age and can be readily calculated. Regional departures from the expected depth correspond remarkably closely to the gravity variations. Scientists have found many good reasons to believe the variations in gravity and depth are associated with convection currents operating at the base of the plate. Chains of volcanoes, such as the Hawaiian Islands, do not move away from their sources of volcanism as fast as the plates on which they ride. Therefore, the sources must be in motion, indicating that convection currents are distorting the flow of upwelling mantle plumes that contain molten mantle material originating from sources as deep as just above the core.

THE RIFTING OF CONTINENTS

The continental crust and underlying lithosphere are generally between 50 and 100 miles thick. Therefore, cracking open the continental lithosphere would appear to be a formidable task, considering its great thickness. During the rifting of continents into separate plates, thick lithosphere must somehow transform into thin lithosphere. The transition from a continental rift to an oceanic rift is accompanied by block faulting. Huge blocks of continental crust drop down along extensional faults, where the crust is diverging.

As rocks heat up in the asthenosphere, they become plastic, slowly rise by convection, and after millions of years, they reach the topmost layer of the mantle, or lithosphere. When the rising rocks reach the underside of the lithosphere, they spread out laterally, cool, and descend back into the deep interior of the Earth. The constant pressure against the bottom of the lithosphere creates fractures that weaken it. As the convection currents flow out on either side of the fracture, they carry the two separated parts of the lithosphere along with them, and the rift continues to widen.

While the rift proceeds across the continent, large earthquakes strike the region. Volcanic eruptions are also prevalent due to the abundance of molten magma rising from the mantle as it nears the surface. Since the crust beneath a rift is only a fraction of its original thickness, magma finds an easy way out. As the crust continues to thin, magma approaches the surface, causing extensive volcanism. The rifting of continents begins with hot-spot volcanism at rift valleys. This explains the marked increase in volcanic activity during the early stages of continental rifting, producing vast quantities of lava that flood the landscape.

Continental breakup begins with hot spots erupting at triple junctions. The hot spots, which are mantle plumes originating from deep within the mantle, weaken the crust by burning holes through it like geologic blowtorches. When a continent comes to rest over a mantle plume, hot spots in the mantle make the continent dome upward and rupture into a three-branched pattern of rifts, similar to the crust of a pie baking in an oven. The three rifts might then develop into two active rifts and an inactive one (Fig. 47). The hot spots are connected by rifts along which the continent eventually breaks up. More than 2 million cubic miles of molten lava erupt when a continent rifts apart over a hot mantle plume, producing enough basalt to cover the entire United States to a depth of half a mile.

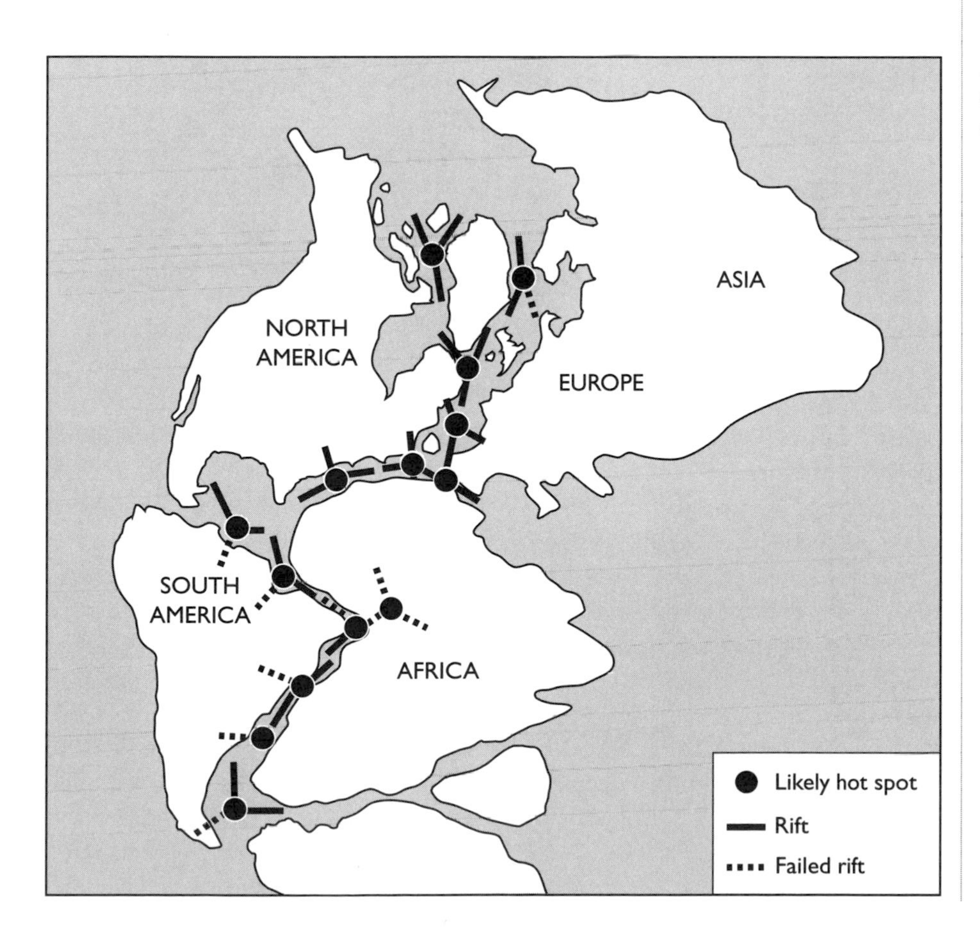

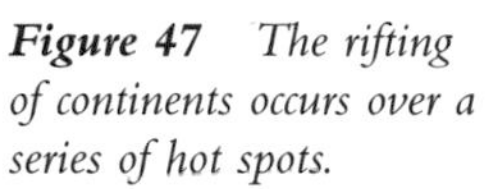
Figure 47 *The rifting of continents occurs over a series of hot spots.*

Figure 48 *The Manzano Mountains in the background border the eastern edge of the Rio Grand Rift, Bernalillo County, New Mexico.*

(Photo courtesy USGS Earthquake Information Bulletin)

Mantle material rises in giant plumes and underplates the crust with basaltic magma, further weakening it and causing huge blocks to drop down to form a series of grabens, from German meaning "ditch." Grabens are long, trenchlike structures formed by the down-dropping of large blocks of crust bounded by normal or gravity, faults. Some grabens are expressed in the surface topography as linear structural depressions considerably longer than wide. The highland areas flanking the grabens often consist of horsts, from German meaning "ridge." They are long, ridgelike structures produced when large blocks of crust bounded by reverse, or thrust, faults are upraised with little or no tilting (see Fig. 4). Horsts and grabens combine to form long parallel mountain ranges and deep valleys such as the Great East African Rift, the Rhine Valley in Germany, the Dead Sea Valley in Israel, the Baikal Rift in southern Russia, and the Rio Grande Rift in the American Southwest (Fig. 48), which slices northward through central New Mexico into Colorado.

The Basin and Range Province of North America is a 600-mile-wide region of fault-block mountain ranges bounded by high-angle normal faults. The province covers southern Oregon, Nevada, western Utah, southeastern California, the southern portions of Arizona and New Mexico, and northern Mexico. The crust has been broken, tilted, and upraised nearly a mile above the basin, radically transforming a land of once gently curving hills into nearly parallel ranges of high mountains and deep basins. The mountains formed at the tops of the fault blocks, while V-shaped valleys formed at the bases. The valley floors flattened as they filled with sediments washed down from the mountains. The stretching and faulting gradually moved westward, finally peaking in Death Valley, California about 3 million years ago.

The Great Basin centered over Nevada and Utah is a 300-mile-wide closed depression created by the stretching and thinning of the crust. The crust in the entire Basin and Range Province is actively spreading apart by forces originating in the mantle. Most of the deformation results from extension along a line running approximately northwest to southeast. The tectonic changes ushered in volcanism in places where the crust was weakening, allowing magma to well up to the surface. As the crust continues to separate, some blocks sink, forming grabens separated by horsts. About 20 horst-and-graben structures extend from the 2-mile-high scarp of the Sierra Nevada in California to the Wasatch Front, a major fault system running roughly north-south through Salt Lake City, Utah. The horst-and-graben structures trend northwestward, roughly perpendicular to the movement of the blocks of crust that are spreading apart.

Basins lying between the ranges contain dry lake bed deposits. This indicates the low-lying areas once contained lakes. Utah's Great Salt Lake and the Bonneville Salt Flats are good examples. The region is literally stretching apart as the crust is weakened by a series of down-dropped blocks. Consequently, 15 million years ago, the present sites of Reno, Nevada and Salt Lake City, Utah were 200 to 300 miles closer together, and have since been spreading apart about an inch per year.

At 280 feet below sea level, Death Valley (Fig. 49) is the lowest point on the North American continent. It was originally elevated several thousand feet but collapsed when the continental crust thinned from extensive block faulting in the region. The Great Basin is now only a remnant of a broad belt of mountains and high plateaus down-dropped after the crust pulled apart during the growth of the Rocky Mountains. The Andes Mountains could suffer a similar fate as the plate upon which they stand thins out and collapses. Without buoyant support from below, the mountains will cease rising, only to be torn down by erosion.

Figure 49 *A view southwest across Death Valley, Inyo County, California.*

(Photo by W. B. Hamilton, courtesy USGS)

The rift valleys in Africa are literally tearing the continent apart. They are a complex system of parallel horsts, grabens, and tilted fault blocks with a net slip on the border faults as much as 8,000 feet. The eastern rift zone lies east of Lake Victoria and extends 3,000 miles from Mozambique to the Red Sea. The western rift zone lies west of Lake Victoria and extends 1,000 miles northward. The rift just north of Lake Victoria holds Lake Tanganyika, the second deepest lake in the world. Russia's Lake Baikal, at about 5,700 feet deep and 375 miles long, is the record holder. It fills the Baikal rift zone, a crack in the crust similar to the East African Rift.

The East African Rift Valley (Fig. 50) marks the boundary between the Nubian plate to the west and the Somalian plate to the east. Much of the East African Rift system has been uplifted thousands of feet by an expanding mass of molten magma lying just beneath the crust. It has not yet fully ruptured and therefore provides some of the best evidence for the rifting of continents. The region extends from the shores of Mozambique to the Red Sea, where it splits to form the Afar Triangle in Ethiopia. Afar is considered a classic example of a triple junction caused by a hot spot. The rift is a complex system of tensional faults, indicating the continent is in the initial stages of rupture.

Once the region finally breaks up, the continental rift will be replaced by an oceanic rift when the area is flooded with seawater. The Gulf of California is an example of a rift propagating through a continent and breaking it apart. As the rift spreads farther apart, the ocean floods in. Eventually, a new sea forms. As the rift continues to widen and deepen, it is replaced by a spreading ridge system. Hot material from the mantle wells up through the rift to form new oceanic crust between the two separated segments of continental crust.

This transition is presently occurring in the Red Sea, which is rifting from south to north (Fig. 51). The Gulf of Aden is a young oceanic rift between the fractured continental blocks of Arabia and Africa, which have been diverging for more than 10 million years. The breakup of North America, Eurasia, and Africa beginning about 170 million years ago was probably initiated by a similar type of upwelling of basaltic magma that is presently taking place under the Red Sea and the East African Rifts.

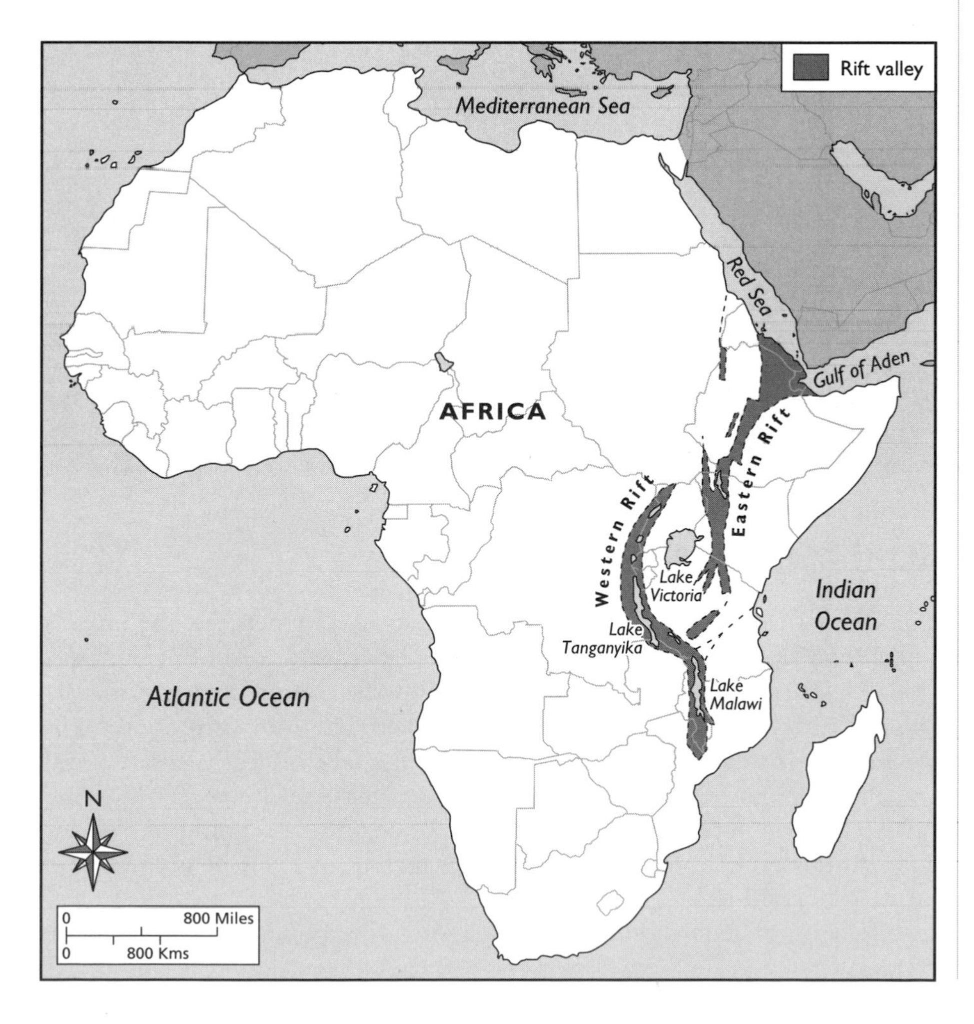

Figure 50 The rifting of the African continent is occurring in the Red Sea, the Gulf of Aden, the Ethiopian rift valley, and the East African Rift.

Figure 51 *The Red Sea and the Gulf of Aden are prototype seas created by seafloor spreading.*

(Photo courtesy USGS Earthquake Information Bulletin)

SUPERCONTINENTS

Throughout the Earth's history, the continents have undergone the process of collision and rifting on a grand scale. Separate blocks of continental crust collide and merge into larger continents. Later, the continents are torn apart by deep rifts that eventually become new oceans. The convective motions in the mantle that drive the continents around the face of the Earth are powered by heat generated by the decay of radioactive elements. The Earth's internally generated radiogenic heat is steadily declining, however. Therefore, it cannot be responsible for the alternating cycle of assembly and breakup of continents.

The key phenomenon in this process is not the production of heat but the conduction and loss of heat through the crust. Continental crust is only half as efficient as oceanic crust at conducting heat and acts as an insulating thermal cap.

A supercontinent covering a portion of the Earth's surface, where heat from the mantle can accumulate under the crust, causes it to dome upward, creating a superswell. New continents eventually rift apart, slide off the superswell, and move toward colder sinking regions in the mantle. Individual continents then become trapped over separate, cool downflows of mantle rock.

After the continents have been widely dispersed, heat is more easily conducted through the newly formed ocean basins. When a certain amount of heat has escaped, the continents halt their progress and start to move toward each other. When the present continents have reached their maximum dispersal, the crust of the Atlantic will age and become dense enough to sink under the surrounding landmasses, thus beginning the process of closing the Atlantic Basin. As the convection patterns shift, all continents will rejoin into a supercontinent over a large downflow, and the cycle begins anew.

The rifting of a continent caused by hot-spot volcanism at rift valleys takes place on average about once every 30 million years. The rising of mantle material in giant plumes underplates the crust with basaltic magma. The magma heats and weakens the crust, causing huge blocks to down-drop and form a series of grabens, or down-faulted blocks. Convection currents in the mantle pull the thinning crust apart. This forms a deep rift valley similar to the East African Rift system. As the crust thins, volcanic eruptions become more prevalent, further weakening the crust.

The rift then fills with seawater after a breach has opened to the ocean. Eventually, the continental rift is replaced by a midocean ridge spreading system. The creation of new ocean floor between the two pieces of crust pushes them farther apart. In this manner, more than 2 million cubic miles of molten rock are released every time a supercontinent rifts apart near a hot plume. After the breakup, the continents travel in spurts rather than drift at a constant speed.

The regions bordering the Pacific Ocean have apparently not participated in continental collisions. In effect, the Pacific is a remnant of an ancient sea that has narrowed and widened in response to the drifting of the continents. Oceans have repeatedly opened and closed in the vicinity of the North Atlantic. In contrast, a single ocean has existed continuously in the vicinity of the Pacific. The Pacific plate is being subducted under all the continents that surround it. The oceanic crust in the Atlantic, however, butts against the surrounding continents.

As the material on the ocean floor ages, it cools, becomes denser, and subsides, increasing the depth of the ocean. Eventually, about 200 million years after the first rift formed, the oldest part of the new ocean floor adjacent to the continents becomes so dense it sinks under the continental crust and is subducted into the mantle. The process of subduction closes the ocean, bringing the continents back together.

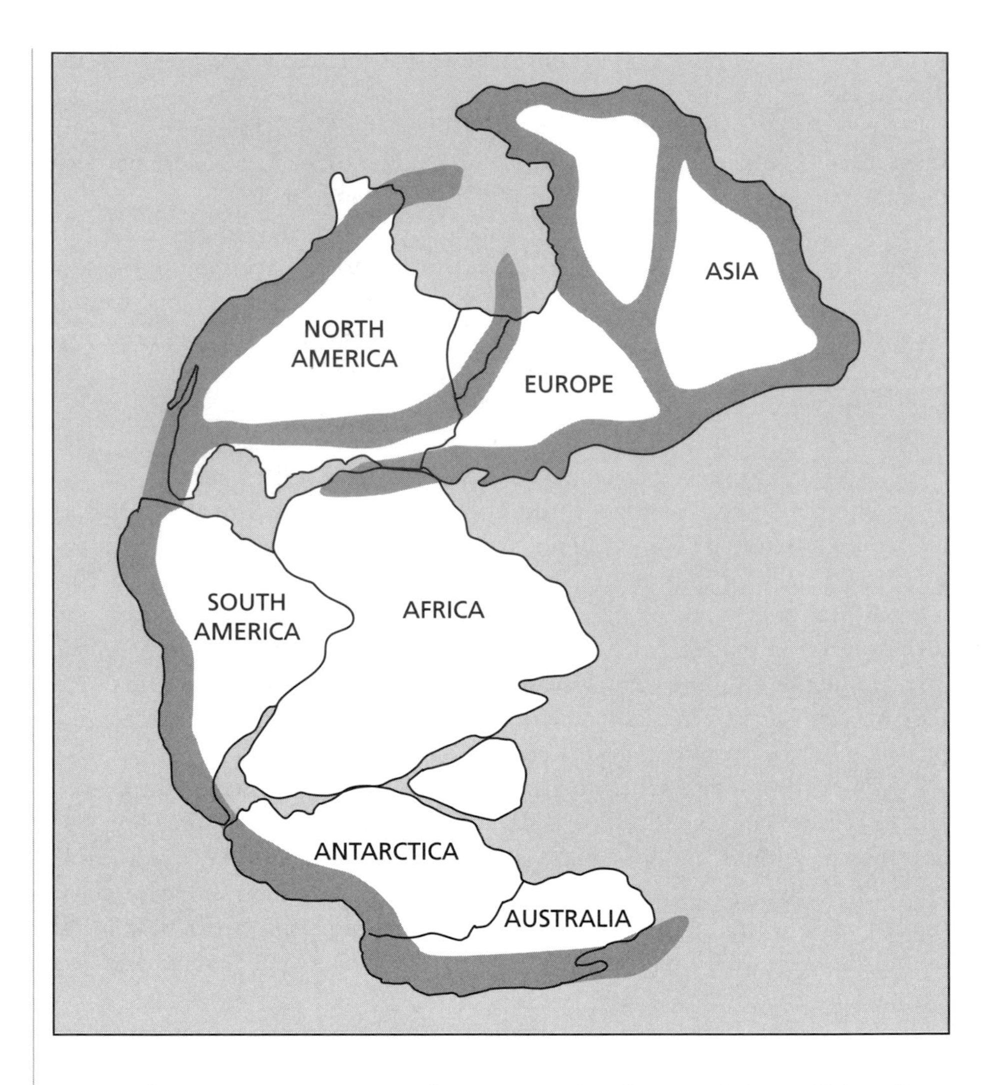

Figure 52 *Pangaea showing mountain belts formed by continental collisions.*

When continents rejoin, the compressive forces of the collision form mountain belts (Fig. 52). The ages of mountain ranges produced by continental collisions are remarkable in their regularity. Intense mountain building occurred about 2.6 billion years ago, 2.1 billion years ago, 1.7 billion years ago, 1.1 billion years ago, 650 million years ago, and 250 million years ago. The timing of these episodes shows an apparent periodicity of between 400 million and 500 million years.

About 100 million years after each of these events, a period of rifting recurred about 2.5 billion years ago, 2 billion years ago, 1.6 billion years ago, 1 billion years ago, 600 million years ago, and 150 million years ago. Continents separate some 40 million years after rifting. They take about 160 million years to reach their greatest dispersal and for subduction to begin in the new oceans. After the continents begin to move back toward each other, they might take another 160 million years before reforming a new supercontinent. The supercontinent might survive for about 80 million years before it rifts apart, changing again the face of the Earth.

After discussing the driving force that moves the continents around, the next chapter will examine what happens with these lithospheric plates in motion.

4

CRUSTAL PLATES

THE LITHOSPHERE IN MOTION

This chapter examines the mobile crustal plates, their interactions, and their effects on the planet. Not even the largest and most prominent features on the surface can be regarded as permanent and immovable. The Earth's outer shell comprises rigid lithospheric plates. Because they are constantly in motion, continents and oceans are continuously being reshaped and rearranged.

Plate tectonics and continental drift have been operating since the Earth's early history. They have played a prominent role in climate and life. Changes in continental configurations brought on by movable lithospheric plates have had a profound effect on global temperatures, ocean currents, biologic productivity, and many other factors of fundamental importance to the living Earth.

CONTINENTAL CRUST

The crust comprises less than 1 percent of the Earth's radius and about 0.3 percent of its mass. It is composed of ancient continental rocks and comparatively young oceanic rocks. The continental crust resembles a layer cake with sedimentary rocks on top, granitic and metamorphic rocks in the middle, and basaltic rocks on the bottom. This gives it a structure somewhat like a jelly sandwich, with a pliable middle layer placed between a solid upper crust and a hard lithosphere. Most of the continental rock originated when volcanoes stretching across the ocean were drawn together by plate tectonics.

By including continental margins and small shallow regions in the ocean, the continental crust covers about 45 percent of the Earth's surface. It varies from 6 to 45 miles thick and rises on average about 4,000 feet above sea level. The average density of continental crust is 2.7 times the density of water, compared with 3.0 for oceanic crust and 3.4 for the mantle, which buoys up the continental and oceanic crust. The thinnest parts of the continental crust lie below sea level on continental margins, and the thickest portions underlie mountain ranges.

The bulk of the Earth's crust is composed of oxygen, silica, and aluminum (Table 5). These form the granitic and metamorphic rocks that constitute most of the continents. The crust and the upper brittle mantle comprise the lithosphere, which averages about 60 miles thick. The lithosphere

TABLE 5 COMPOSITION OF THE EARTH'S CRUST

Crust Type	Shell	Average Thickness (Miles)	Percent Composition of Oxides						
			Silica	Alum	Iron	Magn	Calc	Sodi	Potas
Continental	Sedimentary	2.1	50	13	6	3	12	2	2
	Granitic	12.5	64	15	5	2	4	3	3
	Basaltic	12.5	58	16	8	4	6	3	3
Total		27.1							
Subcontinental	Sedimentary	1.8							
	Granitic	5.6			Same as above				
	Basaltic	7.3							
Total		14.7							
Oceanic	Sedimentary	0.3	41	11	6	3	17	1	2
	Volcanic sedimentary	0.7	46	14	7	5	14	2	1
	Basaltic	3.5	50	17	8	7	12	3	< 1
Total		4.5							
Average		15.4	52	14	7	4	11	2	2

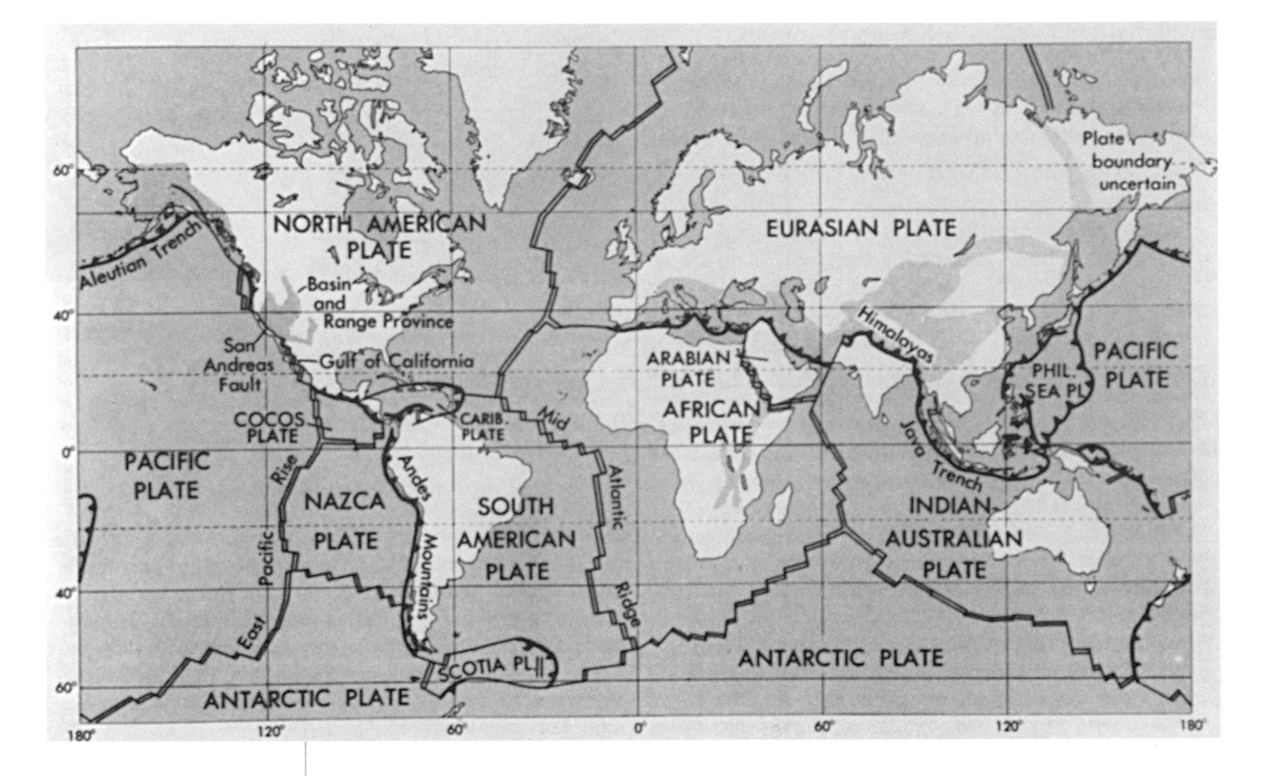

Figure 53 *The Earth's crust is fashioned out of several lithospheric plates that are responsible for the planet's active geology.* (Photo courtesy USGS)

rides freely on the semimolten outer layer of the mantle, the asthenosphere, which lies between about 70 and 150 miles deep. This feature is necessary for the operation of plate tectonics. Otherwise, the crust would be just jumbled up slabs of rock, and the Earth would be an alien place indeed.

The Earth's crust is relatively thin compared with that of its moon and the other terrestrial planets, Mercury, Venus, and Mars. They have thick, buoyant, nonsubducting crusts because they are either too cold or too hot and have been tectonically dead for more than 2 billion years. A thick, buoyant crust could not be easily broken up and subducted into the mantle, which is important for the operation of global tectonics. The lithospheric plates would simply float on the surface like pack ice in the arctic.

The continental crust averages 25 to 30 miles thick and is as much as 45 miles thick in the mountainous regions. The continents also have thick roots of relatively cold mantle material extending down to a depth of about 250 miles. The oceanic crust, by comparison, is considerably thinner. In most places, it is only 3 to 5 miles thick. Like an iceberg, only the tip of the continental crust shows. The rest is out of sight deep below the surface. The continental crust is 20 times older than the oceanic crust. This is because the older oceanic crust has been consumed by the mantle at subduction zones spread around the world (see Chapter 6). Because of plate tectonics, perhaps as many as 20 oceans have come and gone during the last 2 billion years.

Eight major and about a half dozen minor lithospheric plates (Fig. 53) act as rafts that carry the crust around on a sea of molten rock. The plates diverge at midocean ridges and converge at subduction zones. These are expressed on the ocean floor as deep-sea trenches, where the plates are subducted into the mantle and remelted. The plates and oceanic crust are continuously recycled through the mantle. However, because of its greater buoyancy, the continental crust usually remains on the surface.

An interesting feature about the Earth's crust that geologists found quite by accident was that Scandinavia and parts of Canada are slowly rising nearly half an inch a year. Over the centuries, mooring rings on harbor walls in Baltic seaports have risen so far above sea level they could no longer be used to tie up ships. During the last ice age, which ended about 12,000 years ago, the northern landmasses were covered with ice sheets up to two miles thick. Under the weight of the ice, North America and Scandinavia began to sink like an overloaded ship. When the ice melted, the crust became lighter and began to rise (Fig. 54).

Figure 54 *The principle of isostasy. Land covered with ice readjusts to the added weight like a loaded freighter. When the ice melts, the land is buoyed upward as the weight lessens.*

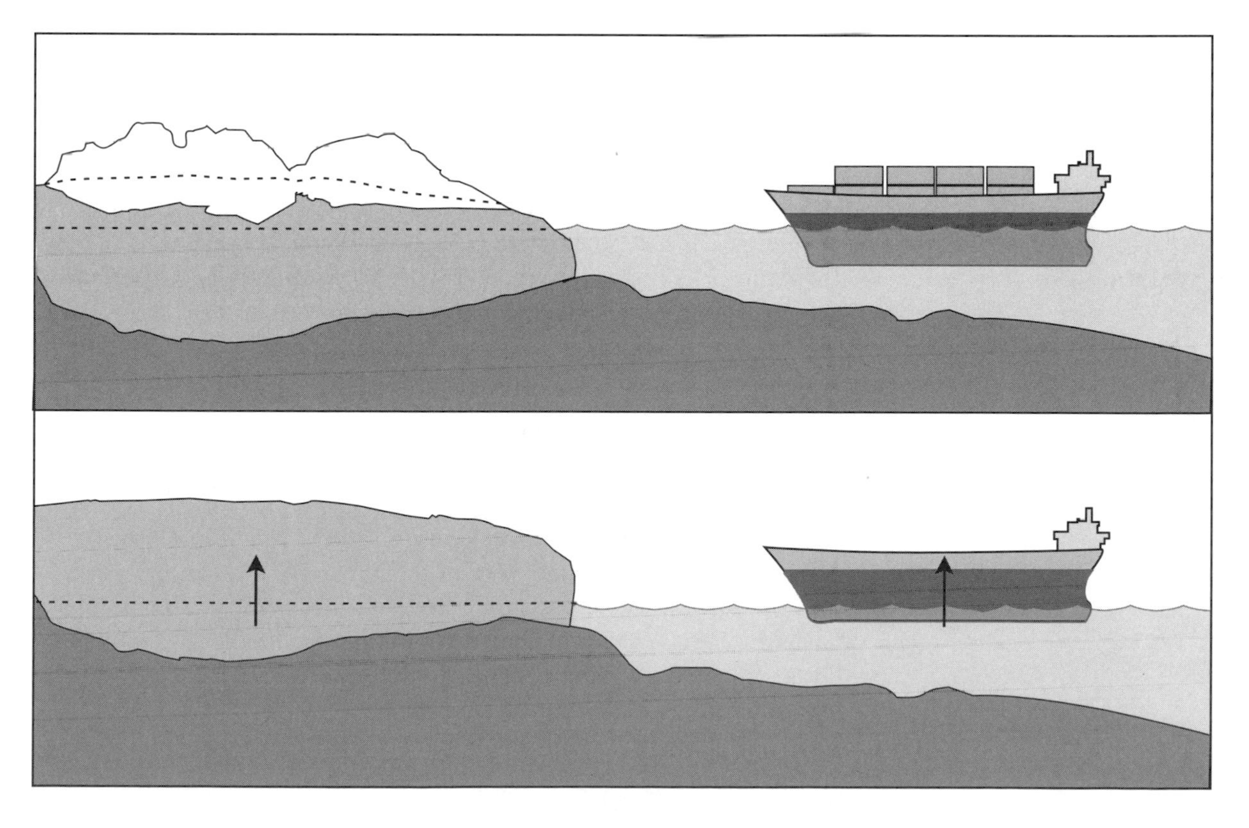

TABLE 6 CLASSIFICATION OF THE EARTH'S CRUST

Environment	Crust Type	Tectonic Character	Thickness (Miles)	Geologic Features
Continental crust overlying stable mantle	Shield	Very stable	22	Little or no sediment, exposed Precambrian rocks
	Midcontinent	Stable	24	
	Basin & Range	Very unstable	20	Recent normal faulting, volcanism, and intrusion; high mean elevation
Continental crust overlying unstable mantle	Alpine	Very unstable	34	Rapid recent uplift, relatively recent intrusion; high mean elevation
	Island arc	Very unstable	20	High volcanism, intense folding and faulting
Oceanic crust overlying stable mantle	Ocean basin	Very stable	7	Very thin sediments overlying basalts, no thick Paleozoic sediments
Oceanic crust overlying unstable mantle	Ocean ridge	Unstable	6	Active basaltic volcanism, little or no sediment

In Scandinavia, marine fossil beds have risen more than 1,000 feet above sea level since the last ice age. The weight of the ice sheets depressed the landmass when the marine deposits were being laid down. When the ice sheets melted, the removal of the weight raised the landmass due to its greater buoyancy. This effect is responsible for maintaining equilibrium in the Earth's crust. Therefore, the lighter continents acted as though they floated on a sea of heavier rocks.

OCEANIC CRUST

The oceanic crust is remarkable for its consistent thickness and temperature, averaging about four miles thick and not varying more than 20 degrees Celsius over most of the globe. Oceanic crust does not form as a single homogeneous mass, however. Instead, it is made in long, narrow ribbons laid side by side with fracture zones in between. The oceanic crust is comparable to a layer cake with four distinct strata (Fig. 55). First, an upper layer of pillow basalts formed when lava extruded undersea at great depths. Next, a second layer of a sheeted-dike complex consists of a tangled mass of feeders that brought

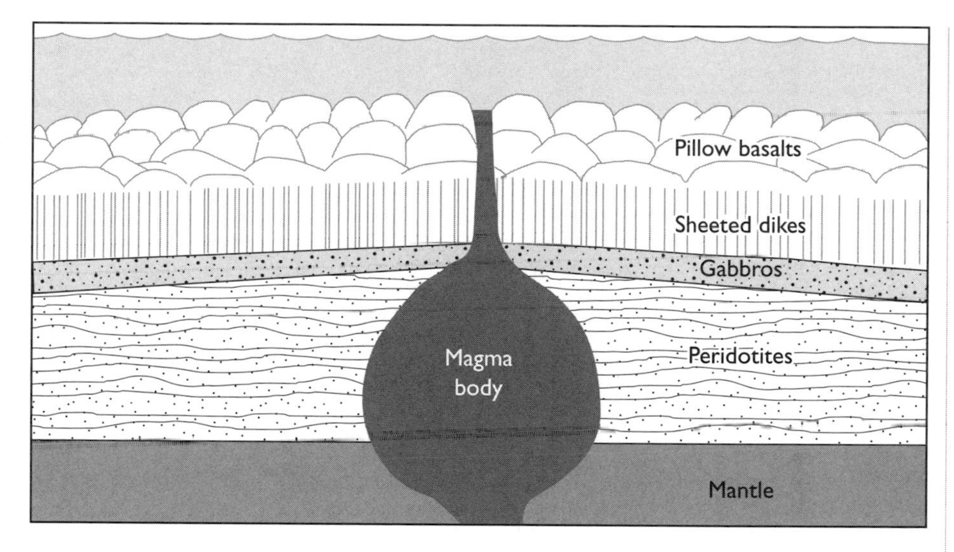

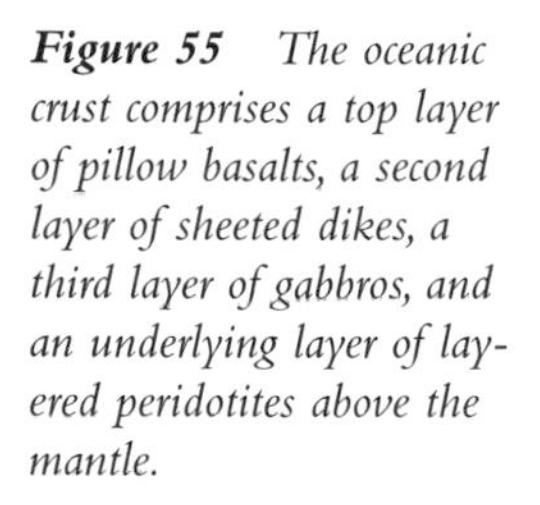
Figure 55 *The oceanic crust comprises a top layer of pillow basalts, a second layer of sheeted dikes, a third layer of gabbros, and an underlying layer of layered peridotites above the mantle.*

magma to the surface. A third layer of gabbros are made up of coarse-grained rocks that crystallized slowly under high pressure in a deep magma chamber. Finally, an underlying layer of peridotites exists above the mantle. Gabbros with higher amounts of silica solidify out of the basaltic melt and accumulate in the lower layer of the oceanic crust. Because this same formation is found on the continents, geologists speculate that these rocks were pieces of ancient oceanic crust called ophiolites.

Most oceanic crust is less than 5 percent of the Earth's age and younger than 170 million years, with a mean age of 100 million years. In comparison, the continental crust is about 4 billion years old. The difference in ages is due to the recycling of oceanic crust into the mantle. Also, almost all the seafloor has since disappeared into the Earth's interior to provide the raw materials for the continued growth of the continents.

New oceanic crust forms at spreading ridges, where basalt oozes out of the mantle through rifts on the ocean floor. This generates about five cubic miles of new oceanic crust every year. Some molten magma erupts as lava on the surface of the ridge through a system of vertical passages. Once at the surface, the liquid rock flows down the ridge and hardens into sheets or rounded forms of pillow lavas, depending on the rate of extrusion and the slope of the ridge.

Magma rising from the upper mantle extrudes onto the ocean floor and bonds to the edges of separating plates. Much of the magma solidifies within the conduits above the magma chamber, forming massive vertical sheets called dikes that resemble a deck of cards standing on end. Individual dikes measure about 10 feet thick, stretch about 1 mile wide, and range about 3 miles long. Periodically, lava overflows onto the ocean floor in gigantic eruptions, providing several square miles of new oceanic crust annually. As the oceanic crust cools and hardens, it contracts, forming fractures through which water circulates.

The oceanic crust thickens with age, from a few miles thick after formation at midocean spreading ridges to more than 50 miles thick in the oldest ocean basins next to the continents. An oceanic plate starts thin and thickens by the underplating of new lithosphere from the upper mantle and the accumulation of overlying sediment layers. The ocean floor at the summit of a midocean ridge consists almost entirely of hard basalt. It acquires a thickening layer of sediments farther outward from the ridge crest. By the time the oceanic plate spreads out as wide as the Atlantic Ocean, the portion near continental margins where the sea is the deepest is about 60 miles thick. Eventually, the oceanic plate becomes so thick and heavy it can no longer remain on the surface, bends downward, and subducts into the Earth's interior.

When an oceanic plate dives into the mantle at a subduction zone, it remelts and acquires new minerals from the mantle. It then provides material for new oceanic crust as molten magma reemerges at volcanic spreading centers along midocean ridges. With reduced pressure, the rocks melt and rise through fractures in the lithosphere. As the molten magma passes through the lithosphere, it reaches the bottom of the oceanic crust. It then forms magma chambers that further press against the crust and continue to widen the rift. Molten lava pouring out of the rift forms ridge crests on both sides and adds new material to the spreading ridge system.

The mantle material below spreading ridges, where new oceanic crust forms, is mostly peridotite. This is a strong, dense rock composed of iron and magnesium silicates. As the peridotite melts on its journey to the base of the oceanic crust, a portion becomes highly fluid basalt. Basalt is the most common magma erupted onto the surface of the Earth. About five cubic miles of basaltic magma is removed from the mantle and added to the crust every year. The oceanic crust gradually increases density, and the added weight causes it to subduct into the mantle. On its way deep into the Earth's interior, the lithosphere melts and the molten magma rises toward the surface to provide new molten rock in a continuous cycle.

TERRANES

The continents are composed of a patchwork of crustal blocks that combined into geologic collages known as terranes (Fig. 56). They are usually bounded by faults and are distinct from their geologic surroundings. The boundaries between two or more terranes, called suture zones, are commonly marked by ophiolite belts. These consist of ancient oceanic crust shoved onto the continents by drifting plates. The composition of terranes generally resembles that

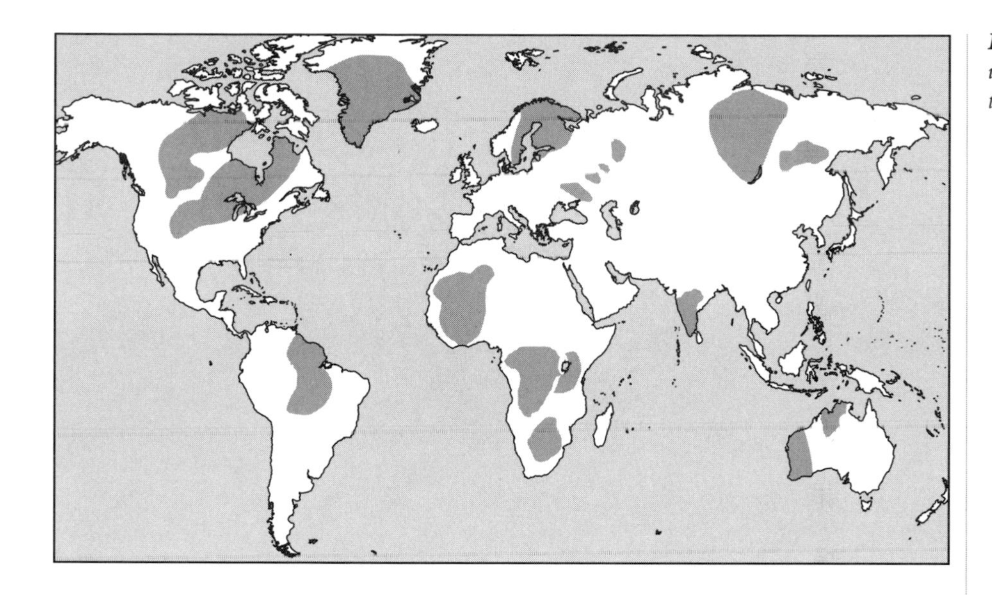

Figure 56 *The distribution of 2-billion-year-old terranes.*

of a volcanic island or an undersea plateau. Some terranes are composed of a consolidated conglomerate of pebbles, sand, and silt that accumulated in an ocean basin between colliding crustal fragments.

Terranes are generally elongated bodies that deformed when colliding and accreting to a continent. For example, the assemblage of terranes in China is being stretched and displaced in an east-west direction as India continues to press against southern Asia after colliding with the mainland some 45 million years ago. The buckling crust raised the Himalaya Mountains and the broad Tibetan Plateau, the largest topographic upland on Earth. Asia was elongated to accommodate the northward advancement of India. A belt of ophiolites marks the boundary between the sutured continents. Eurasia is still accumulating pieces of crust arriving from the south.

Terranes come in a variety of shapes and sizes. They vary from small slices of crust to subcontinents as large as India, itself a single great terrane. They range in age from well over a billion to less than 200 million years old. The ages of the terranes were determined by studying entrained fossil radiolarians (Fig. 57). These are marine protozoans with skeletons made of silica. They were abundant from about 500 million to 160 million years ago. Different species also define specific regions of the ocean where the terranes originated.

Most terranes are fault-bounded blocks with geologic histories apart from those of neighboring terranes and of adjoining continental masses. Many have traveled considerable distances before finally colliding with continental margins. For instance, some North American terranes originated in the western Pacific and traversed thousands of miles eastward. The actual distances terranes travel vary considerably. Basaltic seamounts accreting to the margin of Oregon moved

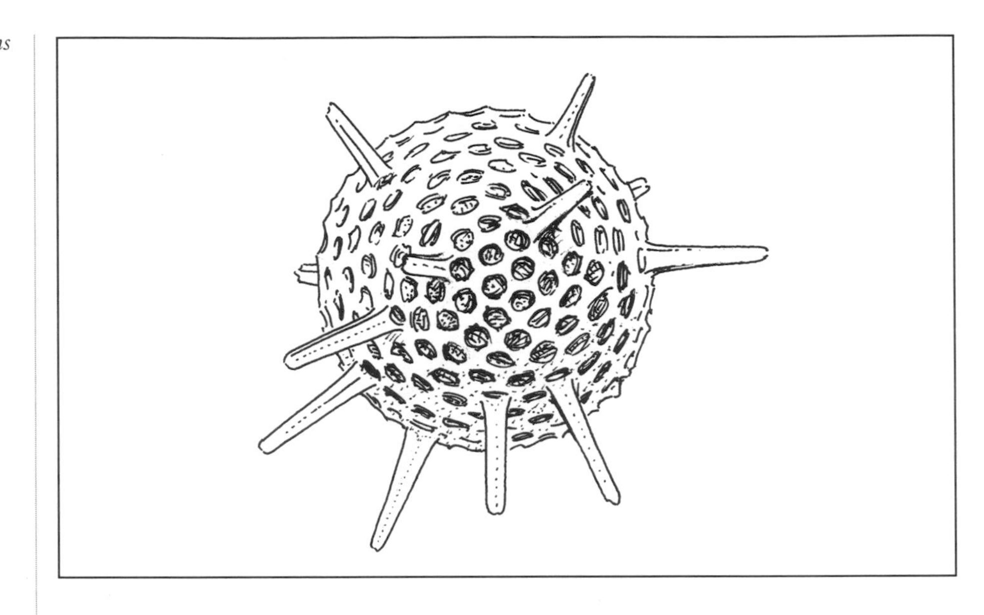

Figure 57 *Radiolarians were marine planktonic protozoans.*

from nearby offshore. In contrast, similar rock formations consisting of three distinct rock units located around San Francisco, California arrived from halfway across the Pacific Ocean. Northern California is a jumble of crust assembled more than 100 million years ago, comprising rock formations that originated as far away as 2,500 miles. At their usual rate of travel, terranes could make a complete circuit of the globe in perhaps a half-billion years.

During the assembly of Pangaea from 360 to 270 million years ago, the western edge of North America ended near present-day Salt Lake City. Over the last 200 million years, North America has expanded some 25 percent during a major pulse of crustal growth following the breakup of Pangaea. The entire Pacific Coast from the Baja California Peninsula to the tip of Alaska was grafted onto the continent by the piecemeal addition of crustal blocks. Much of western North America was assembled from island arcs and other crustal debris skimmed off the 7,500-mile-wide Pacific plate as the North American plate headed westward. Many terranes in western North America have rotated clockwise as much as 70 degrees or more, with the oldest terranes having the most rotation.

The entire state of Alaska is an agglomeration of terranes. Some are well exposed in the Brooks Range (Fig. 58), the spine of northern Alaska, consisting of great sheets of crust stacked one on another aligned in an east-west direction, the only other mountain range in North America to do so. The Alexander terrane, making up a large portion of the Alaskan panhandle, began as part of eastern Australia some 500 million years ago. Beginning about 375 million years ago, it broke free of Australia, traversed the Pacific Ocean, stopped briefly at the coast of Peru, and sliced past California, swiping part of the Mother Lode gold belt. It finally collided with the upper North American continent around 100 million years ago.

Alaska is a unique assemblage of some 50 terranes set adrift over the past 160 million years by the wanderings and collisions of crustal plates, fragments of which are still arriving from the south. As an example, some 70 million years ago, Vancouver Island, British Columbia, was nestled along the coast of what is now Baja California. During the next 50 million years, the Pacific plate carrying California west of the right-lateral-moving San Andreas Fault (Fig. 59) will slide northward relative to the North American plate at a rate of one to two inches per year. Finally, it will come to rest at the continental margin off Alaska, adding another piece to the puzzle.

Along the mountain ranges in western North America, the terranes are elongated bodies due to the slicing of the crust by a network of northwest-trending faults. The most active of these is the San Andreas, which has been displaced some 200 miles over the last 25 million years. The Gulf of California, separating the Baja California Peninsula from mainland Mexico, is a continuation of the San Andreas Fault system. The landscape is literally being torn apart while opening one of the youngest and richest seas on Earth. It

Figure 58 *Steeply dipping Paleozoic rocks of the Brooks Range, near the head of the Itkillik River east of Anaktuvuk Pass, Northern Alaska.*

(Photo by J. C. Reed, courtesy USGS)

Figure 59 *Multiple fault traces and stream offsets on the San Andreas Fault, Carrizo Plains, California.*

(Photo by R. E. Wallace, courtesy USGS)

began rifting some 6 million years ago, offering a new outlet to the sea for the Colorado River, which thereafter began carving out the Grand Canyon.

HOT SPOTS

More than 100 small regions of isolated volcanic activity known as hot spots are found in various parts of the world (Fig. 60). Unlike most other active volcanoes, those created by hot spots rarely exist at plate boundaries but, instead, lie deep in the interior of a plate. They are notable if only for their very isolation, far removed from normal centers of volcanic and earthquake activity. They might be the only distinctive features in an otherwise monotonous landscape.

Hot spots provide a pipeline for transporting heat from the planet's interior to the surface. The magma plumes rise through the mantle as separate giant bubbles of hot rock. When a plume passes the boundary between the lower and upper mantle, some 410 miles below the surface, the bulbous head separates from the tail and rises. This is often followed millions of years later by another similarly created plume.

Almost all hot spot volcanism occurs in regions of broad crustal uplift or swelling. Lavas of hot spot volcanoes differ markedly from those of rift systems and subduction zones. Hot spot lavas are composed of basalts containing larger amounts of alkali minerals such as sodium and potassium, indicating that their source material is not associated with plate margins.

Hot spots appear to result from plumes of hot material rising from deep within the mantle, possibly just above the core. The distinctive composition of hot spot lavas seems to indicate a source outside the general circulation pattern of the mantle. Plumes might also arise from stagnant regions in the center of convection cells or from below the region in the mantle stirred by convection currents. As plumes of mantle material flow upward into the asthenosphere, the portion rich in volatiles rises toward the surface to feed hot-spot volcanoes. The plumes' range of sizes might indicate the depth of their source material. The plumes are not necessarily continuous flows of mantle material, however. Instead, they might consist of molten rock rising in giant blobs or diapirs.

Figure 60 *The world's hot spots, where mantle plumes rise to the surface.*

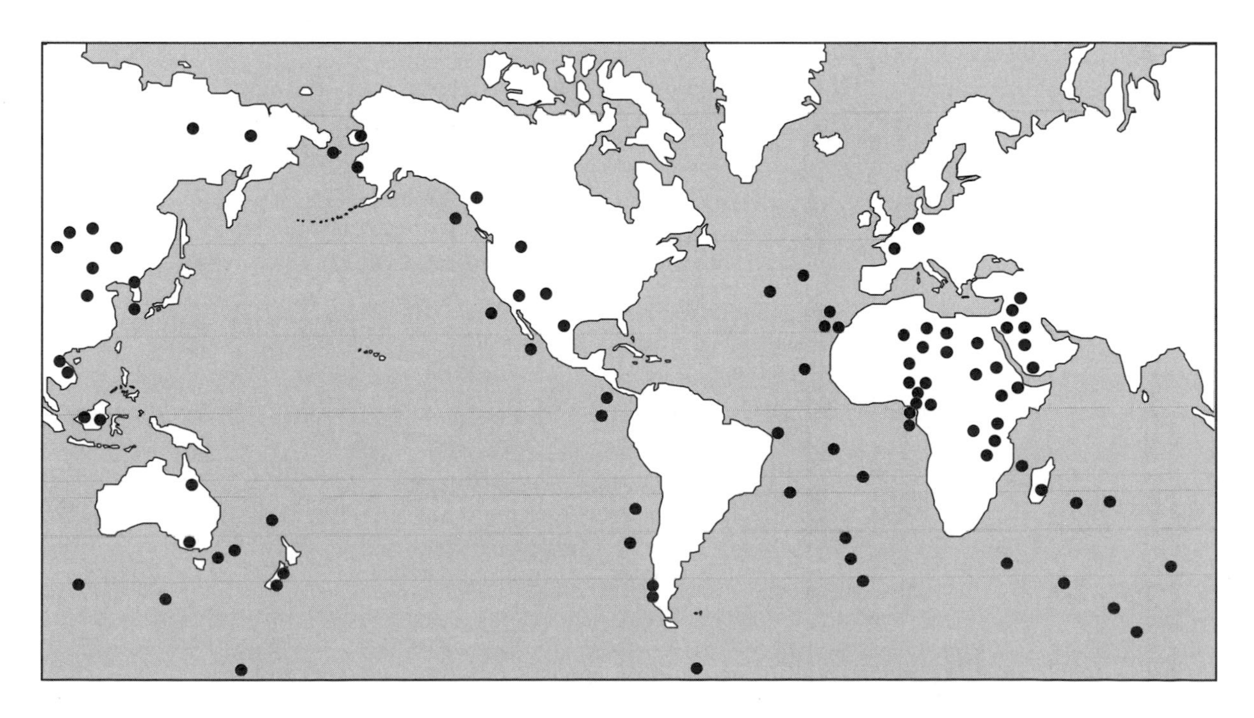

Most hot spots move very slowly, at a lower rate than the oceanic or continental plates above them. When a continental plate hovers over a hot spot, the molten magma welling up from deep below creates a broad, domelike structure in the crust that averages about 125 miles across and accounts for about 10 percent of the Earth's total surface area. As the dome grows, it develops deep fissures through which magma can rise to the surface.

Often the passage of a plate over a hot spot results in a trail of volcanic features whose linear trend reveals the direction of plate motion. This produces volcanic structures aligned in a direction oblique to the adjacent midocean ridge system rather than parallel to it as with rift volcanoes. The hot-spot track might be a continuous volcanic ridge or a chain of volcanic islands and seamounts that rise high above the surrounding seafloor. The hot-spot track might also weaken the crust, cutting through the lithosphere like a hot knife through butter.

The most prominent and easily recognizable hot spot created the Hawaiian Islands (Fig. 61). Apparently all the islands in the Hawaiian chain were produced by a single source of magma over which the Pacific plate had passed, proceeding in a northwesterly direction. The volcanic islands popped out onto the ocean floor conveyor belt fashion, with the oldest trailing off to the northwest farthest away from the hot spot. Similar chains of volcanic islands exist in the Pacific that trend in the same direction as the Hawaiian Islands, including the Line and Marshall-Gilbert Islands and the Austral and Tuamotu Seamounts. This effect indicates that the Pacific plate is moving off in the direction defined by the volcanic chains.

The trail of volcanoes left by the hot spots changes abruptly to the north, where it follows the Emperor Seamounts, an isolated chain of undersea volcanoes strung out across the interior of the Pacific plate. This deflection occurred about 40 million years ago, near the time India rammed into Asia, which might have shifted the Asian plate. A sharp bend in the long Mendocino Fracture Zone jutting out from northern California confirms that the Pacific plate abruptly changed direction at the same time as the India-Asia plate convergence. The timing is also coincident with the collision of the North American plate and the Pacific plate. Therefore, hot spots could be a reliable means for determining the direction of plate motion.

The Bermuda Rise in the western Atlantic appears to be a contradiction to this rule. It is oriented in a roughly northeast direction parallel to the continental margin off the eastern United States. It is nearly 1,000 miles long and rises some 3,000 feet above the surrounding seafloor, where the last of the volcanoes ceased erupting about 25 million years ago. A weak hot spot unable to burn a hole through the North American plate apparently was forced to take advantage of previous structures on the ocean floor. This explains why the volcanoes trend nearly at right angles to the motion of the plate.

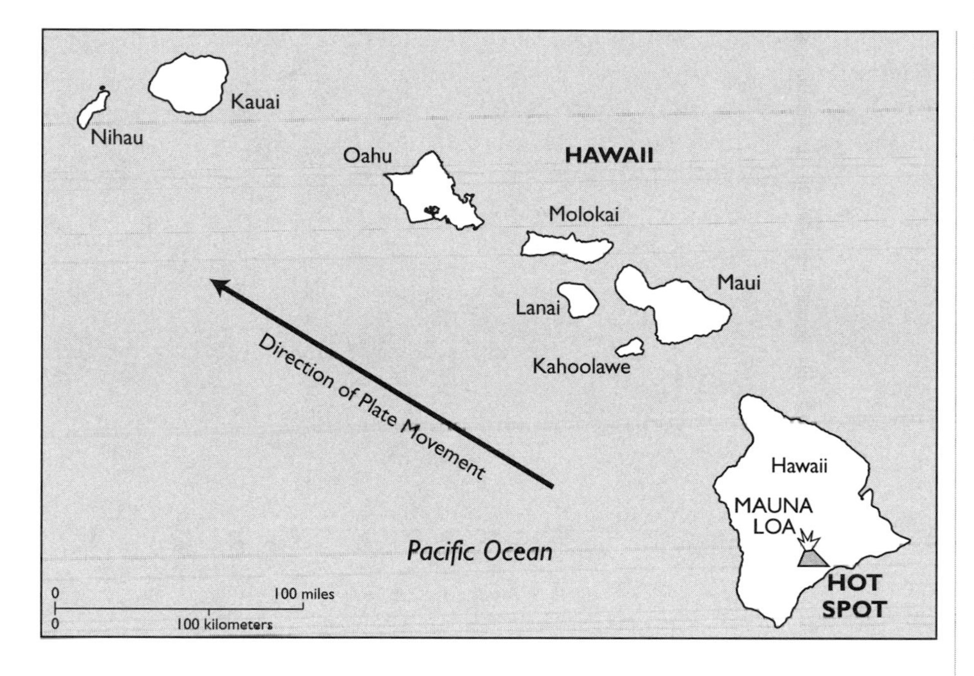

Figure 61 *The Hawaiian Islands formed by the drifting of the Pacific plate over a hot spot.*

The Bowie Seamount is the youngest in a line of submerged volcanoes running toward the northwest off the west coast of Canada. It is fed by a mantle plume nearly 100 miles in diameter and more than 400 miles below the ocean floor. However, rather than lying directly beneath the seamount, as plumes are supposed to do, this one lies about 100 miles east of the volcano. The plume could have taken a tilted path upward, or the seamount somehow moved with respect to the hot spot's position.

More than half the hot spots exist on the continents with the greatest concentration in Africa, which has remained essentially stationary over about 25 hot spots for millions of years. Hot spots might have been responsible for the unusual topography of the African continent, which is characterized by numerous basins, swells, uplifted highlands, and massive basalt flows (Fig. 62). The effect might also indicate that the African plate has come to rest over a population of hot spots. Further evidence that Africa is stationary is that hot spot lavas of several different ages are superimposed on one another. If the continent were drifting, the hot spot lavas would spread laterally in a chronological sequence.

A direct relationship therefore appears to exist between the number of hot spots and the rate of drift of a continent. Besides Africa, hot spots are numerous in Antarctica and Eurasia, thus these regions would appear to be moving at a very slow pace as well. In contrast, on rapidly moving continental plates, such as North and South America, hot-spot volcanism is rare.

Yellowstone National Park is more than 1,000 miles from the nearest plate boundary, yet it is one of several midplate centers of hot spot activity. During the past 2 million years, at least three episodes of intense volcanic activity occurred in the region. An explosive eruption hundreds of times greater than the 1980 Mount St. Helens lateral blast that leveled an entire forest is well overdue. Beneath the park lies a hot spot responsible for the continuous thermal activity giving rise to a number of geysers such as Old Faithful (Fig. 63). The geysers are produced when water seeps into the ground, is heated near a magma chamber, and rises explosively through fissures in the torn crust.

The hot spot was not always under Yellowstone, however. Its positions relative to the North American plate can be traced through volcanic rocks for 400 miles on the Snake River Plain in southern Idaho. Over the past 15 million years, the North American plate slid southwestward across the hot spot,

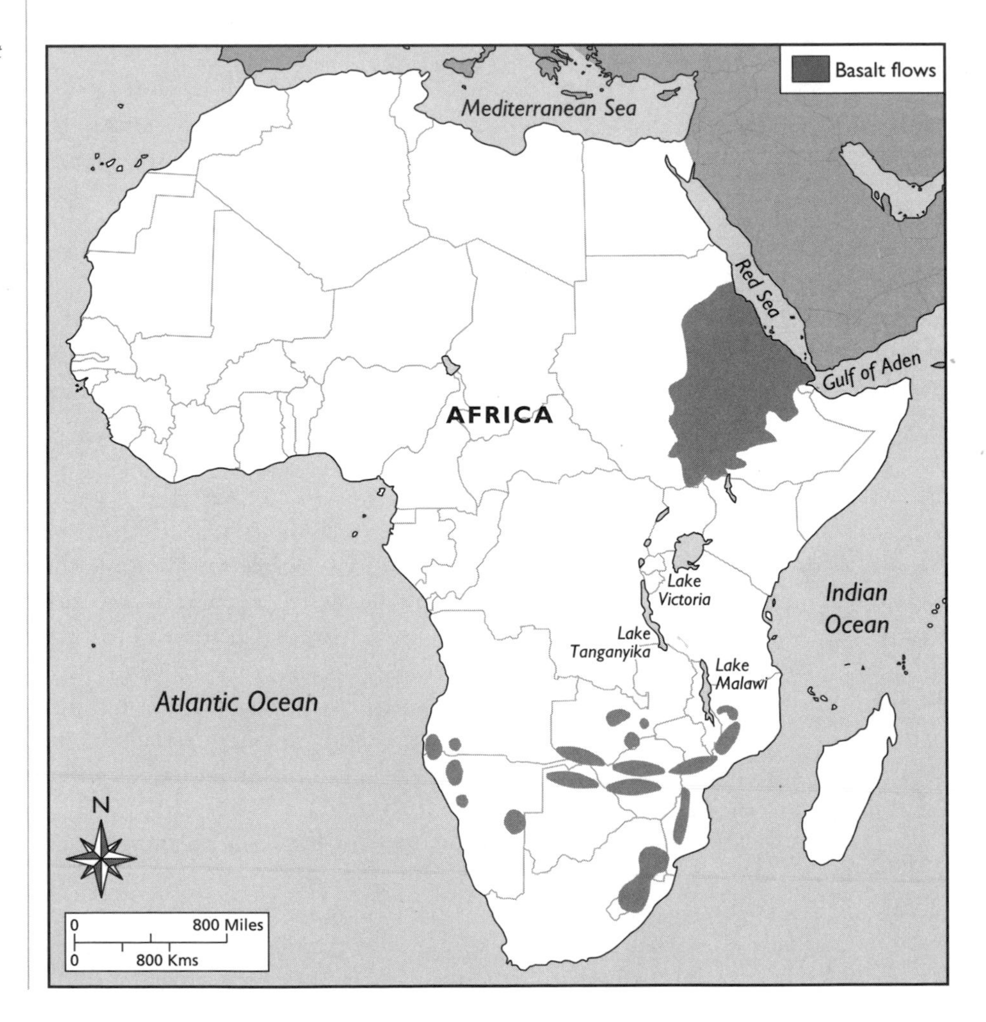

Figure 62 *Major basalt flows in Africa.*

Figure 63 *Eruption of Old Faithful Geyser, Yellowstone National Park, Wyoming.*

(Photo by J. R. Stacy, courtesy USGS)

placing it under its temporary home at Yellowstone. Eventually, as the plate continues in its westerly direction, the relative motion of the hot spot will bring it across Wyoming and Montana.

Sometimes a hot spot fades away entirely, and a new one forms in its place. The typical life span of a plume is on the order of about 100 million years. The position of a hot spot can change slightly as it sways in the convective currents of the mantle. As a result, the tracks on the surface might not always be as linear as those of Hawaii and other volcanic island chains across the Pacific. However, compared with the motion of the plates, the mantle plumes are relatively stationary. Because the motion of the hot spots is only slight, they provide a reference point for determining the direction and rate of plate travel. If the upwelling plumes should cease flowing, the plates would grind to a complete halt due to the loss of internal heat sources.

PLATE MOTIONS

The plates are composed of the lithosphere and the overlying continental or oceanic crusts. The plate boundaries are midocean spreading ridges, subduction zones, and transform faults. The plates ride on the asthenosphere and carry the continents along with them somewhat like ships frozen in floating ice packs. The breakup of a plate results in the formation of a new continent.

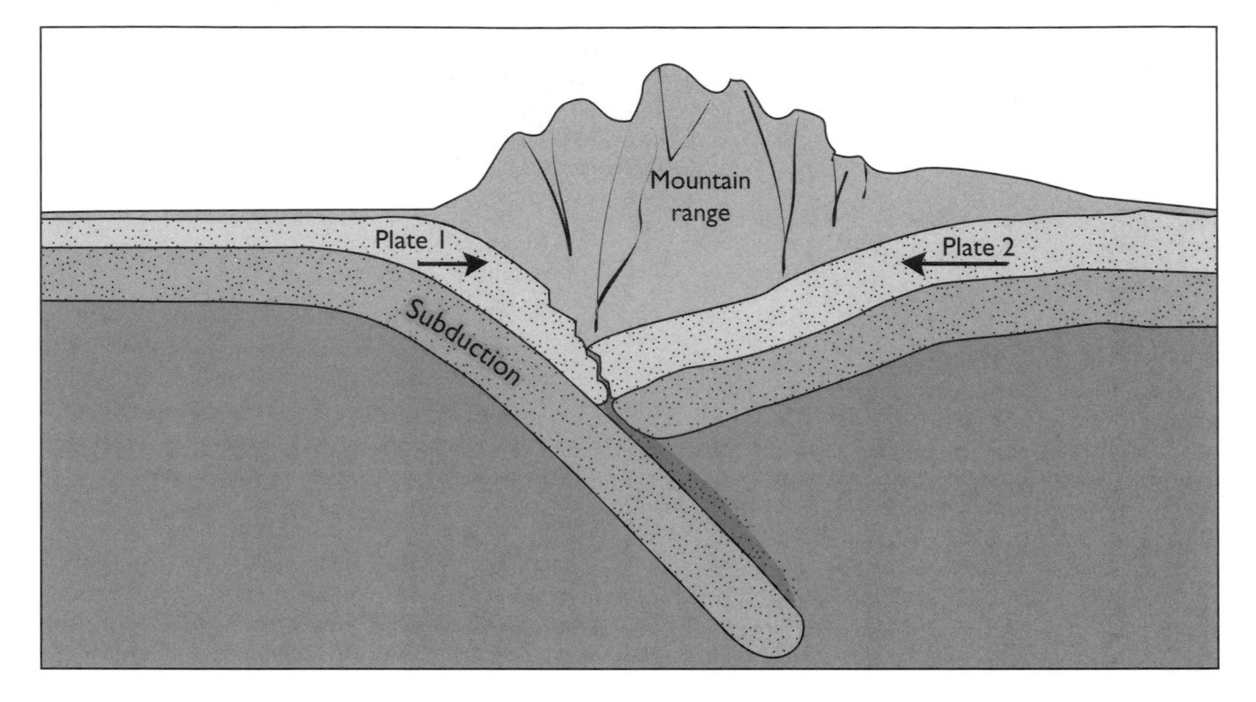

Figure 64 *The formation of mountains by the convergence of two lithospheric plates.*

For example, the breakup of Laurasia and Gondwana at the Mid-Atlantic Ridge created North and South America, Eurasia, and Africa. On the other hand, if two plates collide, the thrust of one under the other uplifts the crust to form mountain ranges (Fig. 64) or long chains of volcanic islands.

Thousands of feet of sediments are deposited along the seaward margin of a continental plate in deep ocean trenches. The increased weight presses downward on the oceanic crust. As the continental and oceanic plates merge, the heavier oceanic plate is subducted or overridden by the lighter continental plate, forcing it further downward. The sedimentary layers of both plates are squeezed and the swelling at the leading edge of the continental crust forms mountain belts. The sediments are faulted at or near the surface, where the rocks are brittle, and folded at a depth where the rocks are more plastic.

As the oceanic crust descends, the topmost layers are scraped off and plastered against the swollen edge of the continental crust, forming an accretionary wedge. In the deepest part of the continental crust, where temperatures and pressures are very high, rocks are partially melted and metamorphosed. As the descending plate dives further under the continent, it reaches depths where the temperatures are extremely high. Part of the plate melts, forming a silica-rich magma that rises because it is lighter than the surrounding rock material. The magma intrudes the overlying metamorphic and sedimentary layers to form large granitic bodies or erupts onto the surface from a volcano.

Continents are neither created nor destroyed by the process of plate tectonics, only the plates on which they ride are absorbed into the mantle. Divergence of lithospheric plates creates new oceanic crust. Convergence destroys oceanic crust in well-developed subduction zones, which are prevalent in the western Pacific and responsible for numerous island arcs. Volcanoes of the island arcs are highly spectacular because their lava is silica-rich, contrasting strongly with the basalt of other volcanoes and midocean ridges. The volcanoes are mostly explosive and build steep-sided cinder cones. Island arcs are also associated with belts of deep-seated earthquakes 200 to 400 miles below the surface.

Rifts open not only in ocean basins but also under continents. This activity is occurring in eastern Africa, creating a great rift valley that will eventually widen and flood with seawater to form a new subcontinent. Old extinct rift systems, where the spreading activity has stopped, or failed rifts, where a full-fledged spreading center did not develop, are overrun by continents. For example, the western edge of North America has overrun the northern part of the now extinct Pacific rift system. The North American continental mass has run into the northern extension of the active Pacific rift system, called the East Pacific Rise, creating the San Andreas Fault in California.

Measuring the rate of plate motions requires extreme accuracy over a distance of thousands of miles. Standard geodetic survey methods cannot provide this accuracy. However, measurements using satellites can. In Satellite Laser Ranging, distances are measured by comparing how long laser pulses take to leave the surface, bounce off a satellite, and return to ground stations. In the satellite-based Global Positioning System (Fig. 65), plate positions can be measured with a precision of about 1 inch over a distance of 300 miles.

In Very Long Baseline Interferometry (VLBI), radio signals from distant quasars (rapidly spinning collapsed stars) are monitored at different stations on Earth. The difference in arrival times for these signals determines the distance between the receiving stations. The VLBI method is the more exact of these techniques, with an accuracy approaching several parts per billion. This is comparable to measuring the length of 100 football fields to within the width of a human hair. These measurements of plate motion are in good agreement with geologic methods, which are based on the spacing of magnetic stripes on the ocean floor produced by seafloor spreading.

EARTHQUAKES

Earthquakes are by far the strongest natural forces on Earth. In a matter of seconds, a large temblor can level an entire city. Every year, a dozen or so major earthquakes strike somewhere in the world. Most are in areas along the rim

Figure 65 *An artist's concept of one of the 24 satellites of the Global Positioning System.*

(Photo courtesy U.S. Air Force)

of the Pacific plate. The Pacific Basin area is also prone to destructive seismic sea waves or tsunamis from undersea earthquakes.

The mechanism for creating earthquakes was poorly understood until after the big quake that struck San Francisco in 1906. The American geologist Harry Reid discovered that for hundreds of miles along the San Andreas Fault, fences and roads crossing the fault had been displaced by as much as 21 feet (Fig. 66). This observation led him to propose the modern elastic-rebound theory of earthquake faulting.

The San Andreas Fault is a fracture zone 650 miles long and 20 miles deep that runs northward from the Mexican border through southern California, plunging into the ocean at Cape Mendocino 100 miles south of the Oregon border. The fault represents the margin between the Pacific plate and the North American plate, which are moving relative to each other in a right lateral direction at a rate of nearly two inches per year. The fault absorbs most of this motion. The rest is dissipated by the spreading of the Basin and Range Province and by the deformation of California's southern Coast Ranges. In the 50 years before the great San Francisco earthquake, land surveys showed displacements as much as 10 feet along the San Andreas Fault. Tectonic forces slowly deformed the crustal rocks on both sides of the fault, causing large displacements. During this time, the rocks were bending and storing up elastic energy, similar to stretching a rubber band.

Eventually, the forces holding the rocks together were overcome, and slippage occurred at the weakest point. As with a rubber band, the rocks snapped back. The point of initial rupture is called the hypocenter. When near the surface, it can cause large displacements in the crust. These displacements can exert strain further along the fault, where additional slippage can occur until most of the built-up strain is released. The slippage allowed the deformed rock to rebound to its original shape elastically, which released heat generated by friction and produced vibrations called seismic waves. The seismic waves radiated outward from the hypocenter in all directions, like the ripples produced when a rock is thrown into a quiet pond.

Figure 66 *A road near Point Reyes Station is offset 20 feet by the San Andreas Fault during the 1906 San Francisco earthquake in California.*

(Photo by G. K. Gilbert, courtesy USGS)

The rocks do not always rebound immediately, however. They might take days or even years, resulting in aseismic slip. The seismic energy thus produced is then quite small. Why seismic energy is released violently in some cases and not in others is still not fully understood. Moreover, some types of shallow earthquakes with magnitudes greater than 5.0 might be triggered by outside events such as large meteorite impacts. Scientists even suspect that when the gravitational attractions of the sun and the moon pull together on the Earth, they can cause some earthquake fault systems to rupture.

The magnitude of an earthquake is recorded by a seismograph and measured on the moment magnitude scale, based on the earlier Richter scale devised by the American seismologist Charles F. Richter. The magnitude scale is logarithmic. An increase of 1 magnitude signifies a 10-fold increase in ground motion and about 30 times the energy. A magnitude of 3 on the scale is barely perceptible, whereas an 8 or higher can be catastrophic.

The San Andreas is perhaps the most heavily instrumented and the best-studied fault system in the world. Various remote-sensing techniques are employed for making earthquake predictions. Laser-ranging devices measure the amount of crustal strain along the fault with an accuracy of 0.5 inches over a distance of about 20 miles. Faults give out several precursory signals that might also aid in earthquake prediction. These include changes in the tilt of the ground, magnetic anomalies, increased radon gas content in nearby water wells, and swarms of microearthquakes.

Faults also produce a phenomenon known as earthquake lights (Fig. 67) before and during rupture. Apparently, the strain on the rocks in the vicinity of the fault causes them to emit energy, producing a faint atmospheric glow at night. Seismic activity is associated with a variety of other electrical effects that might aid in predicting earthquakes. A system of radio wave monitors dis-

Figure 67 *Earthquake lights during the Matsushiro earthquake swarm in Japan, which lasted from 1965 to 1967.*

(Photo by T. Kuribayashi, courtesy USGS)

Figure 68 *Collapsed bridge on Interstate 880 in San Francisco from the October 17, 1989 Loma Prieta earthquake, Alameda County, California.*

(Photo by G. Plafker, courtesy USGS)

tributed along the San Andreas Fault have recorded changes in atmospheric radio waves prior to several earthquakes, including the 7.1-magnitude Loma Prieta earthquake on October 17, 1989 (Fig. 68). The increased electrical conductivity of rocks under stress near the fault apparently causes radio waves to be absorbed by the ground 1 to 6 days before an earthquake. Moreover, investigators have observed short pulses of increased radio interference caused by the release of electromagnetic energy by cracking rocks.

Over the past few years, geodesists (scientists who measure the Earth) have significantly improved the precision with which they can determine positions on the Earth's surface by using the satellite-based Global Positioning System. With this system, investigators can monitor the strain accumulation on the San Andreas Fault more efficiently than with standard geodetic methods. Positions are determined with an accuracy of about 1 part in 10 million over a distance of up to 300 miles. The orbit of the satellite is calculated with great precision by comparing satellite signals received at two stations on the ground. This, in turn, provides data about their relative positions.

In addition to monitoring the San Andreas Fault, the satellites can measure the rate of seafloor spreading in Iceland, subsidence of the crust due to the removal of groundwater, and the bulging crust above the magma chamber under Long Valley, California. Slight bulges in the crust detected by the Global Positioning System might signal when a volcano is poised for eruption.

FAULT ZONES

Roughly 95 percent of the seismic energy released by earthquakes is concentrated in broad zones that wind around the globe and are associated with plate boundaries (Fig. 69). A continuous belt extends for thousands of miles through the world's oceans and coincides with the midocean rift systems. Earthquakes are also associated with terrestrial rift zones such as the 3,600-mile East African Rift.

The greatest amount of energy is released along a path located near the outer edge of the Pacific Ocean, known as the circum-Pacific belt. This zone includes the San Andreas Fault in southern California, which was responsible for numerous powerful earthquakes. Another belt runs through the folded mountainous regions that flank the Mediterranean Sea. It continues through Iran and past the Himalaya Mountains into China. A massive earthquake in the Caspian Sea just north of Iran killed 100,000 people and left half a million homeless in June 1990. A huge earthquake of 7.9 magnitude struck northwest India, causing enormous destruction and loss of a great many lives on January 26, 2001.

In the eastern Himalaya range lies perhaps the most seismically active region of the world. An immense seismic belt some 2,500 miles long stretches across Tibet and much of China, where the 1976 Tangshan earthquake killed nearly half a million people. For centuries, this area has been shaken by catastrophic earthquakes responsible for the deaths of millions. In the last cen-

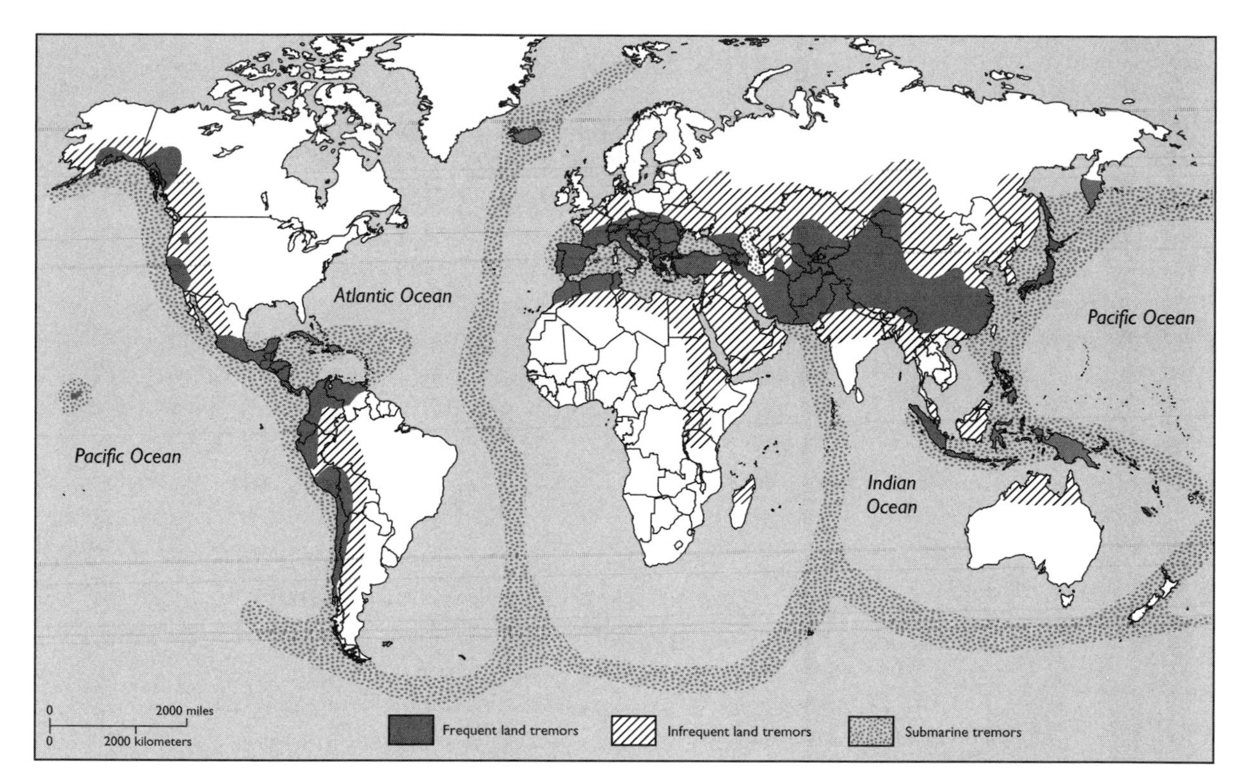

Figure 69 Most earthquakes occur in broad zones associated with plate boundaries.

tury, more than a dozen earthquakes of 8.0 magnitude or greater have been recorded in this region.

West of this belt, in the Hindu Kush range of north Afghanistan and the nearby Russian Republic of Tadzhikstan, is the seat of many earthquakes. Three earthquakes in the last century have had magnitudes of 8.0 or over. The great 1988 Armenian earthquake killed 25,000 people and left a million more homeless. This is a notoriously active seismic belt, with some 2,000 minor earthquakes registered annually.

From there, the Persian arc spreads in a wide sweep through the Pamir and Caucasus Mountains and on to Turkey. The August 17, 1999 earthquake of 7.4 magnitude killed more than 17,000 people in the industrial heartland of Turkey. The eastern end of the Mediterranean is a jumbled region of colliding plates, providing highly unstable ground. The whole of the Near East is inherently unstable, attesting to the many earthquakes reported in biblical times. The remaining regions surrounding the Mediterranean have been devastated by earthquakes throughout history.

The circum-Pacific belt coincides with the Ring of Fire because the same tectonic forces that produce earthquakes are also responsible for volcanic activity (see Chapter 8). Earthquakes in this region also produce most of the

TABLE 7 SUMMARY OF EARTHQUAKE PARAMETERS

Magnitude	Surface Wave Height (Feet)	Length of Fault Affected (Miles)	Diameter Area Quake Is Felt (Miles)	Number of Quakes per Year
9	Largest earthquakes ever recorded—between 8 and 9			
8	300	500	750	1.5
7	30	25	500	15
6	3	5	280	150
5	0.3	1.9	190	1,500
4	0.03	0.8	100	15,000
3	0.003	0.3	20	150,000

world's seismic sea waves (Fig. 70). The area of greatest seismicity is on the plate boundaries associated with deep trenches and volcanic island arcs, where an oceanic plate is thrust under a continental plate. Japan, which is in the process of being plastered against Asia, is a constant reminder of powerful earthquakes associated with subduction zones. The great 1923 Tokyo earthquake of 8.3 magnitude took some 140,000 lives. The January 17, 1995 Kobe earthquake of 7.2 magnitude killed more than 5,500 people and caused more than $100 billion in property damage.

The Andes Mountain regions of Central and South America, especially in Chile and Peru, are known for some of the largest and most destructive earthquakes in historic times. In the last century, nearly two dozen earthquakes of 7.5 magnitude or greater have taken place in Central and South America, including the largest ever recorded, the great 1960 Chilean earthquake of estimated 9.5 magnitude.

The whole western seaboard of South America is affected by an immense subduction zone just off the coast. The lithospheric plate on which the South American continent rides is forcing the Nazca plate to buckle under, causing great tensions to build up deep within the crust. The plate is being consumed by the Peru-Chile trench at a rate of about 50 miles every million years. While some rocks are being forced deep down, others are pushed upward toward the surface, raising the Andean mountain chain. The resulting forces are building great stresses into the entire region. When the stresses become large enough, earthquakes crack open the crust, forming tall scarps that slice across the countryside.

Even in the so-called stable zones, earthquakes occur, although not nearly as frequently as in the earthquake-prone areas. The stable zones are generally associated with continental shields, comprising ancient granitic rocks in

the interior of the continents. When earthquakes strike these regions, they might be due to the weakening of the crust by compressive forces that originate at plate edges. The underlying crust might also have been weakened by previous tectonic activity, resulting in the sudden release of pent-up stresses.

After learning about the Earth's outer layer of lithospheric plates and their interactions, the next chapter will examine perhaps the most important aspects of plate tectonics—seafloor spreading.

Figure 70 *Tsunamis washed many vessels into the heart of Kodiak from the March 27, 1964 Alaska earthquake.*

(Photo courtesy USGS)

5

SEAFLOOR SPREADING

THE MIDOCEAN RIDGES

This chapter reveals how the best evidence for plate tectonics was found lying on the bottom of the ocean. Geologists once thought the ocean floor was barren and featureless, covered by thick, muddy sediments washed off the continents and by debris from dead marine organisms piled up several miles thick after billions of years of accumulation. During the mid-1800s, soundings were made of the ocean floor in preparation for laying the first transcontinental telegraph cable to link the United States with Europe. The depth recordings indicated hills, valleys, and a mid-Atlantic rise, named Telegraph Plateau, where the ocean was supposed to be the deepest.

In 1874, the British cable-laying ship HMS *Faraday* was attempting to mend a telegraph cable in the North Atlantic. The cable had broken at a depth of 2.5 miles, where it passed over a large rise in the ocean floor. While grappling for the cable, the ship caught the strong claws of its grapnel onto a rock. With the winch straining to free the grapnel, it finally came loose and was brought to the surface. Clutched in one of its claws was a large chunk of black basalt, a volcanic rock found where volcanoes should not have been.

THE MYSTERIOUS OCEAN

The ocean spans some 70 percent of the Earth's surface. It covers an area of about 140 million square miles with more than 300 million cubic miles of seawater. About 60 percent of the planet is covered by water no less than 1 mile deep, with an average depth of about 2.3 miles. The ocean floor lies much deeper below sea level than the continents rise above it. The midocean spreading ridges lie at an average depth of 1.5 miles. The ocean bottom slopes away on both sides to a depth of about 3.5 miles. In the Pacific Basin, the ocean depth is as much as 7 miles. If Mount Everest, the world's tallest terrestrial mountain, were placed there, the sea would still extend over a mile above it.

If the oceans were completely drained of water, the planet would look much like the rugged surface of Venus (Fig. 71), which lost its oceans eons ago. The deepest parts of the dry seabed would lie several miles below the surrounding continental margins. The floor of the desiccated ocean would be traversed by the longest mountain ranges and fringed in places by the deepest trenches. Vast empty basins would divide the continents, which would stand out like thick slabs of rock.

Most seawater surrounding the continents lies in a single great basin in the Southern Hemisphere, which is nine-tenths ocean. It branches northward into the Atlantic, Pacific, and Indian basins in the Northern Hemisphere, which contains most of the continents. The Arctic Ocean is a nearly landlocked sea connected to the Atlantic and Pacific only by narrow straits. About

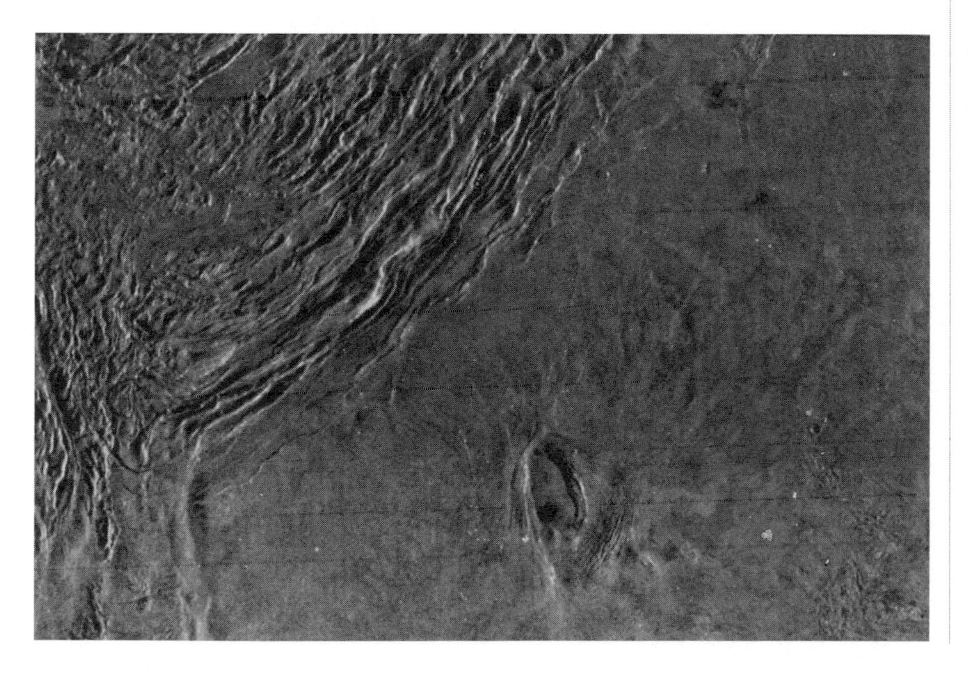

Figure 71 *Radar image of Venus from Vernera spacecraft.*

(Photo courtesy NASA)

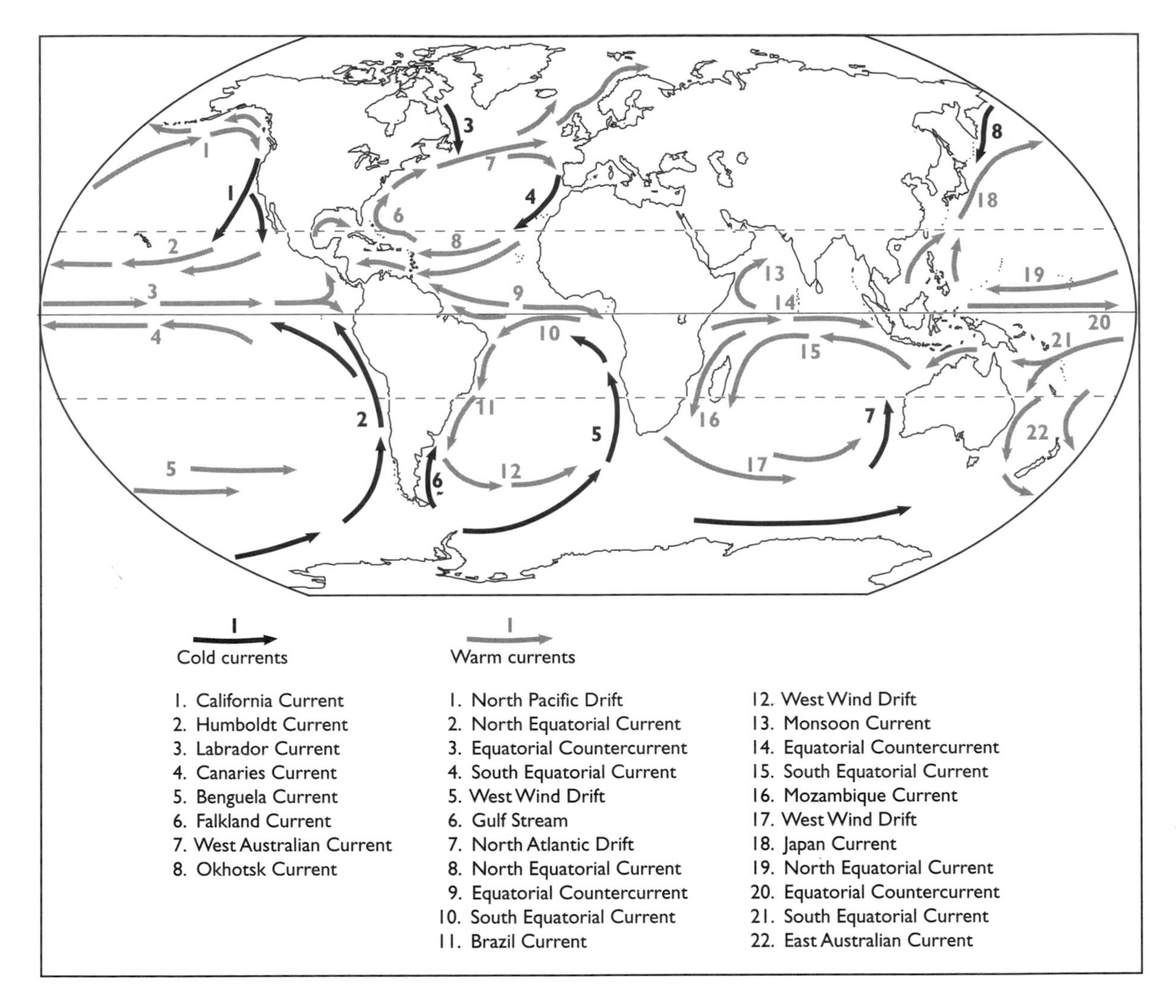

Figure 72 *The major ocean currents.*

20 million years ago, a ridge near Iceland subsided, allowing cold water from the recently formed Arctic Ocean to surge into the Atlantic, giving rise to the oceanic circulation system in existence today (Fig. 72).

If only marine-born sediments settled onto the ocean floor and no bottom currents stirred up the seabed, an even blanket of material would settle onto the original volcanic ocean crust. Instead, the rivers of the world contribute a substantial amount of the sediment deposited onto the deep ocean floor. The largest rivers of North and South America empty into the Atlantic, which receives considerably more river-borne sediment than the Pacific.

The Atlantic is also a smaller and shallower ocean than the Pacific. Therefore, its marine sediments are buried more rapidly and are more likely to survive than those in the Pacific. The deep-ocean trenches around the Pacific trap much of the material reaching its western edge, where it subducts into the mantle. Thus, on average, the floor of the Atlantic receives considerably more sediment than the floor of the Pacific, accumulating at a rate of about an inch every 2,500 years.

Bottom currents redistribute sediments in the Atlantic on a greater scale than in the Pacific. Abyssal storms with powerful currents occasionally sweep patches of ocean floor clean of sediments and deposit the debris elsewhere. On the western side of the ocean basins, periodic undersea storms skirt the foot of the continental rise and transport huge loads of sediment, dramatically modifying the seafloor. The scouring of the seabed and deposition of thick layers of fine sediment result in much more complex marine geology than that developed simply from a constant rain of sediments from above.

The ocean floor presents a rugged landscape unmatched elsewhere on Earth. Chasms dwarfing even the largest continental canyons plunge to great depths. Submarine canyons carved into bedrock 200 feet below sea level can be traced to rivers on land. Submarine canyons on continental shelves and slopes possess many identical features as river canyons, and some rival even the largest on the continents. They are characterized by high, steep walls and an irregular floor that slopes continually outward. The canyons range upward of 30 miles and more in length, with an average wall height of about 3,000 feet. The Great Bahamas Canyon is one of the largest submarine canyons, with a wall height of 14,000 feet, over twice as deep as the Grand Canyon.

THE MID-ATLANTIC RIDGE

In 1872, the British corvette HMS *Challenger,* the world's first fully equipped oceanographic vessel, was commissioned to explore the oceans. Scientists took soundings, water samples, and temperature readings, and they dredged bottom sediments for evidence of animal life living on the deep ocean floor. Hundreds of species never encountered before were brought to the surface. After nearly four years, the research ship charted 140 square miles of ocean bottom and sounded every ocean except the Arctic. The deepest sounding was taken off the Mariana Islands in the Pacific, reaching a depth of five miles.

In the early days of sampling sediments on the ocean floor, scientists used a dredge similar to a bucket tied to the end of a steel cable. The major problem with this technique was that it sampled only the topmost layers of the ocean floor, which could not be recovered in the order they were laid down. In the early 1940s, scientists in Sweden invented a piston corer designed to retrieve a vertical section of the ocean floor intact. The piston corer consisted of a long barrel that plunged into the bottom mud under its

Figure 73 *Piston coring in the Gulf of Alaska.*

(Photo by P. R. Carlson, courtesy USGS)

own weight (Figs. 73 and 74). A piston was fired upward from the lower end of the barrel, sucking up sediments into a pipe. Scientists were thus able to bring up long, cylindrical cores of the ocean floor that dated millions of years old.

The oceanographic research vessel *Glomar Challenger,* commissioned for the Deep Sea Drilling Project in 1968, was developed by a consortium of American oceanographic institutions. Its primary purpose was to drill the ocean floor and take rotary core samples at hundreds of sites scattered around the world. The drill ship dangled beneath the hull a string of drill pipe as much as four miles long. When the drill bit reached the ocean bottom, it bored through the sediments under its own weight. The core was then retrieved through the drill string and brought to the surface.

After dating several cores taken from around various midocean ridges, the scientists discovered something truly remarkable. The sediments were found to be older and thicker the farther the ship drilled away from the deep-sea ridges. What was even more surprising was that the thickest and oldest sediments were not billions of years old as expected but less than 200 million years old.

In order to measure these sediments, scientists invented a seismic device for underwater use. Seismic waves, which are similar to sound waves, are used to locate sedimentary structures such as those that trap oil. An explosive charge or an air gun is set off below the sea surface, and the seismic waves are picked up by an array of hydrophones. Because seismic waves travel slower in soft sediments than they do in hard rock, the data could be used to calculate the thickness of different rock layers.

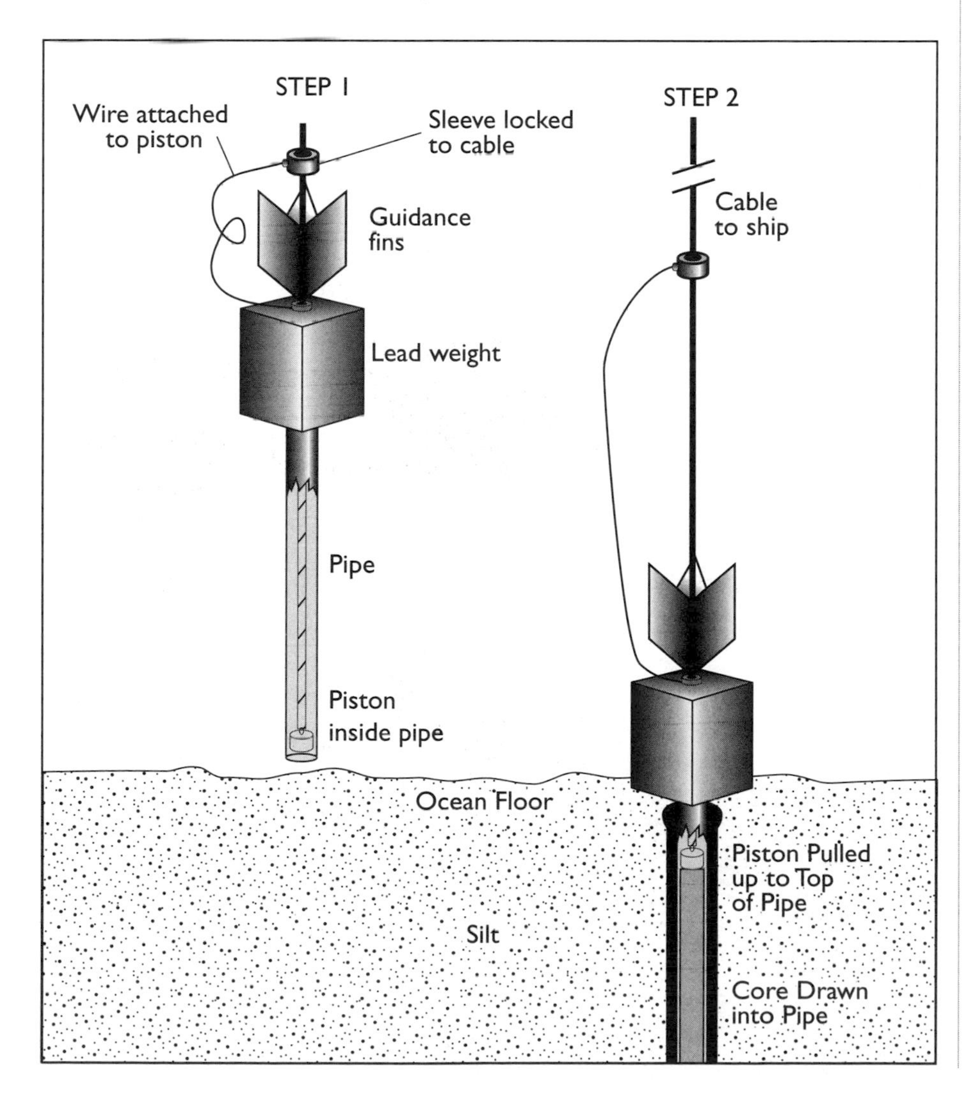

Figure 74 *A piston corer on the ocean floor.*

An ocean bottom seismograph was also lowered to the ocean floor, where it could record microearthquakes in the oceanic crust and then automatically rise to the surface for recovery (Fig. 75). These geophysical methods provided scientists with information about the ocean floor that could not be

Figure 75 *An ocean bottom seismograph provides direct observations of earthquakes on midocean ridges.*

(Photo courtesy USGS)

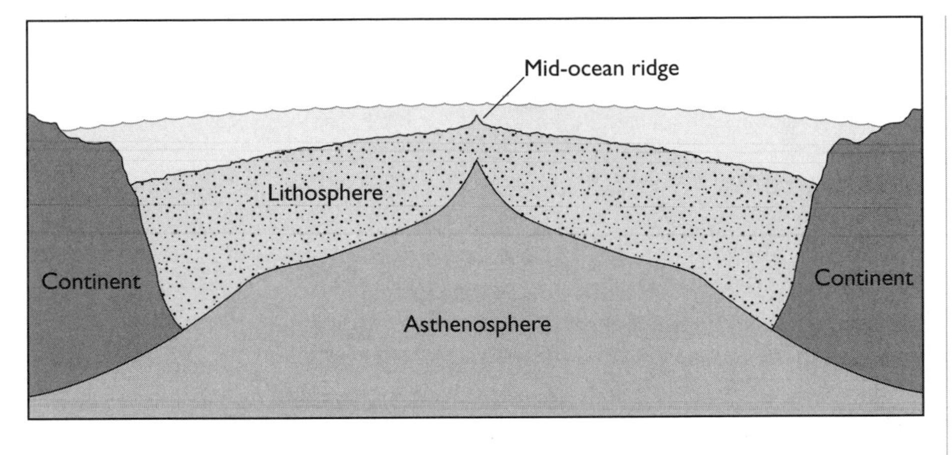

Figure 76 *Cross section of the Earth beneath the spreading Mid-Atlantic Ridge, which separated the New World from the Old World.*

obtained by direct means. Some of their findings, however, came as a complete surprise. Instead of sediments piled miles thick, they found on average sediments only a few thousand feet in thickness, as though some sort of vacuum cleaner had swept the sediments off the ocean floor.

Sonar, a device that bounces sound waves off the ocean bottom, gave scientists an important tool for mapping undersea terrain. As ships crisscrossed the Atlantic Ocean, onboard sonographs painted a remarkable picture of the ocean floor. Lying in the middle of the ocean was a huge submarine mountain range, surpassing in scale the Alps and Himalayas combined. However, what was truly remarkable about this unusual mountain range was that it bisected the Atlantic almost exactly down the middle, weaving halfway between the continents (Fig. 76).

As more detailed maps of the ocean floor were made, the Mid-Atlantic Ridge, as it was later named, became the most peculiar mountain range ever known. It runs from Iceland in the north to Bovet Island about 1,000 miles off Antarctica. The midocean ridge is a string of volcanic seamounts created by molten magma upwelling from within the mantle. Through the middle of the 10,000-foot-high ridge crest ran a deep trough as though it were a giant crack in the Earth's crust. It was 4 miles deep in places and up to 15 miles wide, making it the largest canyon in the world.

A set of closely spaced fracture zones dissects the Mid-Atlantic Ridge in the equatorial Atlantic. The largest of these structures is the Romanche Fracture Zone (Fig. 77), which offsets the axis of the ridge by nearly 600 miles. The floor of the Romanche Trench is as much as 5 miles below sea level. The highest parts of the ridges on either side of the trench are less than a mile below sea level, giving a vertical relief four times that of the Grand Canyon. The shallowest portion of the ridge is capped with a fossil coral reef, suggesting it was above sea level some 5 million years ago. Many similar fracture zones equally impressive span the area, culminating in a sequence of

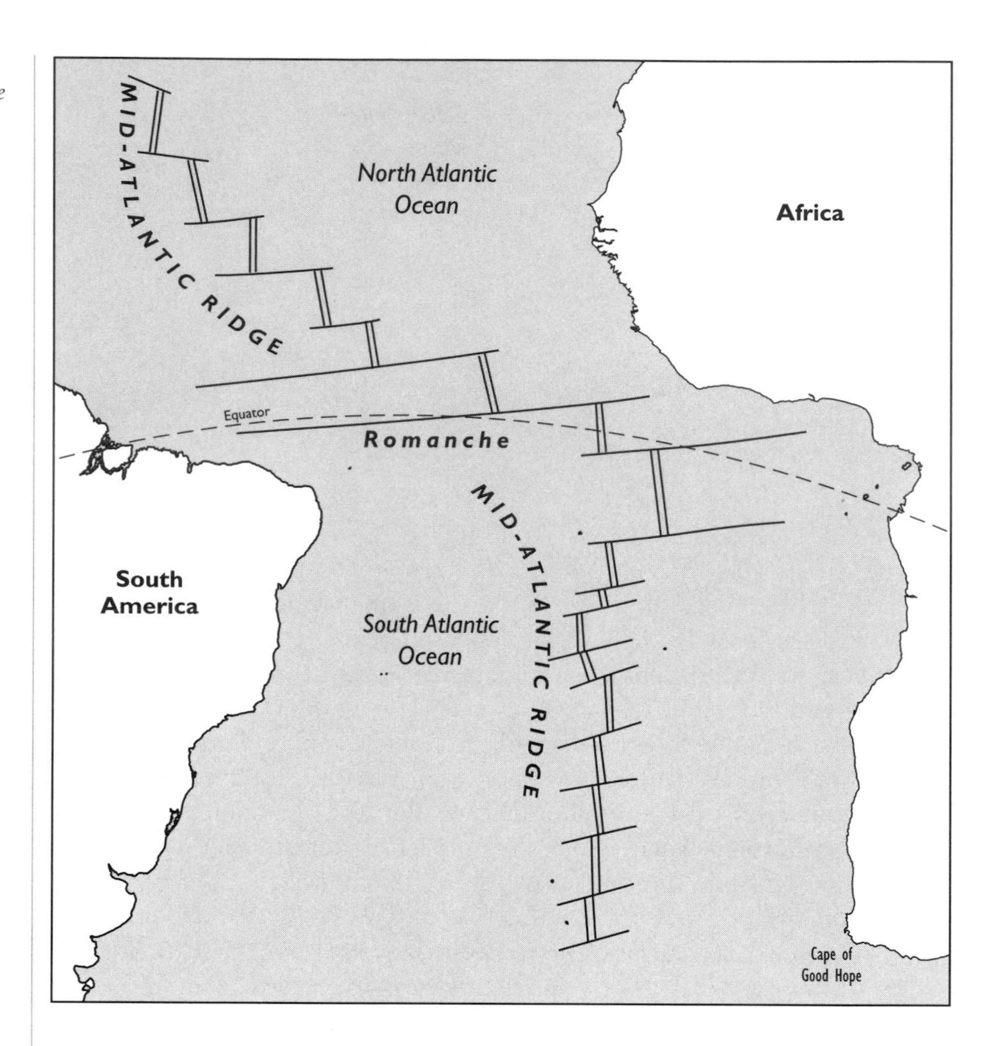

Figure 77 *The Romanche Fracture Zone is the largest offset of the Mid-Atlantic Ridge.*

troughs and transverse ridges that are several hundred miles wide. The resulting terrain is unmatched in size and ruggedness anywhere else on Earth.

The counterpart of the Mid-Atlantic Ridge is the East Pacific Rise, which stretches 6,000 miles from the Antarctic Circle to the Gulf of California and is a member of the world's largest undersea mountain chain. It lies on the eastern edge of the Pacific plate, marking the boundary between the Pacific and Cocos plates. The rift system is a network of midocean ridges, which lie mostly at a depth of about 1.5 miles. Each rift is a narrow fracture zone. Plates of the oceanic crust diverge at an average rate of about five inches a year in these rifts, resulting in less topographical relief on the ocean floor. The active tectonic zone of a fast spreading ridge is usually quite narrow, usually less than four miles wide.

Massive undersea volcanic eruptions from fissures on the ocean floor at spreading centers along the East Pacific Rise create large megaplumes of hot water. The megaplumes are produced by short periods of intense volcanic activity and can measure up to 50 to 60 miles wide. When the ridge splits open, it releases vast quantities of hot water held under great pressure beneath the surface. At the same time, the lava erupts in an act of catastrophic seafloor spreading. In a matter of a few hours, or at most a few days, up to 100 million tons of superheated water gushes from a large crack in the ocean crust up to several miles long.

A small lithospheric plate, called a microplate, about the size of Ohio sits at the junction of the Pacific, Nazca, and Antarctic plates in the South Pacific about 2,000 miles west of South America. Seafloor spreading along the border between the plates adds new ocean crust onto their edges, causing the plates to diverge. The different rates of seafloor spreading have caused the microplate at the hub of the spreading ridges, which fan out like spokes of a bicycle wheel, to rotate one-quarter turn clockwise in the last 4 million years. A similar microplate near Easter Island to the north has spun nearly 90 degrees over the last 3 to 4 million years, which suggests that most microplates behave in this manner.

Several times in the past 200 million years, other plates and their associated spreading centers have vanished beneath the continents that surround the Pacific Basin, which had a substantial impact on the coastal geology. Three plates bordering the Pacific Ocean—the Nazca, Antarctic, and South American plates—come together in an unusual triple junction. The first two plates spread apart along a boundary called the Chile Ridge off the west coast of South America similar to the way the Americas drift away from Eurasia and Africa along the Mid-Atlantic Ridge. The Chile Ridge lies off the Chilean continental shelf at a depth of more than 10,000 feet. Along its axis, magma rises from deep within the Earth and piles up into mounds forming undersea volcanoes.

The Nazca plate moving northeast subducts beneath the westward-moving South American plate at the Peru-Chile Trench. The eastern edge of the Nazca plate is subducting at a rate of about 50 miles every million years, faster than its western edge is growing. An analogy would be an escalator whose top is moving toward the bottom, consuming the rising steps in between. In essence, the Chile Trench is consuming the Chile Ridge, which will eventually disappear altogether.

THE SPREADING SEAFLOOR

In the late 1950s, during the height of the cold war, American and Russian oceanographic vessels crisscrossed the seas, mapping the ocean floor so deep-diving ballistic missile submarines could navigate without crashing into uncharted seamounts. After all the data had been compiled, the maps showed something that was entirely unexpected. The submerged mountains and undersea ridges formed a continuous 40,000-mile-long chain (Fig. 78) several hundred miles wide, up to 10,000 feet high, and circling the globe like the stitching on a baseball.

Even though it was deep beneath the sea, this midocean ridge system easily became the dominant feature on the face of the planet, extending over an area greater than all the major terrestrial mountain ranges combined. Moreover, it exhibits many unusual features including massive peaks, sawtooth ridges, earthquake-fractured cliffs, deep valleys, and lava formations of every description (Fig. 79). Along much of its length, the ridge system is carved down the middle by a sharp break or rift that is the center of an intense heat flow. Furthermore, the midocean ridges are the sites of frequent earthquakes and volcanic eruptions, as though the entire system is a series of giant cracks in the Earth's crust.

The midocean ridge does not form a continuous line but is broken into small, straight sections called spreading centers. The axis of the midocean ridge is offset laterally in a roughly east-west direction by transform faults (Fig. 80). They range from a few miles to a few hundred miles long and are encountered every 20 to 60 miles along the ridge system. The transform faults were created when pieces of oceanic crust slid past each other, resulting in a series of fracture zones. These are long, narrow, linear regions up to about 40 miles wide and comprise irregular ridges and valleys aligned in a stairstep fashion.

Friction between segments produces strong shearing forces, wrenching the ocean floor into steep canyons. Transform faults dissecting the Mid-Atlantic

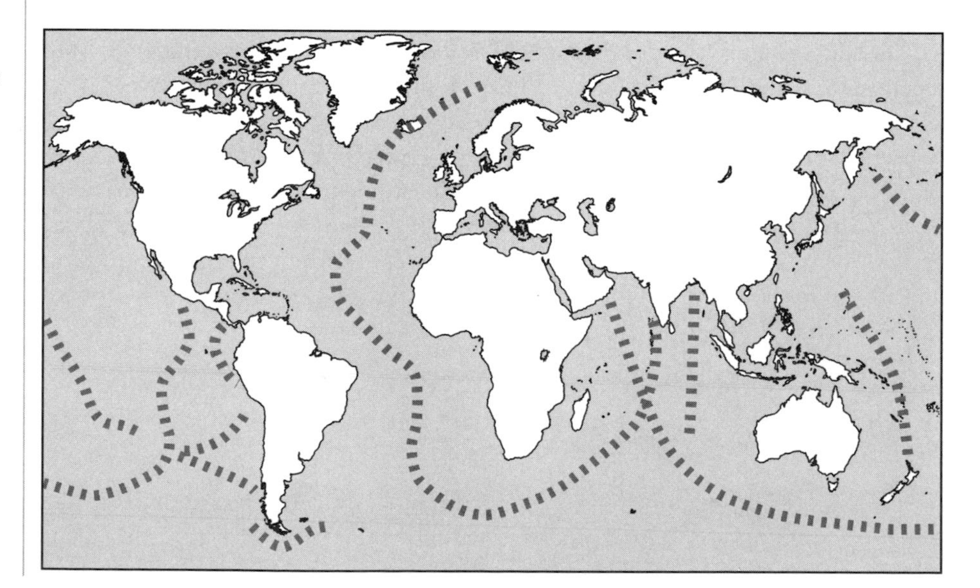

Figure 78 *Midocean ridges, where crustal plates are spreading apart, comprise the most extensive mountain chains in the world and are centers of intense volcanic activity.*

Figure 79 *The rim of a lava lake collapse pit on the Juan de Fuca Ridge in the east Pacific.*
(Photo courtesy USGS)

Ridge are generally more rugged than those of the East Pacific Rise. Moreover, fewer widely spaced transform faults exist along the East Pacific Rise, where the rate of seafloor spreading is 5 to 10 times faster than the Mid-Atlantic Ridge. Therefore, the crust affected by transform faults is younger, hotter, and less rigid in the Pacific than in the Atlantic, giving the undersea terrain much less relief.

Several spreading centers, 20 to 30 miles long, are separated by non-transform offsets that are up to 15 miles wide. The end of one spreading center might run past the end of another, or the tips of the segments might bend toward each other. The friction between the plates gives rise to strong shearing forces, wrenching the ocean floor into steep canyons. These faults appear to have resulted from lateral strain, which is the way movable plates react on the surface of a sphere.

This activity appears to be more intense in the Atlantic. There, the mid-ocean ridge is steeper and more jagged than in the Pacific or the Indian Oceans, where branches of oceanic ridges actually dive under continents. The ocean floor is indeed far more active and younger than had been imagined. Temperature surveys showed heat seeping out of the Earth in the mountain-

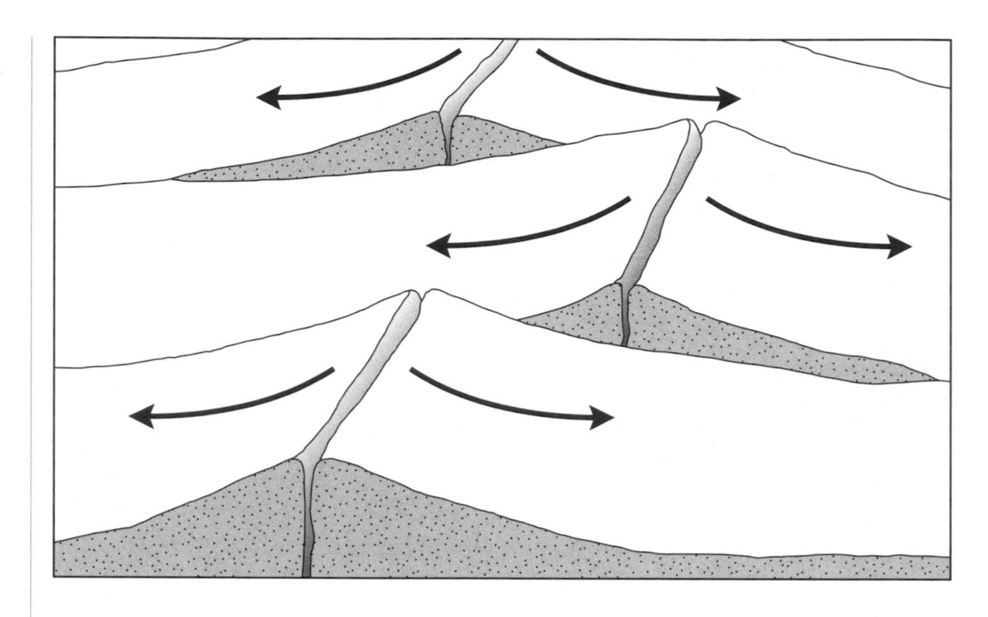

Figure 80 *Transform faults at spreading centers on the ocean floor.*

ous regions of the middle Atlantic. Volcanic activity in the ridges indicate they were adding new material to the ocean floor.

In contrast, the trenches off the continents suggest they are the sites where old oceanic crust disappeared into the Earth's interior. Gravitational surveys in these regions were highly confusing, however. In the deep trenches off the edges of certain continents, gravity was found to be much too weak to be responsible for the downward pull on the sediments in vast geosynclines, which are elongated basins in the crust caused by the weight of sediments washed off the continents.

Scientists explained the cracks in the ocean floor by envisioning an expanding Earth due to an increase in the internally generated heat or a decrease in gravity. The weakening of the gravitational field could cause the Earth to bulge out, forming cracks along the crust similar to cracks on a boiled egg. Most scientists, however, rejected the expansion hypothesis because the force of gravity is considered a constant and has never been known to change. Furthermore, if the Earth of today was significantly larger than in the past, obvious defects would be found in the shapes of the continents. Their continental margins would not fit as well as they do during reconstructions.

Observation of these and other interesting features on the ocean floor led the American geologist Harry Hess to propose a process called seafloor spreading in 1962. His hypothesis described the creation and destruction of the ocean floor. However, it did not specify rigid lithospheric plates. Hess devised a solution that called on an outward expansion of a different sort, which relied on the Earth's internal convection currents to raise material to the surface.

According to Hess's theory, the process of seafloor spreading first begins with hot rocks rising toward the surface by convection currents in the upper mantle. Upon reaching the underside of the lithosphere, comprising the solid portion of the upper mantle and the overriding crust, the mantle rocks spread out laterally, cool, and descend back into the Earth's interior. The mantle completes a single convection loop in perhaps 100 million years or so. The constant pressure against the bottom of the lithosphere creates fractures, which cause it to rift apart.

As the convection currents flow out on either side of the fracture, they carry the now separated parts of the lithosphere along with them. The opening continues to widen. This process reduces the pressure. The mantle rocks melt and ascend through the fractured zone, where the molten rock finds easy passage through the 60 miles or so of lithosphere until it reaches the bottom of the crust. There, it forms magma chambers that press against the crust and widen it further. Meanwhile, magma pours out onto the ocean floor from the trough between ridge crests, adding layer upon layer of basalt to both sides of the spreading ridge (Fig. 81). The pressure of the upwelling magma forces both sides of the ridge farther apart and pushes the ocean floor away from the midocean ridge.

Any new material added to the ocean floor at midocean spreading ridges must be subtracted somewhere else. Therefore, Hess suggested that the old seafloor and the lithosphere upon which it rides were destroyed in the deep-sea trenches at the edges of continents or along volcanic island arcs. The rocks dive back into the Earth, where they are broken up, remelted, and reabsorbed into the mantle to be used again in a continuous cycle.

The theory cleared up many problems connected with the mysterious features on the ocean floor, including the midocean ridges, the relatively

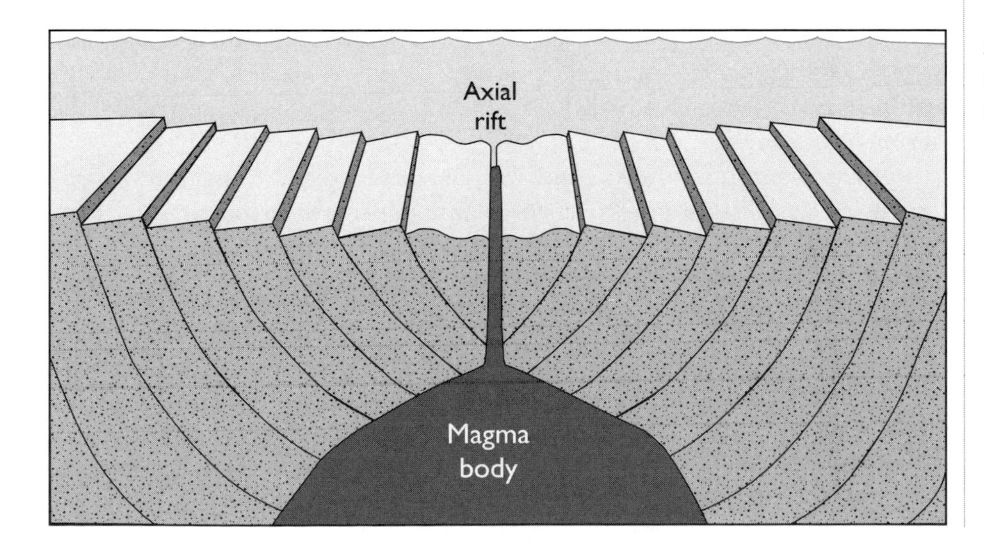

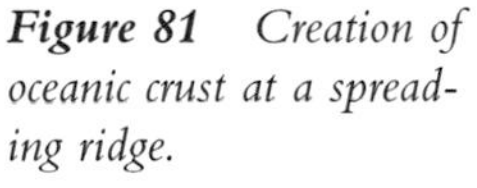
Figure 81 *Creation of oceanic crust at a spreading ridge.*

young ages of rocks in the oceanic crust, and the formation of island arcs. More importantly, though, here at last was the long-sought mechanism for continental drift. The continents do not plow through the ocean crust like an icebreaker slices through frozen seas, as Alfred Wegener envisioned. Instead, they ride like ships frozen in ice floes.

BASALTIC MAGMA

Most of the Earth's surface above and below the sea is of volcanic origin. The shifting of lithospheric plates on the surface of the planet generates new crust in a continuing process of crustal rejuvenation. About 80 percent of oceanic volcanism occurs along spreading ridges, where magma wells up from the mantle and spews out onto the ocean floor. The spreading crustal plates grow by the steady accretion of solidifying magma to their edges. More than one square mile of new oceanic crust, comprising some five cubic miles of new basalt, is generated in this manner each year.

Seafloor spreading has been described as a wound that never heals as magma slowly oozes out of the mantle. However, sometimes gigantic flows erupt on the ocean floor with enough new basalt to pave the entire U.S. interstate highway system 10 times over. The magma also flows from isolated volcanic structures, called seamounts, strung out in chains across the interior of plates.

Mantle material that extrudes onto the surface is black basalt, which is rich in silicates of iron and magnesium. Most of the world's nearly 600 active volcanoes are entirely or predominately basaltic. The magma from which basalt is formed originated in a zone of partial melting in the upper mantle more than 60 miles below the surface. The semimolten rock at this depth is less dense than the surrounding mantle material and rises slowly toward the surface. As the magma ascends, the pressure decreases and more mantle material melts. Volatiles such as dissolved water and gases also aid in making the magma flow easily.

Magma rising toward the surface contributes to the formation of shallow reservoirs or feeder pipes that are the immediate source for volcanic activity. The magma chambers closest to the surface are under spreading ridges, where the crust is only six miles or less thick. Large magma chambers lie under fast spreading ridges where the lithosphere is being created at a high rate such as those in the Pacific. Narrow magma chambers lie under slow spreading ridges such as those in the Atlantic. As the magma chamber swells with molten rock and begins to expand, the crest of the spreading ridge is pushed upward by the buoyant forces generated by the magma.

The magma rises in narrow plumes that mushroom out along the spreading ridge. It wells up as a passive response to plate divergence, somewhat like having the lid taken off a pressure cooker. Only the center of the plume is hot enough to rise all the way to the surface, however. If the entire plume were to erupt, it could build a massive volcano several miles high. Not all magma is extruded onto the ocean floor. Some solidifies within the conduits above the magma chamber and forms massive vertical sheets called dikes.

The composition of magmas varies according to the source materials as well as the depth within the mantle from which they originated. Degrees of partial melting of mantle rocks, partial crystallization that enriches the melt with silica, and assimilation of a variety of crustal rocks in the mantle affect the ultimate composition of the magma. When the erupting magma rises toward the surface, it incorporates a variety of rock types along the way and changes composition, which is the major controlling factor determining the type of eruption.

As the magma reaches the surface, it erupts a variety of gases, liquids, and solids. Volcanic gases mostly consist of steam, carbon dioxide, sulfur dioxide, and hydrochloric acid. The gases are dissolved in the magma and released as it rises toward the surface and pressures decrease. The composition of the magma determines its viscosity and type of eruption, whether mild or explosive. If the magma is highly fluid and contains little dissolved gas when reaching the surface, it flows from a volcanic vent or fissure as basaltic lava, and the eruption is usually quite mild, as with Hawaiian Island volcanoes (Fig. 82).

If the rising magma contains a large quantity of dissolved gases, it suddenly separates into liquid and bubbles. With decreasing pressure, the bubbles expand explosively and burst the surrounding fluid, fracturing the magma into fragments. The fragments are driven upward by the force of the expansion and hurled far above the volcano. The fragments cool and solidify during their flight through the air. They can range in size from fine dust-size particles to large blocks weighing several tons.

MAGNETIC STRIPES

The more scientists probed the ocean floor, the more complex it became. Sensitive magnetic recording instruments called magnetometers were towed behind ships over the midocean ridges, and revealed that the magnetic patterns locked in the volcanic rocks on the ocean floor alternated from north to south and were mirror images of each other on both sides of the ridge crest. The magnetic fields captured in the rocks showed not only the past position of the magnetic poles but their polarity as well.

Recognition of the reversal of the geomagnetic field began in the early 1950s. Over the last 170 million years, the Earth's magnetic field has reversed 300 times. No reversals occurred during long stretches of the Permian and Cretaceous periods. Furthermore, a sudden polar shift of 10 to 15 degrees

Figure 82 *The development of a 1,900-foot-high fountain during the December 17, 1959 eruption of Kilauea, Hawaii.*

(Photo by D. H. Richter, courtesy USGS)

occurred between 100 million and 70 million years ago. Since about 90 million years ago, reversals have steadily become more frequent, and the polar wandering has decreased to only about 5 degrees.

The last time the geomagnetic field reversed was about 780,000 years ago, and the Earth appears to be well overdo for another one. Two thousand years ago, the magnetic field was considerably stronger than it is today. The Earth's magnetic field seems to have weakened over the past 150 years, amounting to a loss of about 1 percent per decade. If the present rate of decay

TABLE 8 COMPARISON OF MAGNETIC REVERSALS WITH OTHER PHENOMENA (IN MILLION YEARS AGO)

Magnetic Reversal	Unusual Cold	Meteorite Activity	Sea Level Drops	Mass Extinctions
0.7	0.7	0.7		
1.9	1.9	1.9		
2.0	2.0			
10				11
40			37–20	37
70			70–60	65
130			132–125	137
160			165–140	173

continues, the field could reach zero and go into another reversal within the next 1,000 years or so.

In 1963, the British geologists Fred Vine and Drummond Mathews thought that magnetic reversal would be a decisive test for seafloor spreading. As the basalts of the midocean ridges cool, the magnetic fields of their iron molecules line up in the direction of the Earth's magnetic field. When the ocean floor spreads out on both sides of the ridge, the basalts record the Earth's magnetic field during each successive reversal somewhat like a magnetic tape recording of the history of the geomagnetic field. Normal polarities in the rocks are reinforced by the present magnetic field, while reversed polarities are weakened by it. This produces parallel bands of magnetic rocks of varying width and magnitude on both sides of the ridge that are mirror images of each other (Fig. 83).

Magnetic reversals also provide a means of dating practically the entire ocean floor because the reversals occur randomly and any set of patterns are unique in geologic history. The Canadian geophysicist J. Tuzo Wilson calculated the age of a number of magnetic stripes in selected parts of the ocean floor. Calculating the rate of spreading was a simple matter of determining the age of the magnetic stripes using radiometric dating techniques and measuring the distance from their points of origin at the midocean ridges. The rate of seafloor spreading has changed little over the past 100 million years. Periods of increased acceleration in the past, however, have been accompanied by an increase in volcanic activity. During the past 10 to 20 million years, a progressive acceleration has occurred, reaching a peak about 2 million years ago.

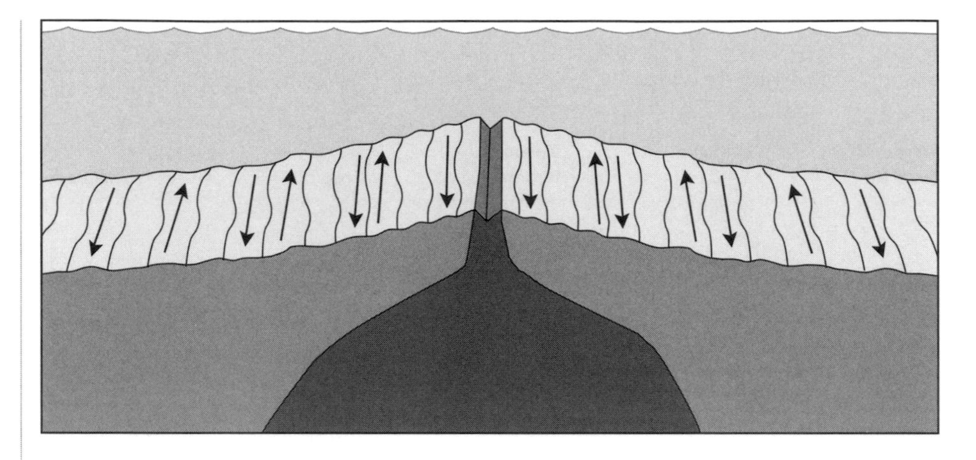

Figure 83 *As volcanic rock cools at midocean ridges, it polarizes in the direction of the Earth's magnetic field, providing a series of magnetic stripes on the ocean floor.*

In the Pacific Ocean, the spreading rates are upward of six inches per year, which accounts for less topographical relief at midocean ridges. The active tectonic zone of a fast spreading ridge is usually quite narrow, generally less than four miles across. In the Atlantic, the rates are much slower, only about one inch per year. This allows taller ridges to form. The Atlantic rift appeared to have opened sometime between 200 and 150 million years ago. This date is remarkably concurrent with Wegener's estimates for the breakup of the continents and the age of the oldest part of the ocean floor, which globally averages only 100 million years old.

SEAMOUNTS AND GUYOTS

Volcanic eruptions associated with midocean ridges are either fissure eruptions, the most common type, or those that build typical conical volcanic structures. Volcanic structures formed on or near the midocean ridges can develop into isolated peaks as they move away from the axis of seafloor spreading. During fissure eruptions, the magma oozes onto the ocean floor in the form of lava that bleeds through fissures in the trough between ridge crests and along lateral faults. Magma welling up along the entire length of the fissure forms large lava pools, similar to those of broad shield volcanoes such as the Hawaiian volcano Mauna Loa, the largest of its kind in the world (Fig. 84).

The two main types of lava formations in the midocean ridges are sheet flows and pillow, or tube, flows. Sheet flows are more prevalent in the active volcanic zone of fast spreading ridge segments, such as those of the East Pacific Rise. They consist of flat slabs of basalt usually less than eight inches thick. The lava that forms sheet flows is much more fluid than that responsible for pillow formations.

Pillow lavas (Fig. 85) appear as though basalt were squeezed out of a giant toothpaste tube. They are mostly found in slow spreading ridges such as the Mid-Atlantic Ridge, where the lava is much more viscous. The surface of the pillows often has corrugations or small ridges pointing in the direction of flow. The pillow lavas typically form small, elongated hills pointing downslope from the crest of the ridge. Below the pillow lavas is a middle layer composed of a sheeted-dike complex. Within this structure is a tangled mass of feeders, which bring magma to the surface. Beneath this is a layer composed of gabbros, which are coarse-grained basalts that crystallized slowly under high pressure.

Upwelling magma from the upper mantle at depths of more than 60 miles below the surface is concentrated in comparatively narrow conduits that lead to the main feeder column. When erupted onto the ocean floor, this forms elevated volcanic structures called seamounts, which are isolated and generally strung out in chains across the interior of a plate. Some seamounts are associated with extended fissures, along which magma wells up through a main conduit, piling successive lava flows on top of one another.

Seamounts associated with midocean ridges that grow tall enough to break the surface of the ocean become volcanic islands. Undersea volcanoes called guyots (pronounced "ghee-ohs") located in the Pacific once towered

Figure 84 *The Mauna Loa Volcano, Hawaii.* (Photo courtesy USGS)

Figure 85 *Pillow lava on the south bank of Webber Creek, Eagle District, Alaska.*

(Photo by E. Blackwelder, courtesy USGS)

above the sea. However, constant wave action eroded them below the sea surface as though the tops of the cones were sawed off (Fig. 86). The remarkable feature about these volcanoes is that the farther away they were from volcanically active regions of the ocean, the older and flatter they became. This observation suggests that the guyots wandered across the ocean floor away from their places of origin. In this respect, the islands appeared to have been pro-

Figure 86 *Guyots were once active volcanoes that moved away from their magma source and have since disappeared beneath the sea.*

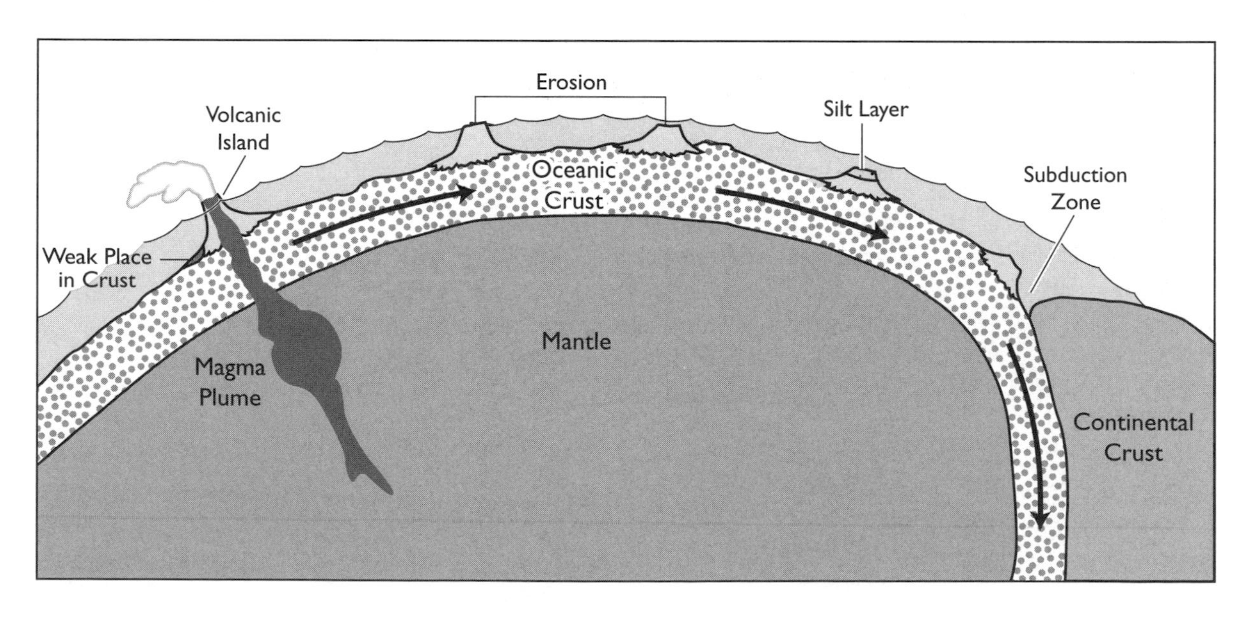

duced assembly line fashion. Each moves away in succession from a magma chamber, called a hot spot, lying beneath the ocean crust.

The most prominent island chain is the Hawaiian Islands. The youngest and most volcanically active island is Hawaii to the southeast, with progressively older islands having extinct volcanoes, trailing off to the northwest. By continuing from there, coral atolls and shoals formed when successive layers of coral grew on the flattened tops of long extinct volcanoes worn down below sea level. From these islands extend an associated chain of undersea volcanoes known as the Emperor Seamounts (Fig. 87). The Hawaiian Islands also lie parallel to two other island chains, the Austral Ridge and the Tuamotu Ridge. The volcanic islands associated with the Mid-Atlantic Ridge system include Iceland, the Azores, the Canary and Cape Verde Islands off West Africa, Ascension Island, and Tristan de Cunha. The volcanic

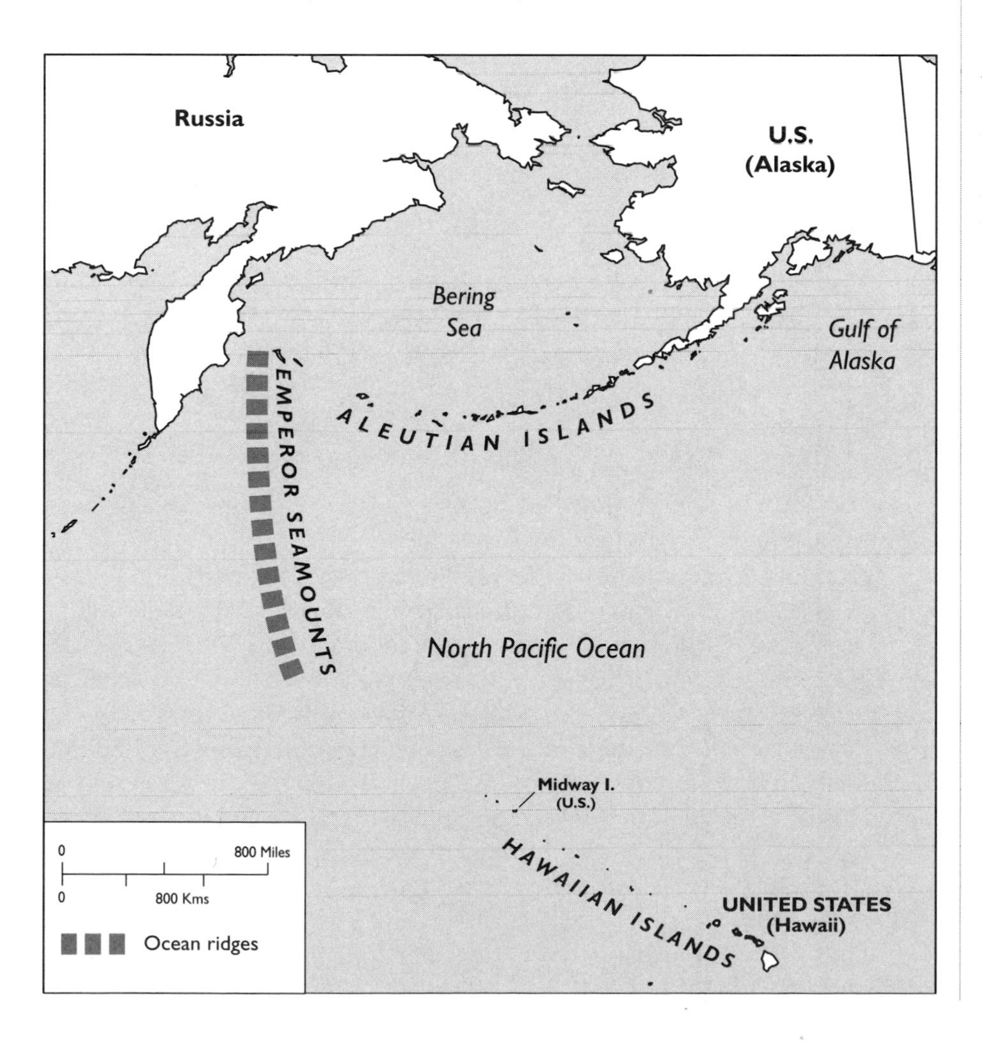

Figure 87 *The Emperor Seamounts and Hawaiian Islands in the North Pacific represent motions in the Pacific plate over a volcanic hot spot.*

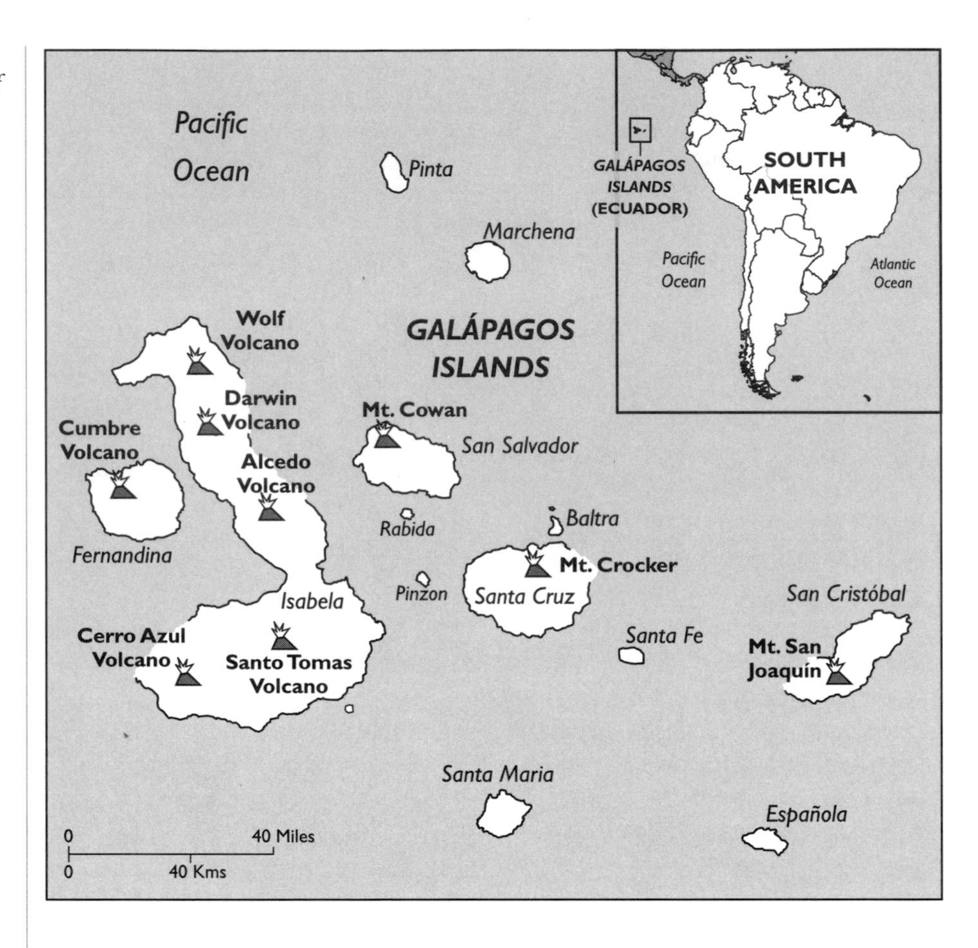

Figure 88 The Galápagos Islands west of Ecuador.

islands associated with the East Pacific Rise are the Galapagos Islands west of Ecuador (Fig. 88).

Volcanoes formed on or near the midocean ridges can develop into isolated peaks as they move away from the ridge axis during seafloor spreading. The ocean floor thickens as it leaves the midocean ridge. This process can influence a volcano's height because as it moves away from the spreading ridge axis, the thicker crust can support a greater mass on the ocean floor. The ocean crust can also bend like a rubber mat under the massive weight of a seamount. For instance, the crust beneath Hawaii bulges by as much as six miles. A volcano formed at a midocean ridge cannot increase its mass unless it continues to be supplied with magma after it leaves the vicinity of the ridge. When its source of magma is cut off, erosion begins to wear the volcano down until it finally sinks below the sea.

If a midocean ridge passes over a hot spot, the plume directly under the spreading center augments the flow of molten rock welling up from the asthenosphere to form new crust. The crust over the hot spot is therefore thicker than it is along the rest of the ridge, resulting in a plateau rising above the surrounding seafloor. The Ninety East Ridge is a ruler-straight undersea mountain chain that runs 3,000 miles south of the Bay of Bengal, India and was formed when the Indian plate passed over a hot spot on its way to Asia.

The most striking example of rift zone hot-spot volcanism is found on Iceland (Fig. 89). It straddles the Mid-Atlantic Ridge and was raised above the sea about 16 million years ago. Along the ridge, the abnormally elevated topography extends in either direction for a distance of about 900 miles, of which 350 miles lies above sea level. South of Iceland, the broad plateau tapers off to form the typical Mid-Atlantic Ridge.

The powerful upwelling currents also produce glacier-covered volcanic peaks up to one mile high. In 1918, an eruption under a glacier unleashed a

Figure 89 *Seawater is sprayed onto lava flow from the outer harbor of Vestmannaeyjar, Iceland, from the May 1973 eruption on Heimaey.* (Photo courtesy USGS)

flood of meltwater 20 times greater than the flow of the Amazon, the world's largest river. A repeat performance occurred in 1996 when massive floods from gushing meltwaters and icebergs dashed 20 miles to the seacoast. These glacial bursts have been known to Icelanders since the 12th century. In a geologically brief period, the Mid-Atlantic Ridge will move away from the hot spot, carrying Iceland along with it. By lacking a source of magma, the previously active volcanoes will cease erupting, and Iceland will become just a cold, ice-covered, uninhabitable island.

The East African Rift Valley extends from the shores of Mozambique to the Red Sea, where it splits to form the Afar Triangle in Ethiopia. Afar is perhaps one of the best examples of a triple junction created by the doming of the crust over a hot spot. The Red Sea and the Gulf of Aden represent two arms of a three-armed rift, with the third arm heading into Ethiopia. For the past 25 to 30 million years, the Afar Triangle has been stewing with volcanism, alternating between sea and dry land. The tiny African nation of Djibouti offers the unusual phenomenon of oceanic crust being extruded as dry land.

Most volcanoes, however, never make it to the surface of the ocean and remain as isolated, undersea volcanoes. Since the crust under the Pacific Ocean is more volcanically active, it has a higher density of seamounts, some 5,000 in all, than the Atlantic or Indian Oceans. The number of undersea volcanoes increases with increasing age and thickness of the crust. The tallest seamounts, which rise more than 2.5 miles above the seafloor, are located in the western Pacific near the Philippine Islands, where the crust is more than 100 million years old. The average density of Pacific seamounts is between 5 and 10 volcanoes per 5,000 square miles of ocean floor, a considerably higher density than on the continents.

ABYSSAL RED CLAY

Further evidence for seafloor spreading was found with the discovery of abyssal red clay. The floor of the Atlantic resembles a huge conveyor belt, transporting lithosphere from its point of origin at the Mid-Atlantic Ridge to its final destination down the Pacific subduction zones. The ocean floor at the crest of the midocean ridge consists mostly of basalt. By continuing away from the crest, the bare rock is covered with an increasing thickness of sediments, composed mostly of red clay from detritus material washed off the continents. Near the ridge crest, the sediments are predominantly composed of calcareous ooze built up by a rain of decomposed skeletons from microorganisms.

Farther away from the ridge crest, the slope falls below the calcium carbonate compensation zone, where calcareous sediments dissolve in seawater at an average depth of about two miles. Therefore, only red clay should exist in deep water far from the crest of the midocean ridge. However, rock cores taken from the abyssal plains near continental shelves, where the oceanic crust is the oldest and the deepest, indicate thin layers of calcium carbonate below thick beds of red clay and above hard volcanic rock. Apparently, the red clay protects the calcium carbonate from being dissolved in the deep waters of the abyss. This observation implies that the midocean ridge was the source of the calcium carbonate discovered near continental margins and that the seafloor appeared to be moving across the ocean basin, carrying the sediments along with it.

After seeing how new oceanic crust is created at midocean ridges, the next chapter will cover the counterpart of seafloor spreading—the subduction of old oceanic crust into the mantle.

6

SUBDUCTION ZONES

THE DEEP-SEA TRENCHES

This chapter explores the ocean floor for evidence of plate subduction. After observing overwhelming evidence on the bottom of the ocean for seafloor spreading and the origin of oceanic crust, scientists needed a convincing theory to account for the disappearance of old oceanic crust. According to the theory of plate tectonics, the creation of new oceanic crust at midocean spreading ridges is matched by the destruction of old oceanic crust at subduction zones, where lithospheric plates along with their overlying sediments are forced down into the mantle.

Most subduction zones are in the western Pacific (Fig. 90), which explains why nowhere is the oceanic crust older than about 170 million years. As the rigid lithospheric plate carrying the oceanic crust descends into the Earth's interior, it slowly breaks up and melts. Over millions of years, it is absorbed into the general circulation of the mantle. The subducted plate also supplies molten magma for volcanoes, most of which surround the Pacific Basin in a zone called the Ring of Fire.

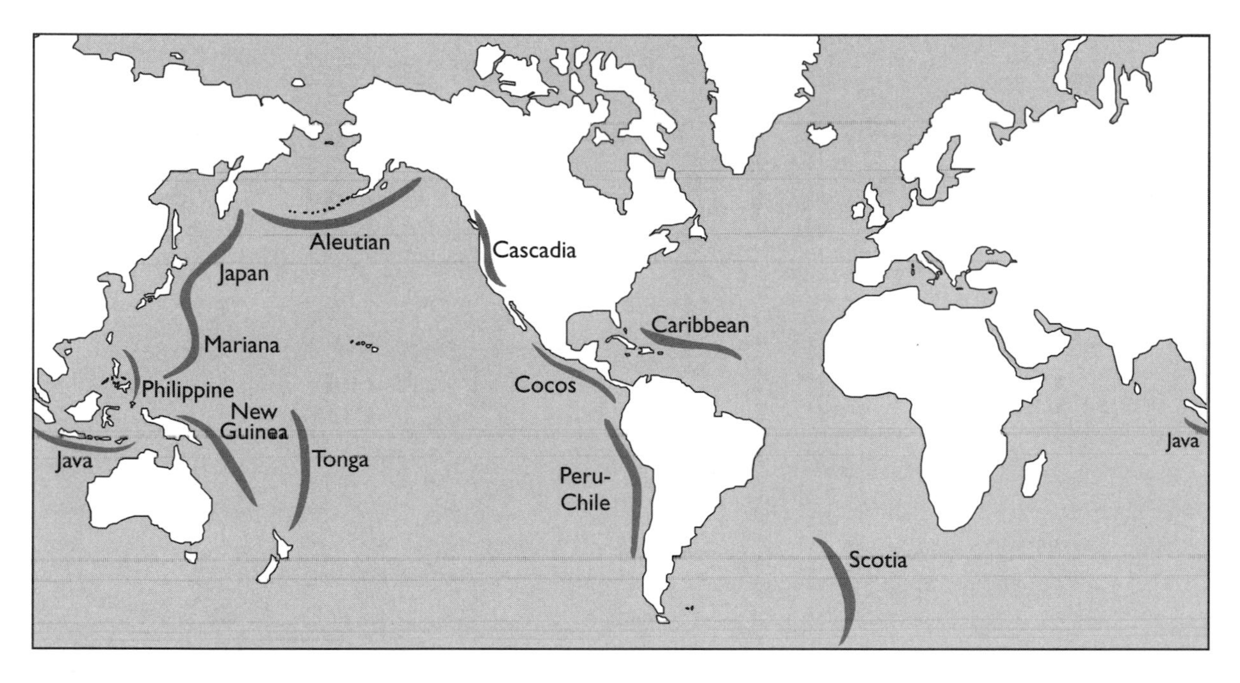

Figure 90 *The major trenches of the world, where crust plates are subducted into the Earth's interior.*

OCEANIC TRENCHES

In the early 1870s, the British oceanographic ship HMS *Challenger* was taking samples in the deep waters off the Mariana Islands in the western Pacific when it encountered a deep trough known as the Mariana Trench, which is, in effect, the lowest point on Earth (Table 9). It forms a long line northward from the island of Guam and is the deepest trench in the world, reaching a depth of about 36,000 feet below sea level.

Since the Earth is not expanding as once thought, new oceanic crust created at midocean spreading ridges has to be displaced somewhere else, and the deep-sea trenches seem to be the likeliest places. Scientists had long thought that the deep-sea trenches off continental margins and island arcs were great bulges in the ocean crust called geosynclines, resulting from the tremendous weight of sediments washed off the continents. Gravity surveys conducted over the trenches, however, indicated that no such material exists. The gravity in the area was found to be much too weak to be responsible for the downward pulling of the seafloor.

New oceanic crust generated by seafloor spreading in the Atlantic and eastern Pacific is offset by the subduction of old oceanic crust along the rim of the Pacific. Most subduction zones are in the western Pacific, which accounts for the oceanic crust being nowhere older than 170 million years. The farther a plate extends from its place of origin at a midocean spreading

TABLE 9 THE WORLD'S OCEAN TRENCHES

Trench	Depth (Miles)	Width (Miles)	Length (Miles)
Peru-Chile	5.0	62	3,700
Java	4.7	50	2,800
Aleutian	4.8	31	2,300
Middle America	4.2	25	1,700
Mariana	6.8	43	1,600
Kuril-Kamchatka	6.5	74	1,400
Puerto Rico	5.2	74	960
South Sandwich	5.2	56	900
Philippine	6.5	37	870
Tonga	6.7	34	870
Japan	5.2	62	500

center, the thicker and colder it becomes. By the time the plate reaches a subduction zone, it has cooled so much since its formation that it begins to thicken as more material from the asthenosphere adheres to its underside. Eventually, the plate becomes so dense it sinks into the mantle, and the line of subduction creates a deep-sea trench. As the subducted portion of the plate dives into the mantle, the rest of the plate, which might be carrying a continent, is pulled along with it, similar to a locomotive pulling a freight train.

Scientists generally thought that the push plates received from the expansion of the ocean floor at spreading ridges would be sufficient to force them into the mantle at subduction zones. However, drag at the base of the plates can greatly resist plate motion. Therefore, an additional source of energy is needed to drive the plates. For this purpose, the force of gravity is called upon to provide the driving mechanism. Pull is favored over push to overcome the resistance caused by plate drag. Therefore, the pull of a sinking slab of ocean crust is the strongest force moving the plates around the surface of the Earth.

An additional force that might help to overcome the resistance caused by plate drag is the pull the sinking plate receives by mantle convection currents. The magnitude of this force depends on the length of the subduction zone, the rate of subduction, and the amount of trench suction. With these forces in place, the plates could practically drive themselves without the aid of seafloor spreading. Therefore, the upwelling of magma at midocean spreading ridges might simply be a passive response to the plates being pulled apart by subduction. The seafloor spreading rate is not always the same as the rate of subduction, however, resulting in the lateral motion of associated midocean ridges.

Subduction zones are regions of low heat flow and high gravity. Conversely, the associated island arcs (Fig. 91) are regions of high heat flow and low gravity due to extensive volcanic activity. The deep-sea trenches are regions of intense volcanism, producing the most explosive volcanoes on Earth. Volcanic island arcs fringe the trenches, and each has similar curves and similar volcanic origins. The trenches produce an arc because this is the geometric figure formed when a plane cuts a sphere, in the same manner that a rigid lithospheric plate subducts into the spherical mantle.

The trenches are also sites of almost continuous earthquake activity deep in the bowels of the Earth. Plate subduction causes stresses to build into the descending lithosphere, producing deep-seated earthquakes that outline the boundaries of the plate. As plates slide past each other along subduction zones, they create highly destructive earthquakes, such as those that have always plagued Japan and the Philippines along with other islands connected with subduction zones.

A band of shallow earthquakes clustered in a line running through Micronesia might mark the earliest stages in the birth of a subduction zone, which is in the process of forming a trench to the north and west of New Guinea in the western Pacific. Gravity in the area is lower than normal, which

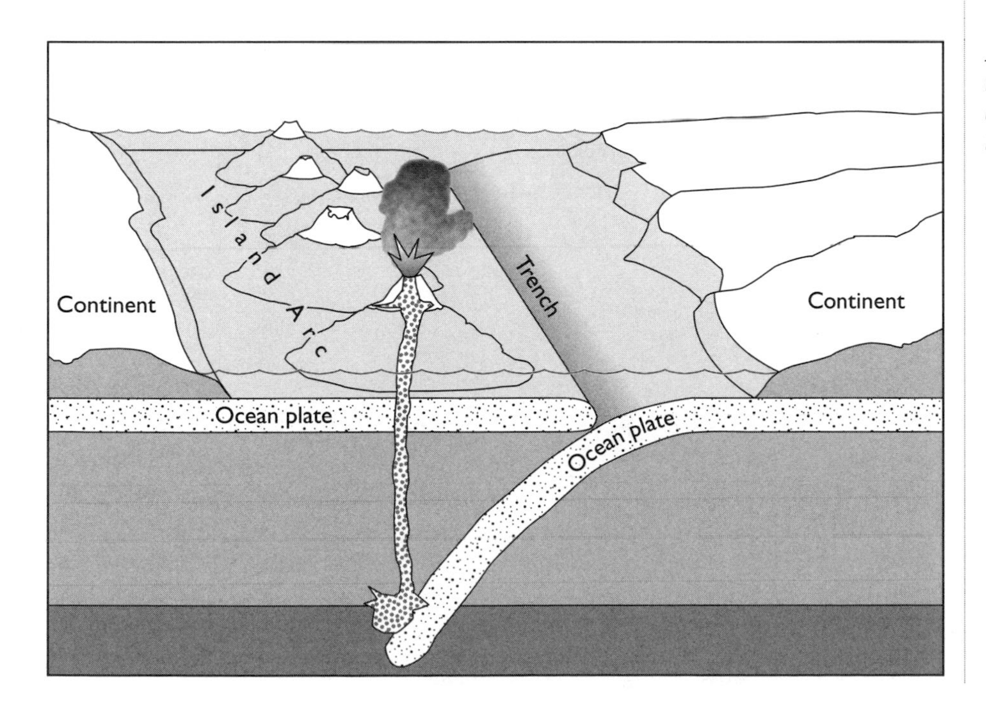

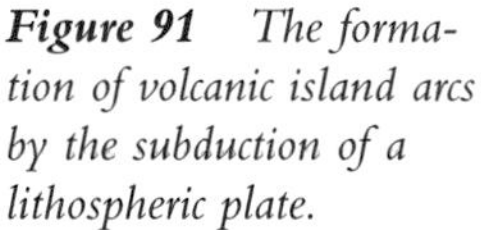
Figure 91 *The formation of volcanic island arcs by the subduction of a lithospheric plate.*

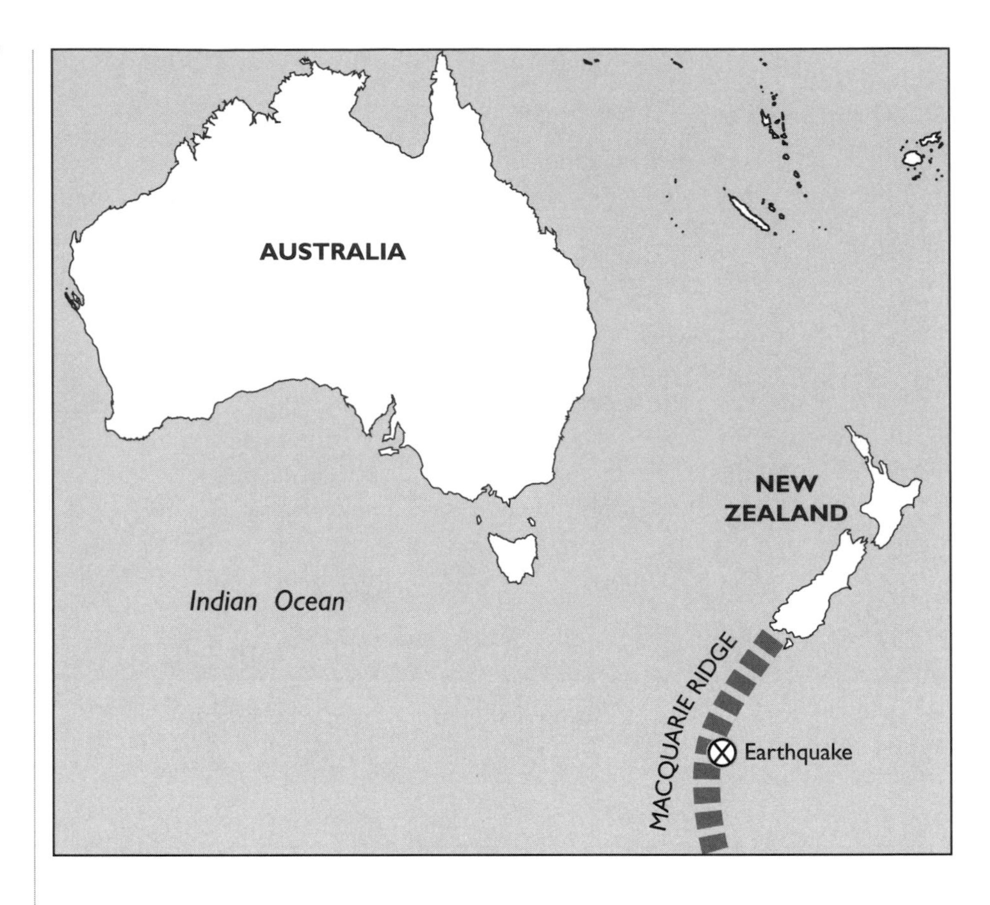

Figure 92 *Location of the Macquarie Ridge south of New Zealand.*

is expected over a trench due to the sagging of the ocean floor. In addition, a bulge in the crust to the south suggests that the edge of a slab of crust is beginning to dive into the Earth. The subduction process might not get fully under way for another 5 or 10 million years as the deep-sea trench nibbles away at the Pacific plate.

The ocean floor south of New Zealand might also be experiencing the early stages of subduction in the process of creating a deep-sea trench. The undersea structure forms a geologic scar on the floor of the Pacific known as the Macquarie Ridge (Fig. 92), an undersea chain of mountains and troughs running south from New Zealand. In 1989, a massive earthquake of magnitude 8.2 struck the ridge, which forms the boundary between the Australian and Pacific plates. The plates pass each other in opposite directions as the Australian plate moves northwest in relation to the Pacific plate.

Ruptures occur along vertical faults that allow the plates to slip past each other, creating large strike-slip earthquakes. The plates are also pressing together along dipping fault planes as they pass one another, creating smaller compressional earthquakes. This suggests that subduction is just beginning along the Macquarie Ridge. However, the separate dipping faults that flank the area have not yet connected to form a single large fault plane, a necessary first step before subduction commences.

The subduction zones are also regions of intense volcanic activity, producing the most explosive volcanoes on the planet. Magma reaching the surface erupts onto the ocean floor, creating new volcanic islands. Most volcanoes do not rise above sea level and become isolated undersea volcanic structures called seamounts. The Pacific Basin is more volcanically active and has a higher density of seamounts than the Atlantic or Indian basins. Subduction-zone volcanoes are highly explosive because their magmas contain large quantities of volatiles and gases that escape violently when reaching the surface.

SEAFLOOR TOPOGRAPHY

In 1978, the radar satellite *Seasat* (Fig. 93) precisely measured the distance to the ocean surface over most of the globe by bouncing radio waves off the ocean surface. Buried structures appeared in full view for the first time. Ridges and trenches on the ocean bottom produced corresponding hills and valleys on the surface of the ocean because of variations in the pull of gravity. The topography of the ocean surface showed bulges and depressions with several hundred feet of relief. However, because these surface variations ranged over a wide area, they are generally unrecognized on the open sea.

The topography of the ocean surface measured by radar altimetry from *Seasat* shows large bulges and depressions with a relief between highs and lows as much as 600 feet. Because these surface variations are spread out over a wide area, they remain undetected by line of sight. The shape of the surface is dictated by the pull of gravity from undersea mountains, ridges, trenches, and other structures of varying mass distributed over the ocean floor. Therefore, the variations in the height of the ocean surface are influenced by variations in the Earth's gravity field.

Massive undersea mountains produce large gravitational forces that make the water pile up around them, resulting in swells on the ocean surface. Because submarine trenches have less mass to attract the water, troughs form in the sea surface over these undersea structures. A trench one mile deep can cause the ocean to drop by several tens of feet. A gravity low, a deviation of the gravity value from the theoretical value, that formed as a plate sinks into the mantle off Somalia in northwest Africa might well be the oldest trench in the world.

The satellite altimetry data was used to produce a map of the entire ocean surface (Fig. 94), representing the seafloor as much as seven miles deep. Chains of midocean ridges and deep-sea trenches were delineated much more

Figure 93 *Artist's concept of the* Seasat *satellite radar mapping the ocean surface.*

(Photo courtesy NASA)

clearly than with any other method of remote sensing of the ocean floor. The maps uncovered many new features such as rifts, ridges, seamounts, and fracture zones and better defined features already known to exist. The maps also provide support for the theory of plate tectonics, which holds that the Earth is broken into several major plates that shift about, crashing into or drifting away from each other. These motions are responsible for all geologic activity taking place on the Earth's surface.

For the first time, geologists could view the ancient midocean ridge that formed when South America, Africa, and Antarctica began separating around 125 million years ago. This particular seafloor spreading center was well concealed due to the buildup of sediment. The boundary between the plates migrated to the west, leaving behind the ancient ridge, which had started to subside. Its discovery might help trace the evolution of the oceans and continents during the last 200 million years.

A newly revealed fracture zone in the southern Indian Ocean might shed some light on India's break from Antarctica around 180 million years ago. The 1,000-mile-long gash, located southwest of the Kerguelen Islands, was gouged out of the ocean floor as the Indian subcontinent inched northward. When India collided with Asia, more than 100 million years after it was set adrift, it pushed up the Himalaya Mountains to great heights like squeezing an accordion. A strange series of east-west wrinkles in the ocean crust just south of India verifies that the Indian plate is still pushing northward, continuously raising the Himalayas and shrinking the Asian continent as much as three inches a year.

The computer-generated satellite imagery also revealed long-buried parallel fracture zones not seen on conventional seafloor maps. The faint lines running like a comb through the central Pacific seafloor might be the result of convection currents in the mantle 30 to 90 miles beneath the ocean crust. Each circulating loop consists of hot material rising and cooler material sinking back into the depths, tugging on the ocean floor as it descends.

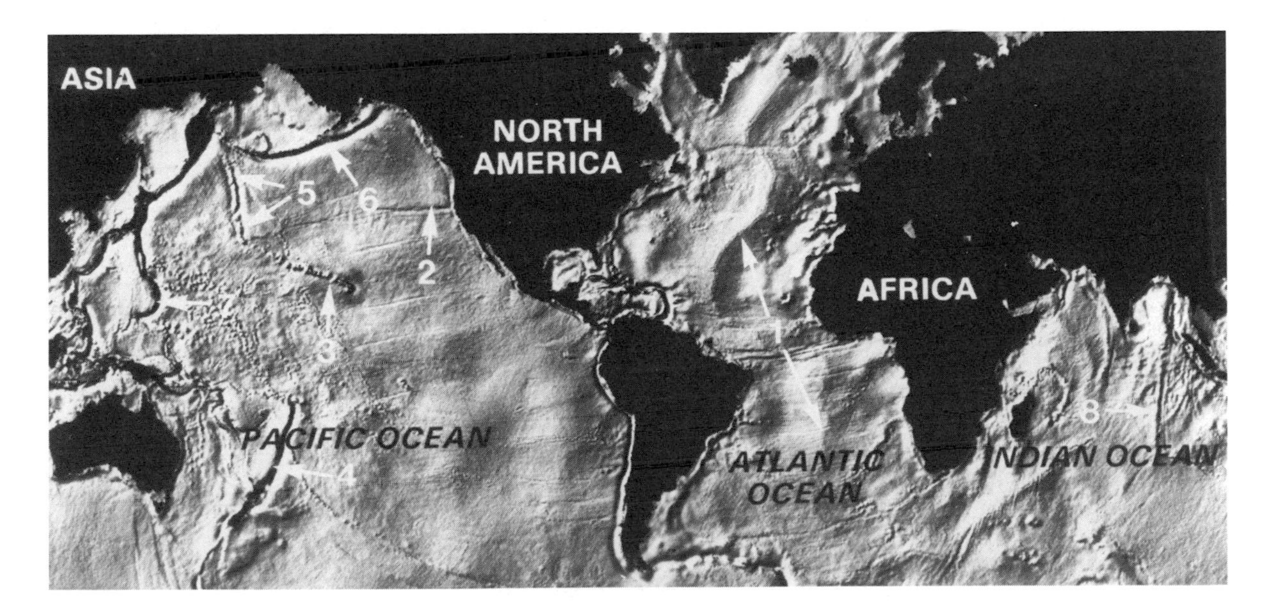

Figure 94 *Radar altimeter data from the* Geodynamic Experimental Ocean Satellite (GEOS-3) *and* Seasat *were used to produce this map of the ocean floor. 1. Mid-Atlantic Ridge, 2. Mendocino Fracture Zone, 3. Hawaiian Island chains, 4. Tonga Trench, 5. Emperor Seamounts, 6. Aleutian Trench, 7. Mariana Trench, 8. Ninety East Ridge.*

(Photo courtesy NASA)

PLATE SUBDUCTION

As a plate extends away from its place of origin at a midocean spreading ridge, it becomes thicker and denser as additional material from the asthenosphere adheres to its underside in a process called underplating. Eventually, the plate becomes so dense it loses buoyancy and sinks into the mantle. The line of subduction creates a deep-sea trench at clearly defined subduction zones where cool, dense lithospheric plates dive into the mantle. The depth at which the oceanic crust sinks as it moves away from the midocean ridges varies with age. Thus, the older the lithosphere, the more basalt that underplates it, making the plate thicker, denser, and deeper. Crust that is 2 million years old lies about 2 miles deep; crust that is 20 million years old lies about 2.5 miles deep; and crust that is 50 million years old lies about 3 miles deep.

The subduction of the lithosphere plays a very significant role in global tectonics and accounts for many geologic processes that shape the surface of the Earth. The seaward boundaries of the subduction zones are marked by the deepest trenches in the world. They are usually found at the edges of continents or along volcanic island arcs. Major mountain ranges and most volcanoes and earthquakes are also associated with the subduction of lithospheric plates.

The amount of subducted plate material is vast. When the Atlantic and Indian Oceans opened and new oceanic crust was created beginning around 125 million years ago, an equal area of oceanic crust had to disappear into the mantle. This meant that 5 billion cubic miles of crustal and lithospheric material were destroyed. At the present rate of subduction, an area equal to the entire surface of the planet will be consumed by the mantle in the next 160 million years.

When two lithospheric plates converge, generally the oceanic plate is bent and pushed under the thicker, more buoyant continental plate. When oceanic plates collide, the oldest and thus more dense plate dives under the youngest plate (Fig. 95). The line of initial subduction is marked by a deep ocean trench. At first the angle of descent is low, but gradually it steepens to about 45 degrees. At this angle, the rate of vertical descent is less than that of the horizontal motion of the plate, typically 2 to 3 inches per year. Upon reaching the boundary between the lower and upper mantle about 410 miles beneath the surface, the plate experiences difficulty in descending any further and is forced to bend to the horizontal. Deep-seated earthquakes are associated with descending plates. The cessation of seismic activity at a depth of about 410 miles indicates that the plate has stop descending. It might also have become too hot and pliable to generate earthquakes.

If continental crust moves into a subduction zone, its lower density and greater buoyancy prevent it from being dragged down into the trench. When two continental plates converge, the crust is scraped off the subducting plate and plastered onto the overriding plate, welding the two pieces of continen-

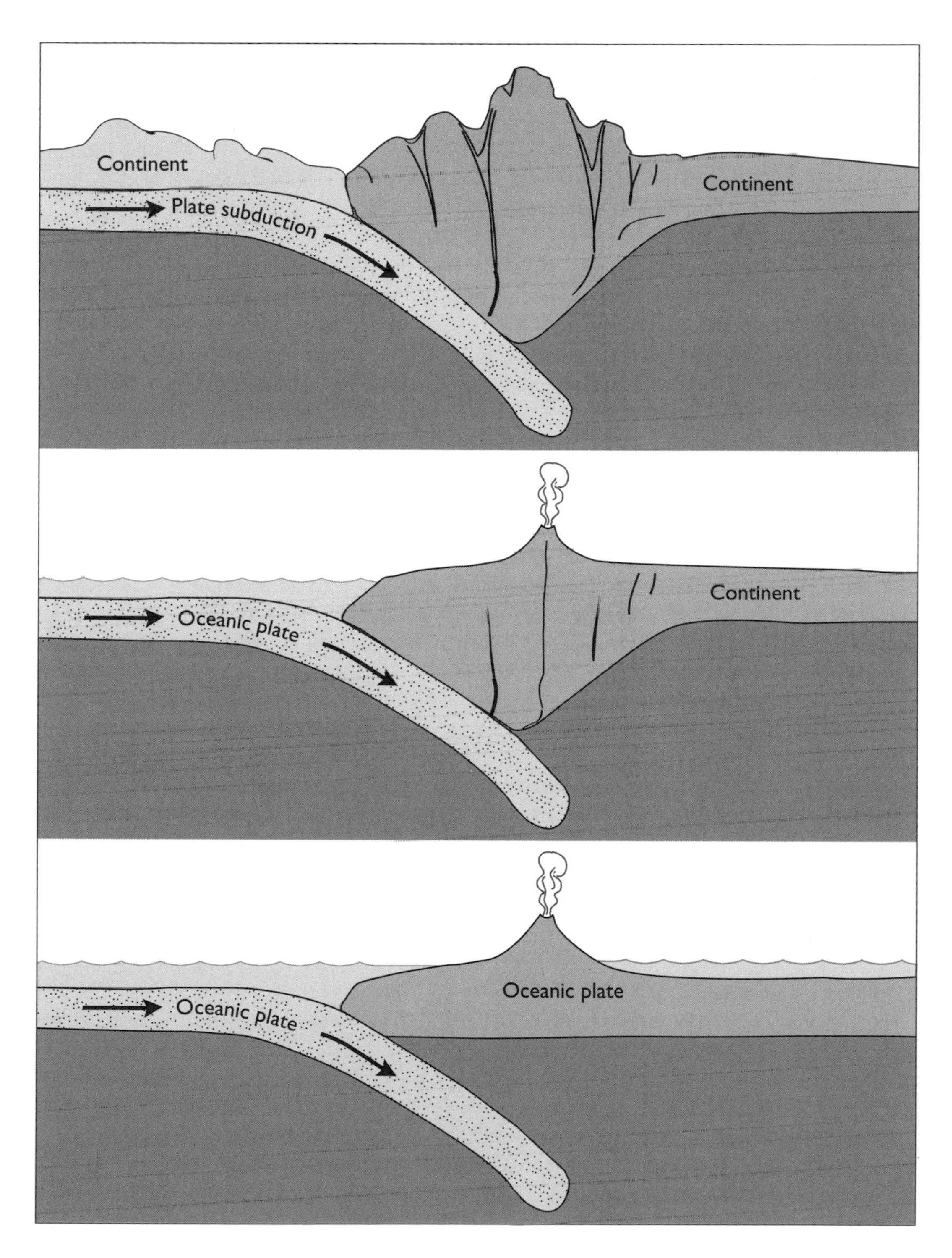

Figure 95 *Collision between two continental plates (top), a continental plate and an oceanic plate (middle), and two oceanic plates (bottom).*

tal crust together. Meanwhile, the subducted lithospheric plate, now without its overlying crust, continues to dive into the mantle.

In some subduction zones, such as the Lesser Antilles, much of the oceanic sediments and their contained fluids are removed by offscraping and underplating in accretionary prisms that form on the overriding plate adjacent to the trench. In other subduction zones, such as the Mariana and Japan trenches, little or no sediment accretion occurs. Thus, subduction zones differ markedly in the amount of sedimentary material removed at the accretionary prism. In most cases, however, at least some sediment and bound fluids appear to be subducted to deeper levels.

As the continental crust is underthrust by additional crustal material, the increase in buoyancy pushes up mountain ranges. This is similar to the building of the Himalayas (Fig. 96) when India collided with Asia around 45 million years ago. Further compression and deformation might take place beyond the line of collision, producing a high plateau with surface volcanoes such as the Tibetan Plateau, the largest in the world. As resisting forces continue to build up, the plate convergence will eventually stop, the Himalayas will cease growing, and erosion will wear the mountains down.

In the ocean, the deep trenches created by descending plates accumulate large deposits of sediments derived from the adjacent continent. When the plate dives into the interior, most of its water goes down with it as well. However, much more water is being subducted into the Earth than was coming out of subduction-zone volcanoes. Heat and pressure act to dehydrate rocks of the descending plate, but just where all the fluid goes remains a mystery. Some fluid expelled from a subducting plate reacts with mantle rocks to produce low-density minerals that slowly rise to the seafloor. There they build mud volcanoes that erupt serpentine, an asbestos mineral formed by the reaction of olivine from the mantle with water.

When the sediments and their content of seawater are caught between a subducting oceanic plate and an overriding continental plate, they are subjected to strong deformation, shearing, heating, and metamorphism. As the rigid lithospheric plate carrying the oceanic crust descends into the Earth's interior, it slowly breaks up and melts. Over a period of millions of years, it is absorbed into the general circulation of the mantle. The subducted plate also supplies molten magma for volcanoes, most of which ring the Pacific Ocean and recycle chemical elements to the Earth. In this manner, the continental crust is rejuvenated, and the total mass of low-density crustal rocks is preserved.

DESCENDING PLATES

When a lithospheric plate descends into the mantle, heat flows into the cooler lithosphere from the surrounding hot mantle. The conductivity of the rocks

Figure 96 *View of the Himalaya Mountains of India and China from the space shuttle.*

(Photo courtesy NASA)

increases with increasing temperature. Therefore, conductive heating becomes more efficient with depth. Heat of compression is introduced into the plate as it continues to descend and is subjected to increasing pressure. Heat is also generated within the plate by the decay of radioactive elements (mainly uranium, thorium, and potassium), by the change in mineral structure of the rocks, and by internal and external friction, especially at the boundaries between the moving plate and the surrounding mantle. Among these heat sources, conductive heating and friction contribute the most toward heating the descending lithosphere.

At the beginning of descent, the interior of the plate remains relatively cool compared with the mantle. The plate is first subjected to internal stresses, faulting, and fracturing as it bends and dives under another plate.

When stresses open a crack in the plate, the weight of the overlying rock layers quickly close the gap. As the plate begins its passage through the mantle, its temperature rises very slowly due mainly to internal heat sources. Since the heat can no longer escape from the plate, as it did when on the surface, temperatures begin to build within the plate. As the plate penetrates to deeper levels, its interior begins to warm more rapidly due to the more efficient transfer of heat by radiation.

As the plate dives deeper into the mantle, heat is eventually conducted from the outside, where the plate is in contact with the mantle. The plate is also subjected to increasing pressure during its journey through the mantle. By the time the plate reaches a depth of several hundred miles, the extreme temperatures and pressures transform it into a highly dense mineral form with a compact crystal structure. As the plate continues downward, the tightly packed crystals begin to melt partially and the plate becomes plastic and is able to flow.

When the subducted segment of the plate reaches the boundary between the upper and lower mantle, at about 410 miles beneath the surface, it is prevented from descending any farther due to a density difference between the two rock layers. It must bend parallel to the boundary and move in the direction of the convective flow. The entire trip from top to bottom might take some 10 million years. In another 50 million years, the plate will have totally lost its identity and be completely assimilated by the mantle.

Not all plates descend to this level, however. The depth a plate reaches before being assimilated by the mantle depends on its rate of descent. A slow-moving plate will attain thermal equilibrium at shallower depths. For example, if a plate were descending at a rate of half an inch per year, it would travel no farther than a depth of about 250 miles before complete assimilation occurs. A fast-moving plate will dive deeper into the mantle before it reaches thermal equilibrium. When the plate reaches the boundary between the upper and lower mantle, it ceases to be thermally distinguishable as a structural unit from the mantle. In effect, it becomes a part of the mantle.

ISLAND ARCS

Deep-ocean trenches are also regions of intense volcanism. Most of the volcanic activity that continually remakes the surface of the Earth takes place at the bottom of the ocean. The trenches are also sites of almost continuous seismic activity. In 1954, the American seismologist Hugo Benioff discovered a descending lithospheric plate by studying deep-seated earthquakes, which act like beacons marking the boundaries of the plate (Fig. 97).

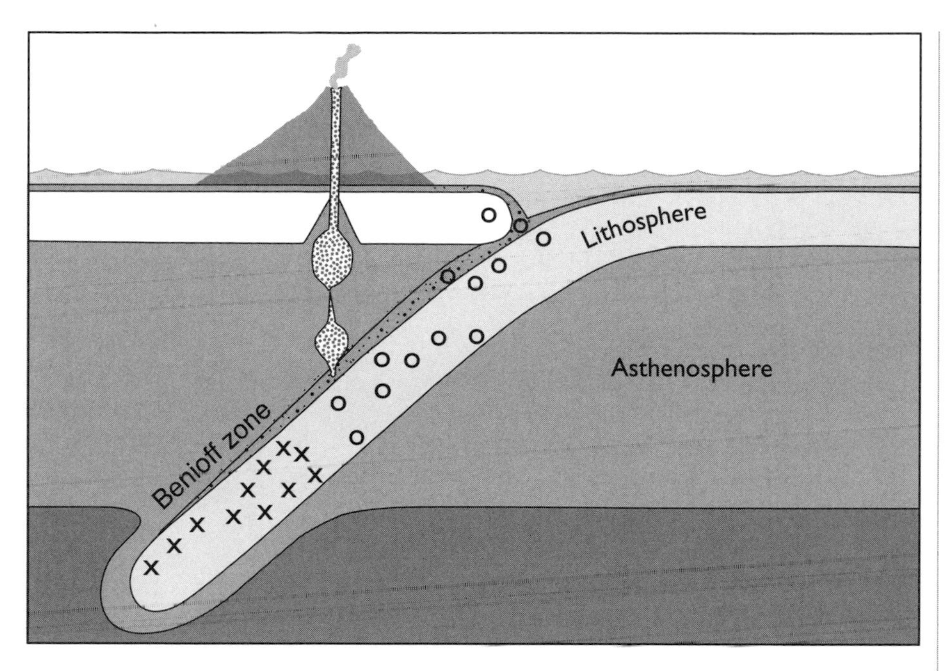

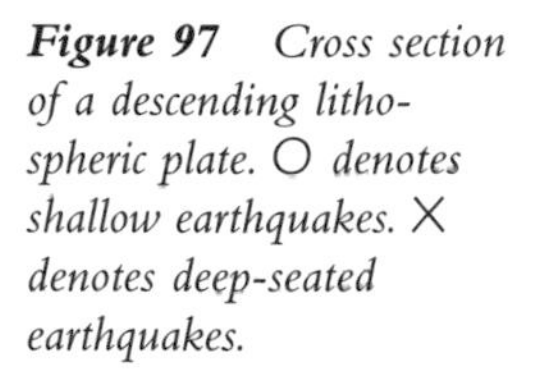
Figure 97 *Cross section of a descending lithospheric plate. O denotes shallow earthquakes. X denotes deep-seated earthquakes.*

Subduction zones noted for their active volcanoes in island arcs accumulate large deposits of sediments, primarily from the adjacent continents. When sediments caught between the subducting oceanic crust and a continental crust are carried deep within the mantle, they are melted in pockets of molten magma called diapirs, which rapidly rise to the surface (Fig. 98). When the diapirs reach the underside of the lithosphere, they burn holes into it as they melt their way upward. After reaching the surface, the magma erupts onto the ocean floor, sometimes explosively. In the process, the magma creates a new volcanic island (Fig. 99), most of which lie in the Pacific. The magma from subduction zones produces a fine-grained, gray rock known as andesite, named for the Andes Mountains, whose volcanoes were created by a zone of subduction beneath the western portion of the South American plate. Andesite is quite different in composition and texture from the upwelling basaltic lavas of spreading midocean rifts. It indicates a deep-seated source, possibly as deep as 70 miles within the mantle.

The longest island arc is the Aleutian Islands, which extend more than 3,000 miles from Alaska to Asia. The Kuril Islands to the south form another long arc. The islands of Japan, the Philippines, Indonesia, New Hebrides, Tonga, and those from Timor to Sumatra also form island arcs. The island arcs have similar, graceful curves, and each is associated with a subduction zone. The curvature of the island arcs is due to the curvature of the Earth; when a plane cuts a sphere, it results in an arc.

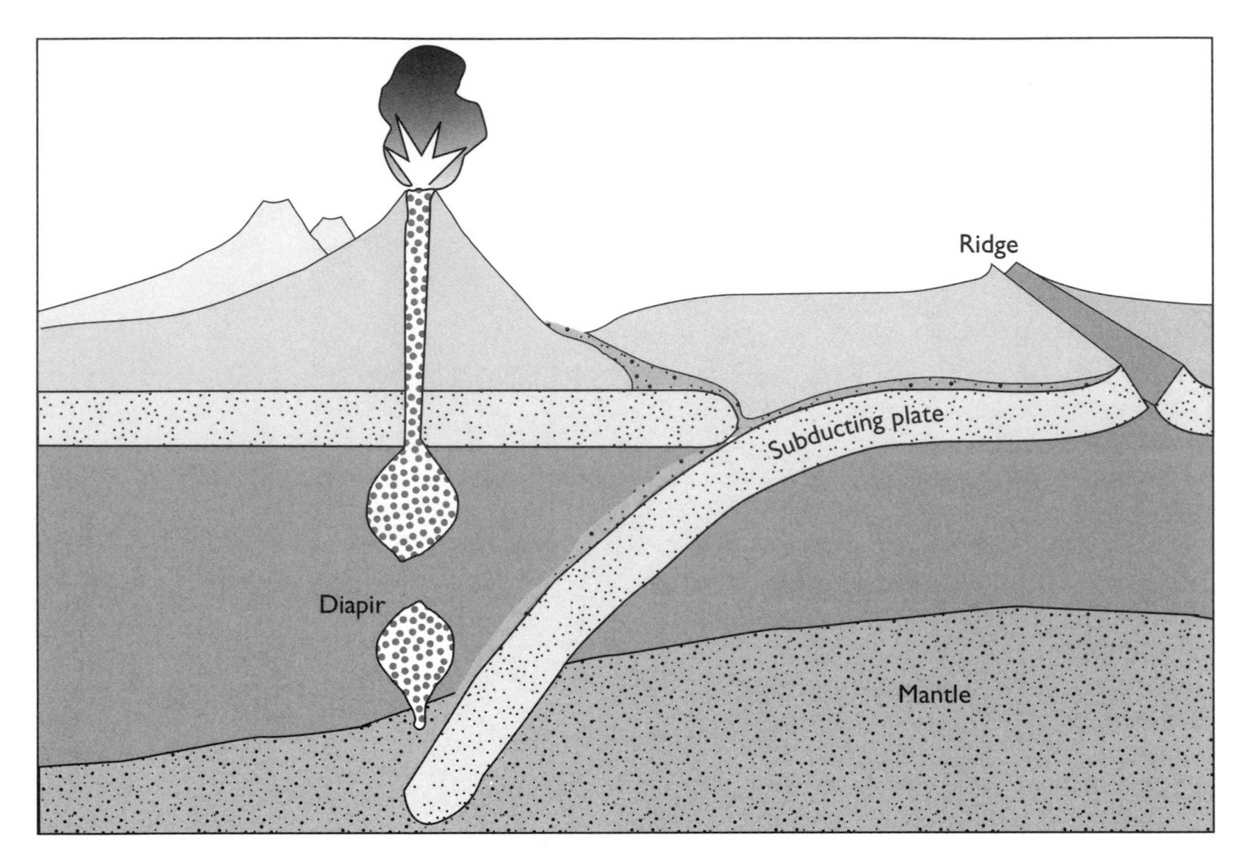

Figure 98 *The subduction of a lithospheric plate into the mantle supplies volcanoes with molten magma, which rises to the surface in blobs called diapirs.*

Behind each island arc lies a marginal or back-arc basin, which is a depression in the ocean crust due to the effects of plate subduction. Steep subduction zones such as the Mariana Trench form back-arc basins, while shallow ones such as the Chilean Trench do not. One classic back-arc basin is the Sea of Japan between China and the Japanese archipelago. Back-arc basins are associated with high heat flow because they are underlain by relatively hot material brought up by convection currents behind the island arcs or by upwelling from deeper regions in the mantle. The trenches have low heat flow because of the subduction of cool, dense lithospheric plates. Their associated island arcs generally have high heat flow due to a high degree of volcanism.

THE RING OF FIRE

A band of subduction zones surrounding the Pacific plate is called the Ring of Fire because of frequent volcanic activity. The subduction zones have devoured almost all the seafloor since the breakup of Pangaea. As a result, the oldest crust lies in a small patch off southeast Japan and dates about 170 million years old. In the process of subduction, the seafloor melts to provide molten magma for volcanoes that fringe the deep-sea trenches. This is why most of the 600 active volcanoes in the world lie in the Pacific Ocean, with nearly half residing in the western Pacific region alone. This produces an almost continuous Ring of Fire along the edges of the Pacific (Fig. 100). In addition, practically all mountain ranges surrounding the Pacific Ocean formed by plate subduction.

Most volcanoes occur on plate boundaries associated with deep-sea trenches along the margins of continents and adjacent to island arcs fringing subduction zones. Subduction-zone volcanoes build volcanic chains on continents when a lithospheric plate dives into a subduction zone and slides beneath the continental crust. As the lithospheric plate subducts into the hot mantle, portions of the descending plate along with the adjacent crustal plate melt under the high temperatures, forming pockets of magma. The molten rock rises toward the surface to feed magma chambers underlying active volcanoes.

Volcanic islands began as undersea volcanoes. At subduction zones where one plate descends beneath another, magma forms when the lighter constituents of the subducted oceanic crust melt. The upwelling magma creates island arcs, mostly in the Pacific. Beginning at the western tip of the

Figure 99 *The September 23, 1952 submarine eruption of Myojin-sho Volcano, Izu Islands, Japan.*

(Photo courtesy USGS)

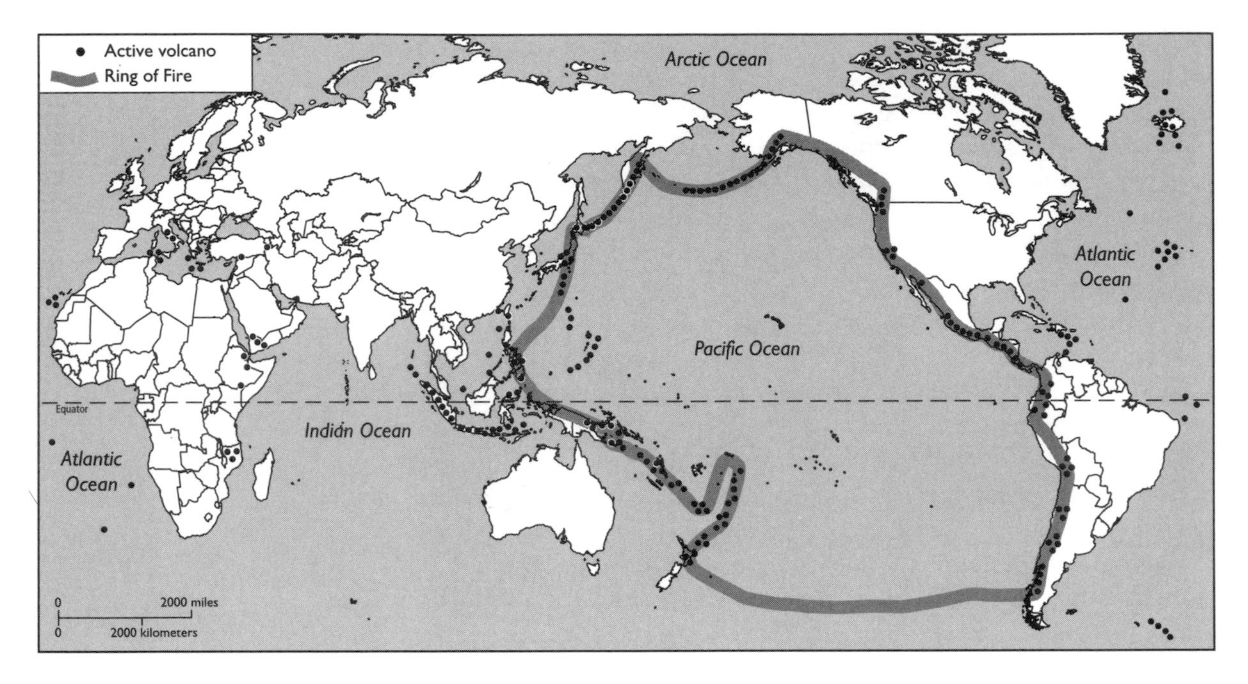

Figure 100 *The Ring of Fire is a band of subduction zones surrounding the Pacific Ocean.*

Aleutian Islands, a string of volcanoes stretches along the Aleutian archipelago, consisting of a series of volcanic islands formed by the subduction of the Pacific plate into the Aleutian Trench. The band of volcanoes turns south across the Cascade Range of British Columbia, Washington, Oregon, and northern California. The volcanic activity in these mountains originates from the subduction of the Juan de Fuca plate into the Cascadia subduction zone. The May 18, 1980 lateral blast from Mount St. Helens caused total destruction along its flanks, providing a vivid reminder of the explosive nature of subduction zone volcanoes.

The chain of volcanoes continues along Baja California and southwest Mexico, where lie the volcanoes Parícutin and El Chichon (Fig. 101). It travels on through western Central America. In this area, many active cones have erupted in recent years, including Nevado del Ruiz of Colombia, whose massive mudflows killed 25,000 people in November 1985. It is among some 20 other volcanoes considered the most dangerous in the world.

The band of volcanoes journeys along the course of the Andes Mountains on the western edge of South America, known for their highly explosive nature from the subduction of the Nazca plate into the Chilean Trench. The volcanic belt then turns toward Antarctica and the islands of New Zealand on the margin of the Pacific and Australian plates, New Guinea in the Malay archipelago, and Indonesia. Here the volcanoes Tambora and Krakatoa

(Fig. 102) have produced the greatest eruptions in modern history from the subduction of the Australian plate into the Java Trench.

The volcanic belt continues across the Philippines, where the subduction of the Pacific plate into the Philippine Trench resulted in the June 1991 Pinatubo eruption, whose huge ash cloud dramatically changed the world's climate. From there, the zone of volcanoes runs across the Japanese Islands. They are built from a combination of rock drawn from the mantle along a volcanic arc and from sediments scraped off the ocean floor. The volcanic belt finally ends on the Kamchatka Peninsula in Northeast Asia, whose volcanoes are renowned for their powerful blasts.

The Ring of Fire coincides with the circum-Pacific belt because the same tectonic forces that produce earthquakes are also responsible for volcanic activity. The area of greatest seismicity is on plate boundaries associated with deep trenches along volcanic island arcs and along the margins of continents. As the Pacific plate inches northwestward, its leading edge dives into the mantle, forming some of the deepest trenches in the world.

Figure 101 *Caldera formed by the March 28, 1982 massive eruption of the El Chichon Volcano, Chiapas, Mexico.*

(Photo courtesy USGS)

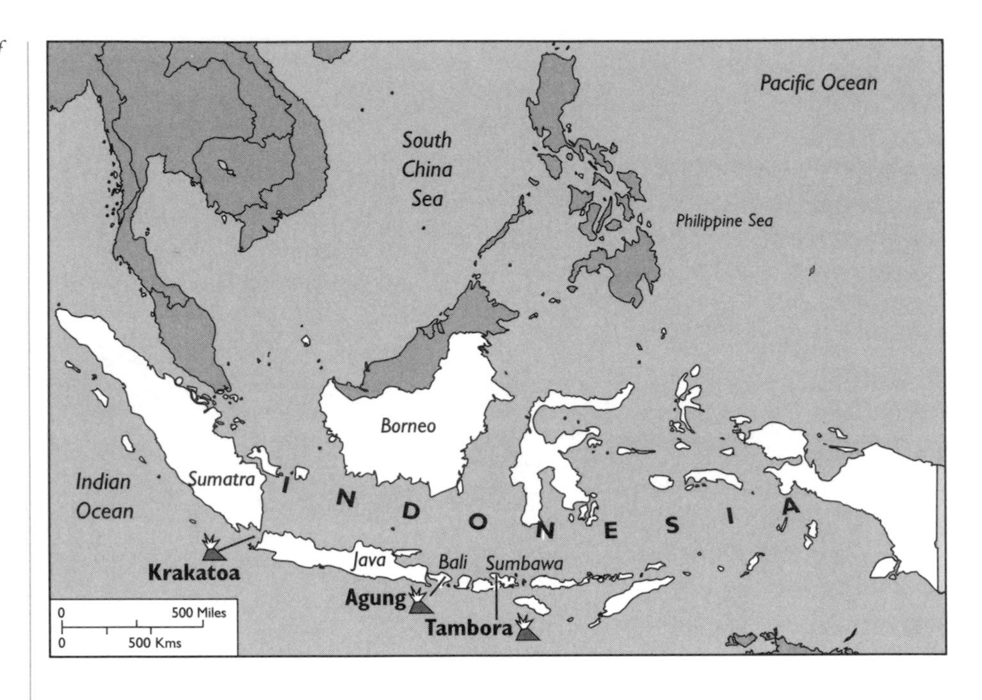

Figure 102 Location of the great Indonesian volcanoes.

The entire western seaboard of South America is affected by an immense subduction zone just off the coast. The lithospheric plate on which the South American continent rides is forcing the Nazca plate to buckle under, causing great tensions to build up deep within the crust. While some rocks are forced deep down, others are buoyed up, raising the Andean Mountain chain. The resulting forces are building great stresses into the entire region. When the stresses become large enough, powerful earthquakes roll across the landscape.

Subduction zone volcanoes such as those in the western Pacific and Indonesia are among the most explosive in the world. They create new islands and destroy old ones, including the near total destruction of the Indonesian island of Krakatoa in 1883. The reason for their explosive nature is that the magma contains large amounts of silica and volatiles, consisting of water and gases derived from sediments on the ocean floor subducted into the mantle and melted. The volatiles lower the melting point of mantle rocks and make them flow easily. When the pressure is lifted as the magma reaches the surface, these volatiles are released explosively, fracturing the magma, which then shoots out of the volcano like pellets from a shotgun.

The Cascade Range in the western United States comprises a chain of volcanoes associated with the Cascadia subduction zone, created when the North American plate overran the northern part of the Pacific plate. As the lithosphere is being forced into the mantle, the tremendous heat melts parts of the descending plate and the adjacent lithospheric plate to form pockets of magma. The magma forces its way to the surface, resulting in explosive volcanic eruptions. The May 18, 1980 eruption of Mount St. Helens, which devastated 200 square miles of national forest (Fig. 103) is a prime example of the explosive nature of subduction-zone volcanoes.

Figure 103 *An entire forest is leveled by the explosive eruption of Mount St. Helens.*

(Photo by Jim Hughes, courtesy USDA Forest Service)

VOLCANIC ERUPTIONS

The vast majority of the world's volcanoes accompany crustal movements at plate margins where lithospheric plates are either diverging or converging. Fissure eruptions on the ocean floor occur at the boundaries between lithospheric plates where the brittle crust pulls apart by the process of seafloor spreading. This creates new oceanic crust. Basaltic magma welling up along the entire length of a fissure forms large lava pools.

Volcanoes erupt a variety of rock types ranging from rhyolite with a high silica content to basalt with a low silica and high iron-magnesium content. Basalt is the heaviest volcanic rock and the most common igneous rock type produced by the extrusion of magma onto the surface. Most of the active volcanoes in the world are completely or predominantly basaltic.

The world's volcanoes come in many shapes and sizes, depending on the composition of the erupted magma. The four main types of volcanoes are cinder cones, composite volcanoes, shield volcanoes, and lava domes. Cinder cones are the simplest volcanic structures, built from particles and congealed lava ejected from a single vent. Explosive eruptions form short, steep slopes usually less than 1,000 feet high. Cinder cones build upward and outward by accumulating layers of pumice, ash, and other volcanic debris falling back onto the volcano's flanks. The general order of events is eruption followed by formation of cone and crater and then lava flow. The bowl-shaped craters are numerous in western North America. Mexico's Parícutin Volcano (Fig. 104), which erupted in a farmer's field in 1943, is a classic example.

Composite volcanoes are constructed from cinder and lava cemented into tall mountains rising several thousand feet. They are generally steep sided and comprise symmetrical lava flows, volcanic ash, cinders, and blocks. The crater at the summit contains a central vent or a cluster of vents. During eruption, the hardened plug in the volcano's throat breaks apart by the buildup of pressure from trapped gases below. Molten rock and fragments shoot high into the air and fall back onto the volcano's flanks as cinder and ash. Layers of lava from milder eruptions reinforce the fragments, forming cones with a steep summit and steeply sloping flanks. Lava also flows through breaks in the crater wall or from fissures on the flanks of the cone, continually building the cone upward. As a result, composite volcanoes are the tallest cones in the world and often end in catastrophic collapse, thus preventing them from becoming the highest mountains.

The broadest and largest volcanoes are called shield volcanoes. The slope along their flanks generally rises only a few degrees and no more than 10 degrees near the summit. They erupt almost entirely basaltic lava from a cen-

Figure 104 *Heavy cinder activity during the July 25, 1943, eruption of Parícutin Volcano, Michoacan, Mexico.*

(Photo by W. F. Foshag, courtesy USGS)

tral vent. Highly fluid molten rock oozes from the vent or violently squirts out, forming fiery fountains of lava from pools within the crater. As the lava builds in the center, it flows outward in all directions, forming a structure similar to an inverted dinner plate. The lava spreads out, covering areas as large as 1,000 square miles. Several of these dome-shaped features in northern California and Oregon such as the Mono-Inyo Craters are 3 or 4 miles wide and 1,500 to 2,000 feet high. Hawaii's Mauna Loa is the largest active shield volcano, creating a huge sloping dome rising 13,675 feet above sea level but still much less than its depth below the sea.

Figure 105 *Ribbon and spindle bombs from the Craters of the Moon National Monument, Idaho.*

(Photo by H. T. Stearns, courtesy USGS)

If lava is too viscous or heavy to flow very far, causing it to pile up around the vent, it forms a lava dome that grows by expansion from within. Lava domes commonly occur in a piggyback fashion within the craters of large composite volcanoes. Good examples are California's Mono domes and Lassen Peak.

If a volcano explosively erupts fluid lava high into the air, it produces particles ranging in size from ash to molten blobs of lava up to several feet wide called volcanic bombs (Fig. 105). They often change shape while flying through the air and flatten or splatter when striking the ground. As volcanic bombs cool in flight, they assume a variety of forms called cannonball, spindle, bread crust, cow dung, ribbon, or fusiform, depending on their shape or surface appearance. Bread crust bombs, which can reach several feet across, are named for their crusty appearance caused by gases escaping from the bomb during the hardening of the outer surface. Some volcanic bombs actually explode when landing due to the rapid expansion of gases in the molten interior as the solid crust cracks open on impact. If the bombs are the size of a nut, they are called lapilli, Latin for "little stones," and form strange, gravel-like deposits along the countryside.

All solid particles ejected into the atmosphere from volcanic eruptions are collectively called tephra, from Greek meaning "ash." Tephra includes an assortment of fragments from large blocks to dust-sized material. It originates from molten rock containing dissolved gases that rises through a conduit and suddenly separates into liquid and bubbles when nearing the surface. With decreasing pressure, the bubbles grow larger. If this event occurs near the orifice, a mass of froth spills out and flows down the sides of the volcano, forming pumice. If the reaction occurs deep down in the throat, the bubbles expand explosively and burst the surrounding liquid, which fractures the magma into fragments. The fragments are then driven upward by the force of the rapid expansion and hurled high above the volcano.

Tephra supported by hot gases originating from a lateral blast of volcanic material is called nuée ardente, French for "glowing cloud." The cloud of ash and pyroclastics flows streamlike near the ground and might follow existing river valleys for tens of miles at speeds upward of 100 miles per hour. The best-known example was the 1902 eruption of Mt. Pelée, Martinique, which produced a 100-mile-per-hour ash flow that, within minutes, killed 30,000 people at St. Pierre. When the tephra cools and solidifies, it forms deposits called ash-flow tuffs that can cover an area up to 1,000 square miles or more.

Lava is molten magma that reaches the throat of a volcano or fissure vent and flows freely onto the surface. The magma that produces lava is much less viscous than that which produces tephra, allowing volatiles and gases to escape with comparative ease. This gives rise to much quieter and milder eruptions. The outpourings of lava take the Hawaiian names aa (pronounced, AH-ah), which is the sound of pain when walking over them barefooted, and pahoehoe (pronounced, pah-HOE-ay-hoe-ay), which means satinlike. Aa or blocky lava forms when viscous, subfluid lavas press forward, carrying along a thick and brittle crust. As the lava flows, it stresses the overriding crust, breaking it into rough, jagged blocks. These are pushed ahead of or dragged along with the flow in a disorganized mass.

Pahoehoe or ropy lavas (Fig. 106) are highly fluid basalt flows produced when the surface of the flow congeals, forming a thin, plastic skin. The melt beneath continues to flow, molding and remolding the skin into billowing or ropy-looking surfaces. When the lavas eventually solidify, the skin retains the appearance of the flow pressures exerted on it from below. If a stream of lava forms a crust and hardens on the surface and the underlying magma continues to flow away, it creates a long cavern or tunnel called a lava tube. Long caverns beneath the surface of a lava flow are created by the withdrawal of lava as the surface hardens. In exceptional cases, they can extend up to several miles within a lava flow.

Figure 106 *Ropy lava surface of pahoehoe near Surprise Cave, the Craters of the Moon National Monument, Idaho.*

(Photo by H. T. Stearns, courtesy USGS)

Undersea volcanic eruptions are responsible for erecting most of the world's islands. Successive eruptions pile up volcanic rocks until they finally break through the surface of the sea. Because these volcanoes rise from the very bottom of the ocean, they make the tallest mountains in the world.

After learning the basics of plate tectonics, the next chapter will apply that knowledge to the subject of mountain building, perhaps the most evident result of crustal plates in motion.

7

MOUNTAIN BUILDING

CRUSTAL UPLIFT AND EROSION

This chapter explores how geologic uplift and erosion sculptures the Earth's surface into mountain ranges. Perhaps the most spectacular result of plate tectonics is mountains. They are the most visible manifestations of the powerful tectonic forces that shape the Earth. Mountains are important landforms in terms of their rugged beauty and their ability to control the climate and the flow of rivers. A mountain is defined as a topographic feature that rises abruptly above the surrounding terrain and involves the massive deformation of the rocks that form the core of the mountain. The usual means of building mountains is for plate motions to shove the crust onto a strong plate or create a deep root of light crustal rock that literally floats the mountain like an iceberg.

Most mountains occur in ranges. Although a few isolated peaks do exist, they are rare. Mountains have complex internal structures formed by folding, faulting, volcanic activity, igneous intrusion, and metamorphism. Mountain building, which provides the forces necessary for folding and faulting rocks at shallow depths, also supplies the stress forces that strongly distort rocks at greater depths. The interiors of mountains contain some of the oldest rocks, which were once buried deep in the crust and subsequently thrust to the surface by tectonic activity.

EPISODES OF OROGENY

During the late Precambrian between 1.3 and 0.9 billion years ago, Laurentia, the ancient North American continent, collided with another large landmass on its southern and eastern borders, creating the supercontinent Rodinia. The collision raised a 3,000-mile-long mountain belt in eastern North America during the Grenville orogeny (mountain building episode). A similar mountain belt occupied parts of western Europe.

When Rodinia rifted apart between 630 and 570 million years ago, the separated continents dispersed around an ancestral Atlantic Ocean called the Iapetus Sea. Near the present Appalachians is a long belt of volcanism as testimony to the breakup of the supercontinent. When all continents reached their maximum dispersal roughly 480 million years ago, subduction of the ocean floor beneath the North American plate initiated a period of volcanic activity and intense mountain building.

Many of today's mountain ranges were uplifted by Paleozoic continental collisions. The collision of North America and Europe in the early Paleozoic was responsible for the Caledonian orogeny, which produced a mountain belt that extended from southern Wales, spanned across Scotland, and ran northeastward through Scandinavia (Fig. 107). In North America, this

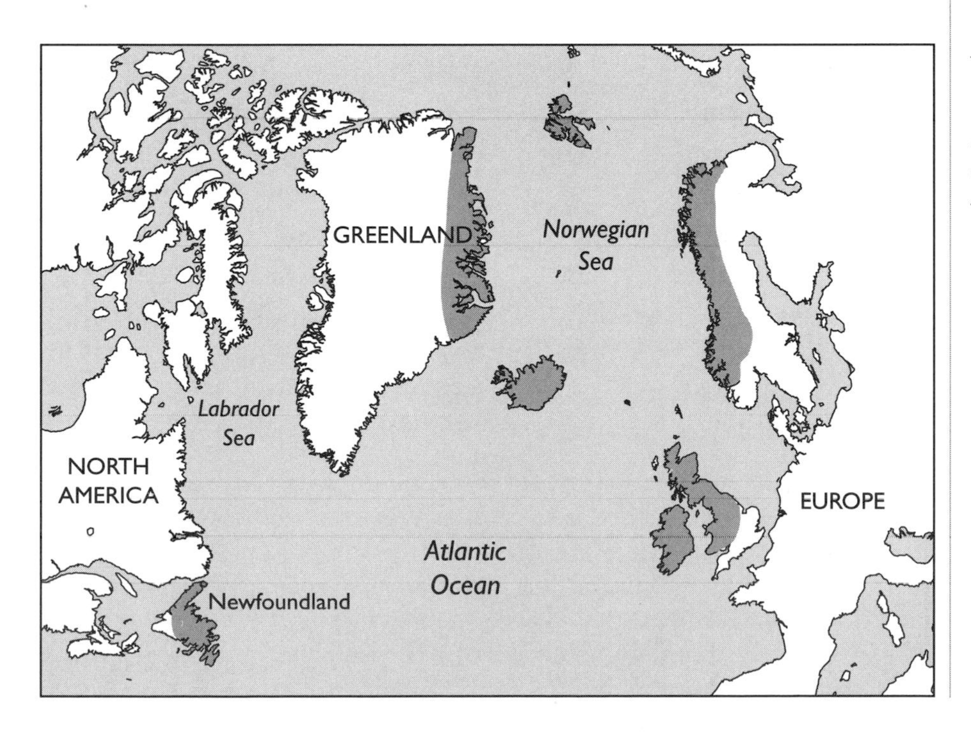

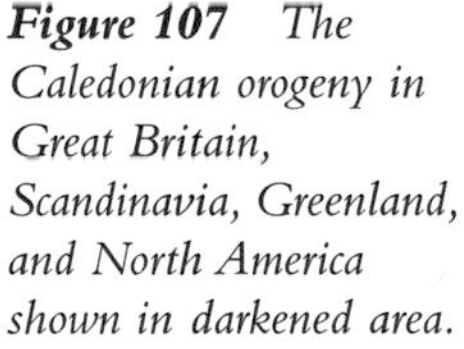
Figure 107 *The Caledonian orogeny in Great Britain, Scandinavia, Greenland, and North America shown in darkened area.*

TABLE 10 THE TALLEST MOUNTAIN PEAKS BY STATE

State	Mountain or Peak	Elevation (Feet)	State	Mountain or Peak	Elevation (Feet)
Alabama	Cheaha Mt.	2,407	Montana	Granite Pk.	12,799
Alaska	McKinley Mt.	20,320	Nebraska	Kimball Co.	5,246
Arizona	Humphreys Pk.	12,633	Nevada	Boundary Pk.	13,143
Arkansas	Magazine Mt.	2,753	New Hampshire	Washington Mt.	6,288
California	Whitney Mt.	14,494	New Jersey	High Point	1,803
Colorado	Elbert Mt.	14,433	New Mexico	Wheeler Pk.	13,161
Connecticut	Frissell Mt.	2,380	New York	Marcy Mt.	5,344
Delaware	Ebright Rd.	442	N. Carolina	Mt. Mitchell	6,684
Florida	Walton Co.	345	North Dakota	White Butte	3,506
Georgia	Brasstown Bald	4,784	Ohio	Campbell Hill	1,550
Hawaii	Mauna Kea Mt.	13,796	Oklahoma	Black Mesa	4,973
Idaho	Borah Pk.	12,662	Oregon	Mt. Hood	11,239
Illinois	Charles Mound	1,235	Pennsylvania	Davis Mt.	3,213
Indiana	Wayne Co.	1,257	Rhode Island	Jerimoth Hill	812
Iowa	Osceola Co.	1,670	S. Carolina	Sassafras Mt.	3,560
Kansas	Sunflower Mt.	4,039	South Dakota	Harney Pk.	7,242
Kentucky	Black Mt.	4,145	Tennessee	Clingmans Dome	6,643
Louisiana	Driskill Mt.	535	Texas	Guadalupe Pk.	8,749
Maine	Katahdin Mt.	5,268	Utah	Kings Pk.	13,528
Maryland	Backbone Mt.	3,360	Vermont	Mansfield Mt.	4,393
Massachusetts	Greylock Mt.	3,491	Virginia	Rogers Mt.	5,729
Michigan	Curwood Mt.	1,980	Washington	Mt. Rainer	14,410
Minnesota	Eagle Mt.	2,301	West Virginia	Spruce Knob	4,863
Mississippi	Woodall Mt.	806	Wisconsin	Timms Hill	1,951
Missouri	Taum Sauk Mt.	1,772	Wyoming	Gannett Pk.	13,804

orogeny built a mountain belt extending from Alabama through Newfoundland and reaching as far west as Wisconsin and Iowa. During the Taconian disturbance, extensive volcanism erupted in this region, culminating in a chain of folded mountains that comprise the Taconic Range of eastern New York State. Vermont still preserves the roots of these ancient mountains, which were shoved upward between 470 and 400 million years ago but have since been planed by erosion.

A collision between present eastern North America and northwestern Europe from 400 million to 350 million years ago raised the Acadian Mountains. The terrestrial redbeds of the Catskills in the Appalachian Mountains of southwestern New York to Virginia are the main expression of this orogeny in North America. Accompanying the mountain building episode at its climax was extensive igneous activity and metamorphism.

The Appalachians and the Ouachitas were formed when North America and Africa slammed into each other in the late Paleozoic from 300 million to 250 million years ago during the formation of Pangaea. The southern Appalachians are underlain by more than 10 miles of sedimentary and metamorphic rocks that are essentially undeformed. In contrast, the surface rocks were highly deformed by the collision. This process suggests that these mountains were the product of horizontal thrusting, in which crustal material was carried great distances. The Appalachians later collapsed when North America separated from Africa, tearing apart the lithosphere upon which the range stood. Today, only remnants remain of once-towering giants (Fig. 108).

The continental collision also raised the Mauritanide mountain chain on western Africa, which is characterized by a series of belts that are similar to the Appalachians. In fact, the two mountain ranges are practically mirror images of each other. This episode of mountain building also uplifted the Hercynian Mountains in Europe, which progressed from England to Ireland and continued through France and Germany. The folding and faulting were accompanied by large-scale igneous activity in Britain and on the European continent.

The Urals resemble no other mountain chain on Earth. They formed during a collision between the Siberian and Russian shields between 600 million and 300 million years ago, roughly the same time the North American plate collided with the African plate to build the Appalachians. The nearly

Figure 108 *The foothills of Blue Ridge Mountains and Piedmont Plateau in the Appalachians, North Carolina.*

(Photo by A. Keith, courtesy USGS)

2,000-mile-long mountain range reaches unprecedented depths of 100 or more miles to the base of the lithosphere underlying central Russia. Unlike most other mountain chains, which have evolved through a similar pattern of growth and collapse, the Urals are frozen in midevolution. The mountains have remained intact because Asia never separated from Europe.

The Transantarctic Range, comprising great belts of folded rocks, was raised when two plates collided to create the continent of Antarctica. Prior to the end of the Permian period, the younger parts of West Antarctica had not yet formed and only East Antarctica was in existence. Then roughly 60 million years ago, two plates merged to construct the crust of West Antarctica.

The Innuitian orogeny resulted from a collision with another crustal plate, which deformed the northern margin of North America. The Old Red Sandstone, a thick sequence of chiefly nonmarine sediments in Great Britain and Northwest Europe, is the main expression of this mountain building episode in Europe. The formation contains great masses of sand and mud that accumulated in the basins lying between the ranges of the Caledonian Mountains.

A mountain building event called the Antler orogeny resulted from a collision of island arcs with the western margin of North America. The island arcs appear to have formed approximately 470 million years ago off the west coast of North America. The orogeny intensely deformed rocks in the Great Basin region from the California-Nevada border to Idaho. The continued clashing of island arcs with North America initiated an episode of mountain building in Nevada called the Sonoma orogeny, which coincided with the final assembly of Pangaea about 250 million years ago.

The mountain belts of the Cordilleran of North America, the Andean of South America, and the Tethyan of Africa-Eurasia arose during plate collisions when one plate slipped under the leading edge of another to increase crustal buoyancy. The Cordilleran and Andean belts were created by the collision of east-Pacific plates with the continental margins of the newly formed American plates when Pangaea rifted apart about 170 million years ago. The Tethyan belt formed when Africa collided with Eurasia around 30 million years ago, raising the Alps and Dolomites of Europe. Huge chunks of rock high in the alps apparently originated hundreds of miles underground, bringing some of the Earth's deepest secrets to the surface.

Since the late Cambrian, the future Rocky Mountain region was near sea level. Farther west about 400 miles from the coast was a mountain belt comparable to the present Andes. This belt developed between 160 and 80 million years ago above a subduction zone possibly responsible for the Cretaceous Sevier orogeny that formed the Overthrust Belt in Utah and Nevada. A region from eastern Utah to the Texas panhandle was deformed during the late Paleozoic Ancestral Rockies orogeny but was completely eroded by the time the present Rocky Mountains formed.

The Laramide orogeny between 80 million and 40 million years ago created the Rocky Mountains, which extend from northern Mexico into Canada. A large part of western North America was uplifted, and the entire Rocky Mountain Region was raised nearly a mile above sea level. The orogeny resulted from an increase in buoyancy of the continental crust due to the subduction of vast areas of oceanic crust and its attached lithosphere beneath the west coast of North America. The Canadian Rockies consist of slices of sedimentary rock that were successively detached from the underlying basement rock and thrust eastward on top of one another.

The Rocky Mountain foreland region subsided by as much as two miles between 85 million and 65 million years ago. It then rose above sea level and acquired its present elevation around 30 million years ago. A region between the Sierra Nevada and the southern Rockies took a spurt of uplift during the past 20 million years, raising the area more than 3,000 feet. The Nevadan orogeny was a surge of volcanic and plutonic activity caused by the subduction of oceanic crust under western North America. It created the Sierra Nevada, which rose an amazing 6,600 feet during the last 10 million years, apparently from an increase in buoyancy by a mass of hot rock in the mantle.

Westward of the Rockies, a large number of parallel faults sliced through the Basin and Range Province (Fig. 109) between the Sierra Nevada and the Wasatch Mountains. This resulted in a series of about 20 peculiar north-south-trending mountain ranges comprising grabens and horsts. The Basin and Range, which includes southern Oregon, Nevada, western Utah, southeastern California, and southern Arizona and New Mexico, consists of numerous fault block mountain ranges bounded by high-angle normal faults called horsts (see Fig. 4). The crust is literally broken into hundreds of blocks that were steeply tilted and raised nearly a mile above the basin, forming nearly parallel mountain ranges up to 50 miles long.

Death Valley, the lowest place on the North American continent at 280 feet below sea level, was once elevated several thousand feet. The area collapsed when the continental crust thinned due to extensive block faulting in the region. The Great Basin area is a remnant of a broad belt of mountains and high plateaus that subsequently collapsed after the crust was pulled apart following the Laramide orogeny, which raised the Rocky Mountains. The Andes Mountains of South America could follow a similar evolution as the Basin and Range sometime in the geologic future.

The still rising Wasatch Range of north-central Utah and south Idaho (Fig. 110) is an excellent example of a north-trending series of normal faults, one below the other. The fault blocks extend for 80 miles, with a probable net

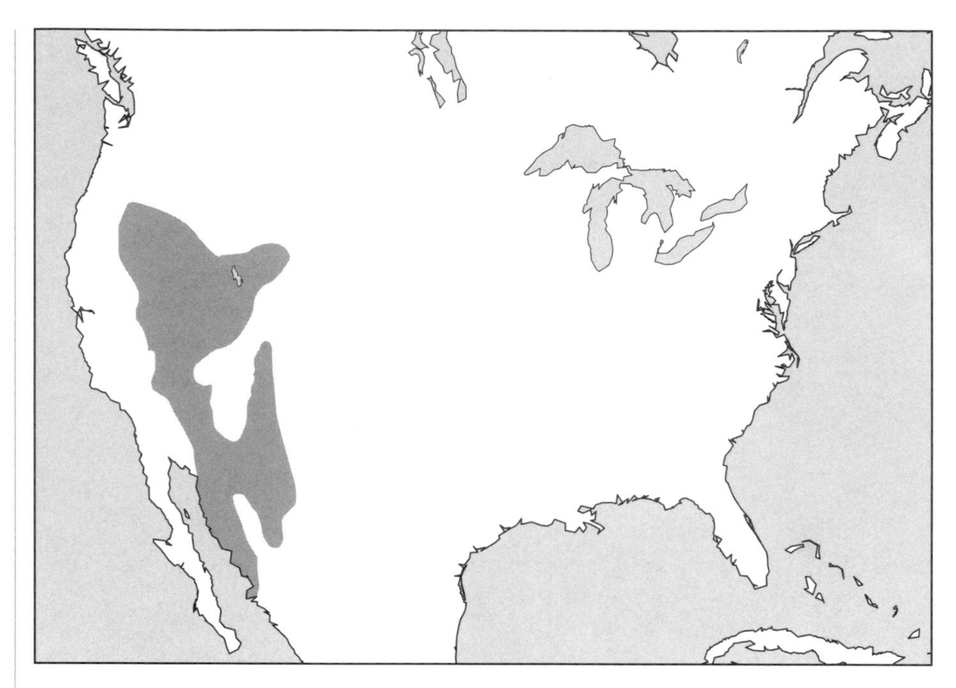

Figure 109 *The Basin and Range Province of the western United States.*

Figure 110 *The Wasatch Range, north-central Utah.*

(Photo by R. R. Woolley, courtesy USGS)

Figure 111 *The Grand Teton Range, Grand Teton National Park, Teton County, Wyoming.* (Photo by I. J. Witkind, courtesy USGS)

slip along the west side of 18,000 feet. The Tetons of western Wyoming (Fig. 111) were upfaulted along the eastern flank and downfaulted to the west. The rest of the Rocky Mountains were created by a process of upthrusting connected with plate collision and subduction similar to that which raised the Andes Mountains of Central and South America. The Andes continue to rise due to an increase in crustal buoyancy resulting from the subduction of the Nazca plate beneath the South American plate. Besides supplying magma for these very powerful volcanoes, the subducting plate also has the potential of generating very strong earthquakes in the region.

India and the rocks that now make up the Himalayas (Fig. 112) broke away from Gondwana early in the Cretaceous, sped across the ancestral Indian Ocean, and slammed into southern Asia about 45 million years ago. In the process, 6,000 miles of subducting plate were destroyed. As the Indian and Asian plates collided, the oceanic lithosphere between them was thrust under Tibet. The increased buoyancy uplifted the Himalaya Mountains and the wide Tibetan Plateau, which is underlain by Indian lithosphere. These regions rose at an incredible rate during the past 5 to 10 million years, when the entire area was uplifted over a mile. So much bare rock jutted skyward that carbon dioxide, an important greenhouse gas, was removed from the atmosphere by the

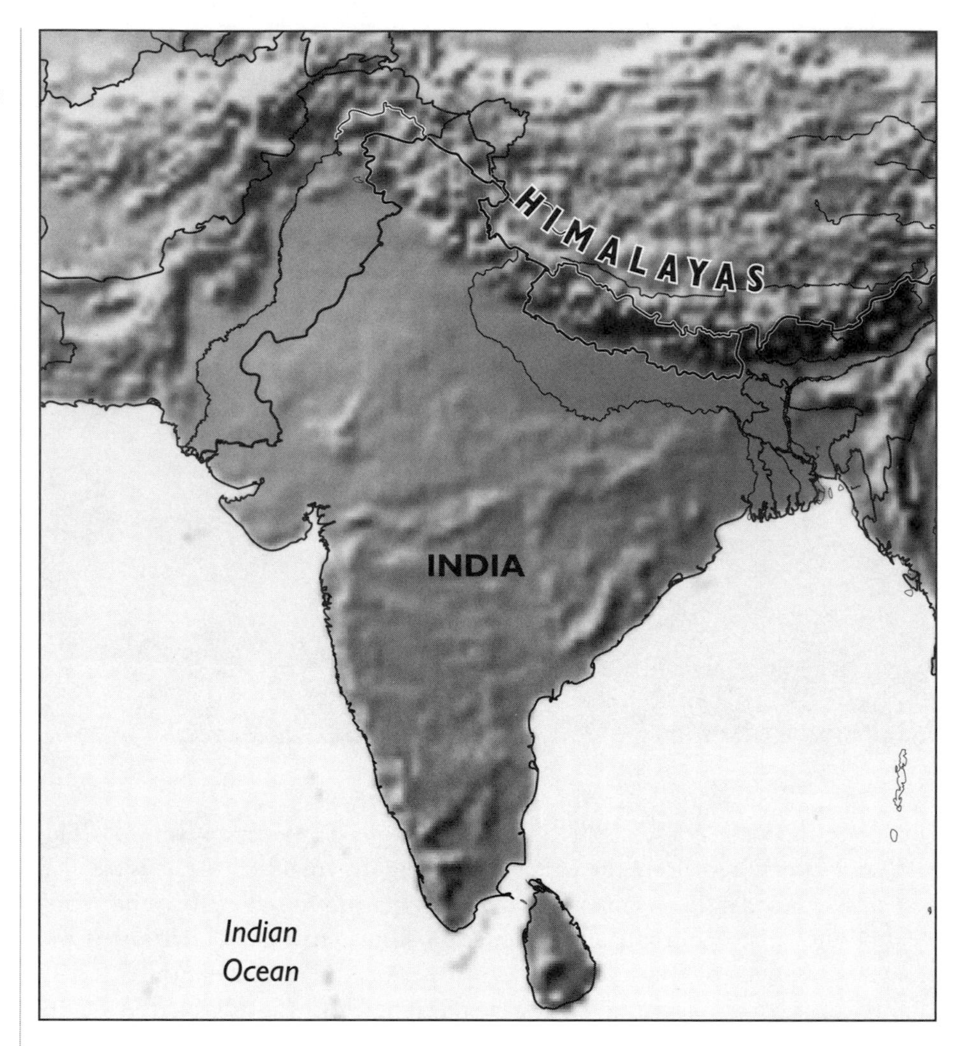

Figure 112 *The formation of the Himalaya Mountains by the collision of the Indian subcontinent with Asia.*

carbon cycle (discussed in more detail in the next chapter). As a result, climatic conditions grew cold, spawning a period of glaciation.

As India continued to slam into Asia, Indochina slid southeastward relative to South China along the 600-mile-long Red River Fault, running from Tibet to the South China Sea. As India plowed into Asia, it pushed Indochina eastward at least 300 miles. In the process of continental escape, Indochina jutted out to sea, rearranging the entire face of Southeast Asia. When the fault locked, the continental escape halted around 20 million years ago. The additional stress on Asia thickened the crust and raised the Himalaya Mountains and the Tibetan Plateau, the largest expanse of land above three miles elevation, half of which appears to have arisen within the last 10 million years.

The Tethys Sea separating Eurasia from Africa filled with thick sediments in a vast geosyncline that formed a huge bulge in the Earth's crust. About 50 million years ago, the Tethys narrowed as Africa approached Eurasia and began to close off entirely some 20 million years ago. Like a rug thrown across a polished floor, the crust crumbled into folds. Thick sediments that had been accumulating for tens of millions of years were compressed into long belts of mountain ranges on the northern and southern continental landmasses.

The entire crusts of both continental plates buckled upward, forming the central portions of the range. This episode of mountain building, called the Alpine orogeny, ended about 26 million years ago. It raised the Pyrenees on the border between Spain and France, the Atlas Mountains of northwest Africa, and the Carpathians in east-central Europe. The Alps of northern Italy formed in much the same manner as the Himalayas, when the Italian prong of the African plate thrust into the European plate. However, because the European plate is only half as thick as the Indian plate, the Alps are not nearly as tall as the Himalayas.

PLATE COLLISIONS

Before the continental drift theory was accepted, geologists had no adequate explanation for the building of mountain ranges. Mountains were thought to have formed early in the Earth's history when the molten crust cooled and solidified, forcing it to contract and shrivel up. However, after further study, geologists were forced to conclude that the folding of rock layers was much too young and intense. Therefore, some other forces had to be at work in raising mountains. Moreover, if mountains had formed in this manner, they would have been scattered evenly throughout the world instead of concentrating in a few mountain ranges.

In his continental drift theory, Alfred Wegener introduced a new idea for mountain building. Although it was convincing because of its elegance and simplicity, the theory was not entirely correct. Wegener thought that as the continents pushed through the ocean floor like an icebreaker plows through arctic ice, the continents encountered increasing resistance, causing the leading edges to crumble, fold back, and thrust upward.

In actual practice, however, when an oceanic and a continental plate collide, thousands of feet of sediments are deposited along the seaward margin of the continental plate in deep-ocean trenches. The increased weight presses downward on the oceanic crust. As the continental and oceanic plates merge, the heavier oceanic plate is subducted under or overridden by the lighter continental plate, forcing it farther downward. As the oceanic crust descends, the topmost layers are scraped off and plastered against the swollen edge of the continental crust, forming an accretionary prism. The sedimentary layers of both plates are then compressed, resulting in a swelling of the leading edge of the continental crust. This forms a mountain belt similar to the Andes of western South America.

A gravity survey in the Andes showed that its gravitational attraction was less than that at sea level. This observation implied that the granites in the mountains were lighter or less dense than the rocks below and thus exerted less gravitational pull on the instruments. From the data, geologists concluded that the continents were composed of lighter granitelike materials called sial (from silica and aluminum) and the ocean floor was composed of heavier basaltlike substances called sima (from silica and magnesium). The difference in density between the two rock types made the continents buoyant, a relationship known as isostasy, from Greek meaning "equal standing."

The buoyant continental crust is generally 25 to 30 miles thick. However, the highest mountain ranges of the world, the Himalayas and the Andes, whose tallest peaks are more than 4 miles high, are supported by deep crustal roots 45 to 50 miles thick. The Sierra Nevada, on the other hand, is underlain by crust only 20 to 25 miles thick, suggesting it has lost elevation over the last 10 million years as the underlying crust thinned. The stable continental roots have remained essentially unchanged since early Precambrian time, especially underneath cratons. These are regions of crust that have remained almost undisturbed for millions of years (Fig. 113). Because continental crust is about 80 to 85

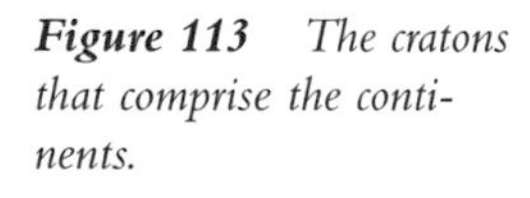
Figure 113 *The cratons that comprise the continents.*

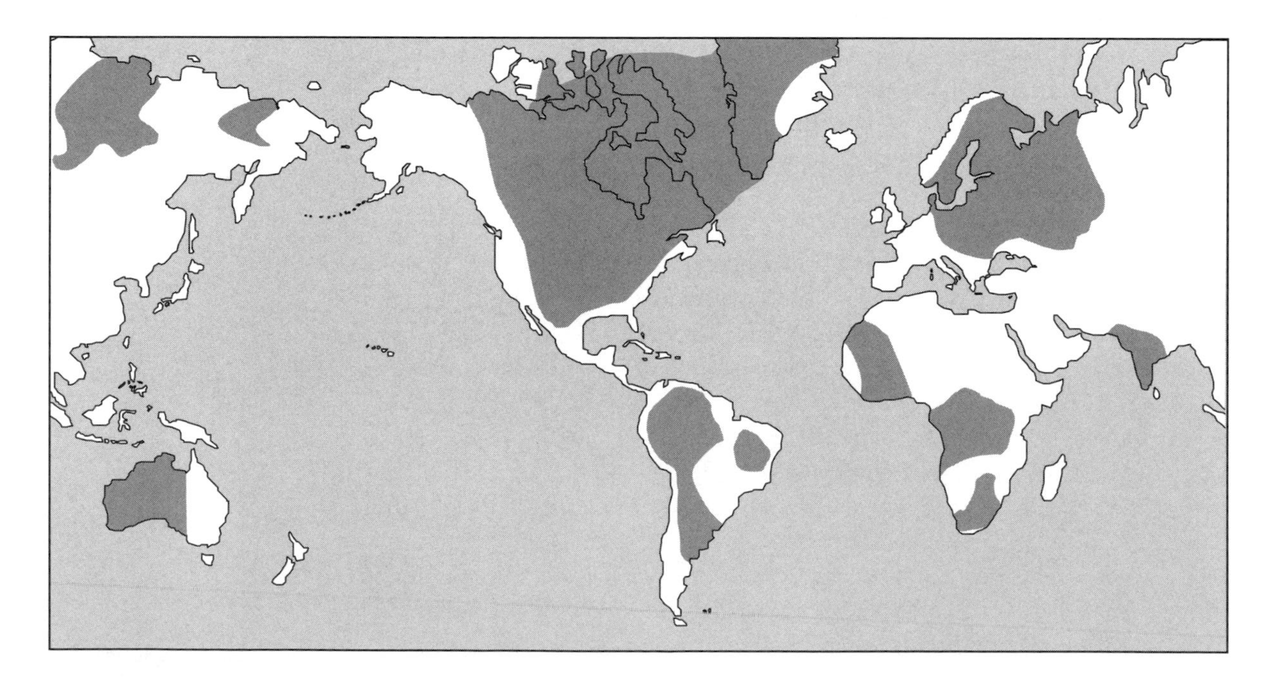

percent as dense as the underlying mantle, deep crustal roots can support mountains several miles high. Like an iceberg, only the tip of the crust shows while the rest is out of sight deep below the surface.

Seismologists using underwater explosions have been trying to get at the roots of a long-standing puzzle about the origin of the southern Andes. Marine seismic survey techniques were used in the waterways that weave through the southern portion of the mountains. Underwater air guns and a string of hydrophones were dragged behind a ship below the water's surface. The hydrophones detected sound waves that reflected off boundaries between rock units in the crust. They produced a sonic picture of the mountains' core like a sort of X-ray image, which might explain why the Andes Mountains are so large.

The Andes Range owes its existence to an increase in crustal buoyancy by the subduction of the Nazca plate beneath the South American plate along the Chile Trench, at a rate of about two inches per year. The interaction of the two plates accounts for the crumpling of the stable continental margin to form belts of folded mountains that constitute the eastern ranges. The western portions of the Andes comprise a chain of active volcanoes and enormous batholiths. The two ranges are separated by a high plateau called the altiplano, the tallest part of the mountains.

The accretion of crustal blocks called terranes played a major role in the formation of mountain chains along convergent continental margins that mark the collision of lithospheric plates. Geologic activity around the Pacific rim was responsible for practically all mountain ranges facing the Pacific Ocean and the island arcs along its perimeter. Therefore, by all accounts, the Andes Mountains should not be there. Most of the world's great mountain ranges were created when plate tectonics slammed two continents together. The Andes, however, appear to have been thrust upward by the accretion of oceanic plateaus along the continental margin of South America as the continent overrode a Pacific plate.

Folded mountain belts (Fig. 114), where strata buckles over, are created by the collision of continental plates. They constitute a massive deformation of rocks in the core of the range. Plate motions built most mountains by shoving the crust of one plate onto another. A gently sloping fault beneath the Wind River Mountains of Wyoming suggests that horizontal squeezing of the continents rather than vertical lifting formed these as well as many other mountain ranges. Mountain building, which provides the forces that fold and fault rocks at shallow depths, also generates stresses that strongly distort rocks deep below.

When two continental plates collide, they crumple the crust, forcing up mountain ranges at the point of impact (Fig. 115). The long-lived continental roots that underlie mountain ranges can extend downward 100 miles or more into the upper mantle. Apparently, through collisions resulting from plate tec-

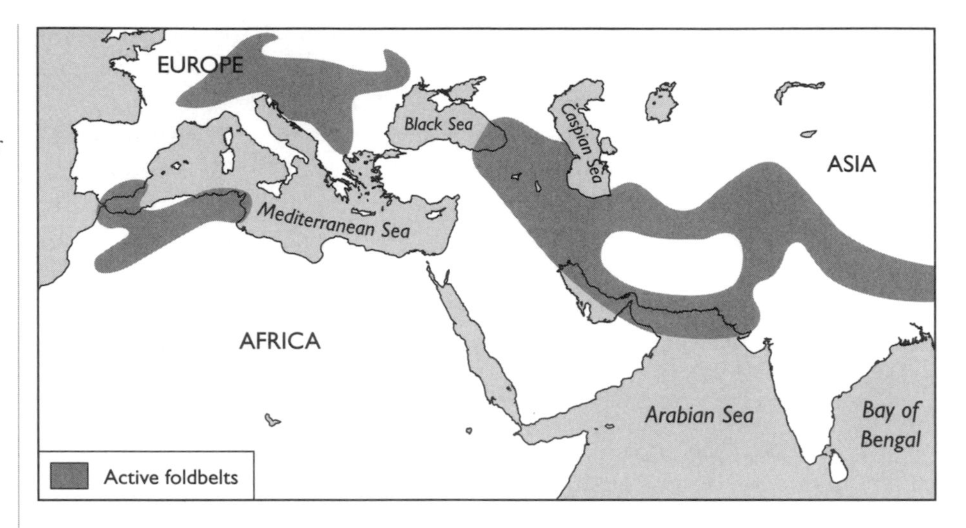

Figure 114 *Active fold belts result from crustal compression where continental tectonic plates collide such as the collision of Africa with Eurasia.*

tonics, continents have stabilized part of the mobile mantle rock below them. The drifting continents carry along with them thick layers of chemically distinct mantle rock. The squeezing of a plate into a thicker one due to continental collision, such as that between the Indian and Asian plates to uplift the Himalayas, might be the very process that forms deep roots.

Continental collisions arising from plate tectonics stabilize a portion of the mobile mantle rock beneath the continents. This causes them to drift

Figure 115 *A wax model illustrating mountain uplift.*

(Photo by J. K. Hillers, courtesy USGS)

along with thick layers of chemically distinct rock as much as 250 miles thick called subcrustal keels. The process that forms deep roots operates by squeezing a plate during continental collision, thus thickening it. This activity was most intense during the collision between the Indian plate and the Eurasian plate, which has shrunk some 1,000 miles while raising the great Himalayas. The Himalaya Mountains stand on a thick shield of strong Precambrian rock. This makes them the tallest range on Earth, comprising all 10 of the world's highest peaks. The orogeny also uplifted the broad, high-rising Tibetan Plateau. Virtually the entire plateau is underlain by Indian lithosphere, the horizontal thrusting of which was responsible for its vast upheaval.

The sutures joining the landmasses are still visible as the eroded cores of ancient mountains known as orogens. Ancient rocks that comprise the interiors of the continents, which were assembled some 2 billion years ago, are called cratons. Caught between the cratons was an assortment of debris swept up by drifting continents, including sediments from continental shelves and the ocean floor, stringers of volcanic rock, and small scraps of continents heavily fractured by faults. Ophiolites, which are pieces of ocean crust thrust up onto land, have been found deep in the interiors of the continents, indicating that they were patched together in the distant past. Blueschists (Fig. 116), which are metamorphosed rocks of subducted ocean crust shoved up onto the continents, might have also been present at a very early age.

Figure 116 *An outcrop of retrograde blueschist rocks in the Seward Peninsula region, Alaska.*

(Photo by C. L. Sainsbury, courtesy USGS)

Many major mountain ranges are associated with the subduction of lithospheric plates. When two continental plates converge, the crust is scraped off the subducting lithospheric plate and plastered onto the overriding plate. The lithospheric plate, now without its overlying crust, continues to dive into the mantle. However, when continental crust moves into a subduction zone, its lower density prevents it from being carried downward into the Earth's interior. The submerged crust is underthrusted by additional crustal material, and the increased buoyancy raises mountain ranges. Additional compression and deformation might take place farther inland beyond the line of collision. This would produce a high plateau with surface volcanoes, similar to those on the wide plateau of Tibet.

Knowledge of the complex tectonic underpinning of the Tibetan Plateau is the key to understanding mountain building. Apparently, the strain of raising the world's highest mountain range by the collision of the Indian plate with Asia has resulted in deformation and powerful earthquakes all along the plate. India is still plowing into Asia at a rate of about two inches a year. As resisting forces continue to build, the plate convergence will eventually stop, the mountains will cease growing, and crustal weakening and erosion will ultimately bring them down to the level of the sea.

UPLIFT AND EROSION

No matter how pervasive the formation of mountain ranges by the convergence of crustal plates, erosion is an equally powerful force wearing them down. Erosion has leveled the tallest mountains, gouged deep ravines into the hardest rock, and obliterated most geologic structures on Earth. Therefore, the shaping of mountains depends as much on the destructive forces of erosion as on the constructive power of plate tectonics. The interactions between tectonic, climatic, and erosional processes exert strong control over the shape and height of mountains as well as the time needed to build or destroy them.

Erosion might actually be the most powerful agent of mountain building by removing mass restored by isostasy, which lifts the entire mountain range to replace the missing mass. If the rate of erosion matches the rate of uplift, the size and shape of mountains can remain stable for millions of years. As the mountains age, the crust supporting them thins out, and erosion takes over to bring down even the most imposing ranges.

The rise of active mountain chains such as the Himalayas is matched by erosion so that their net growth is nearly zero. The world's mountain ranges contain some of the oldest rocks. What was once buried deep below the surface is now thrust high above as huge blocks of granite were pushed up by tectonic forces deep within the Earth and exposed by erosion. The process of erosion is delicately balanced by the forces of buoyancy, which keep the crust afloat. Therefore, erosion can shave off only the top 2.5 miles of continental crust before the mean height of the crust falls below sea level. At that point, erosion ceases and sedimentation begins as the sea inundates the land and sediments accumulate on the seafloor.

Before the spread of vegetation on the land, soil eroded easily because of the lack of plant roots to hold it in place. Therefore, erosion rates were probably much higher than they are today. In the early stages, the relief of the land was not nearly as great as at present. Millions of years of mountain building and erosion provide the current landscape of tall mountains and deep canyons, including the Himalayas, which constitute one of the largest snow fields in the world.

Alpine glaciers (Fig. 117) aggressively attack mountains, becoming perhaps the most potent erosional agents. In addition, mountains evolve their own climate as they grow, increasing rainfall as well as snowfall. Mountains thereby sow the seeds of their own destruction. Therefore, as mountains grow taller, erosion increases, reducing the growth rate. However, some mountains can remain stable for millions of years, with little or no change, because the rate of erosion matches the rate of uplift.

Figure 117 *Chickamin Glacier is a composite valley and slope glacier, Glacier Peak Wilderness, Skagit County, Washington.*

(Photo by A. Post, courtesy USGS)

Water, ice, and wind gradually erase almost all signs of once splendid mountain ranges. However, erosion does not remove everything. Often, the roots of ancient mountains, such as those of the Taconic Range in Eastern New York, survive beneath otherwise unimpressive terrain. Buried deep down are huge faults and folds running through the basement rock. They indicate that long ago, tectonic forces squeezed the crust into mountains now completely leveled by erosion.

Similar folds and faults form the roots of present ranges such as the Appalachian Mountains. The rocks under the northwest Appalachians are riddled with ancient fractures called Iapetan faults that are being forced into earthquake activity by the compression of the crust. The timing of the squeezing, heating, and alteration of rock within mountains is unknown. Yet according to analysis of radioactive elements used in dating rocks, the process of mountain folding apparently spans only a few million years.

IGNEOUS ACTIVITY

Many mountain ranges are associated with volcanic and igneous activity. In the deepest part of the continental crust, where temperatures and pressures are very high, rocks are partially melted and metamorphosed. Most igneous rocks were derived from new material in the mantle, some began when oceanic crust subducted into the mantle, and others formed by the melting of continental crust. Pockets of magma also provide the source material for volcanoes and igneous intrusions, which invade the crust to form large bodies of granitic rocks.

Intrusive magma bodies come in a variety of shapes and sizes. The largest intrusives are called batholiths, the most massive of igneous bodies. These huge granitic structures add significant amounts of new crust to a continent. They are generally greater than 40 square miles in exposed area and are much longer than they are wide. When the overlying sediments are eroded away, the harder, more resistant batholith remains. This results in a major mountain range, such as the Sierra Nevada in California (Fig. 118), which extends for nearly 400 miles but is only about 50 miles wide.

Additional buoyancy might have been provided when the underlying lithosphere dripped away from the crust and was replaced by hotter rock, which shoved the mountain range upward. Globs of relatively cold rock dropping hundreds of miles into the mantle appear to precede this type of mountain building. The 2.5-mile-high southern Sierra Nevada has risen some 7,000 feet over the last 10 million years. However, no plates have converged near the region for more than 70 million years.

Batholiths comprise granitic rocks composed mainly of quartz, feldspar, and mica. The rocks might contain veins, where rich ores have accumulated.

Figure 118 Kern Canyon in Sequoia National Park, Sierra Nevada Mountains, Tulare County, California.
(Photo by F. E. Matthes, courtesy USGS)

The ores formed when metal-rich fluids from a magma chamber migrated into cracks and fractures in the rocks (Fig. 119). Similar to a batholith is a stock, which is less than 40 square miles in exposed area. A stock might be an extension of a larger batholith buried below. Like batholiths, stocks are also composed of coarse-grained granitic rocks.

Another type of granitic rock formation called a dike is an intrusive magma body that is tabular in shape and considerably longer than it is wide. It forms when magma fluids occupy a large crack or fissure in the crust. Because dike rocks are usually harder than the surrounding material, they gen-

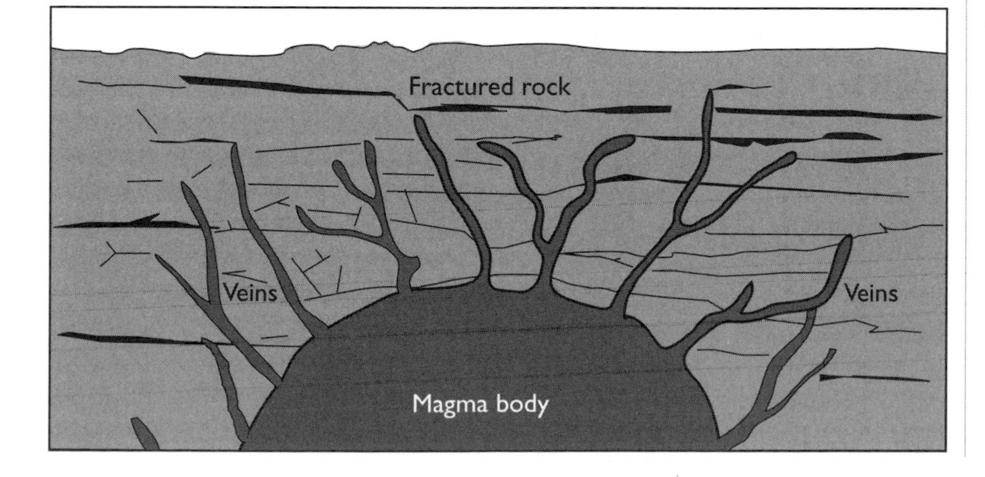

Figure 119 Ore deposits often form when a magma body invades the crust.

Figure 120 *Mount Hillers, Henry Mountains, Garfield County, Utah.*

(Photo by R. G. Luedke, courtesy USGS)

erally form long ridges when exposed by erosion. A sill is similar to a dike in its tabular form except it is produced parallel to planes of weakness such as sedimentary beds. A special type of sill is called a laccolith, which tends to bulge the overlying sediments upward. This sometimes forms solitary peaks such as the Henry Mountains in southern Utah (Fig. 120).

If magma extrudes onto the Earth's surface either by a fissure eruption, the most common type, or a vent eruption, the volcanism builds majestic mountains. The volcanoes of the Cascade Range resulted from the subduction of the Juan de Fuca plate along the Cascadia Subduction Zone beneath the northwestern United States (Figs. 121 & 122). As the plate melts and dives into the mantle, it feeds molten rock to magma chambers that underlie the volcanoes. Much of this activity is accompanied by large-scale intrusions of igneous rocks, some of which provide our present-day world with much of its mineral wealth.

MINERAL DEPOSITS

Throughout geologic history, in many parts of the world, a piece of crust was scraped off an oceanic plate as it plunged under a continental plate and plastered against the leading edge of the continent. These slices of ocean crust are called ophiolites, from the Greek *ophis,* meaning "snake" or "serpent," due to their greenish color. They are associated with an igneous rock known as serpentinite, which has a mottled green appearance like that of a serpent.

When former volcanically active regions of the ocean crust are uplifted onto the continents, they provide rich metal ore deposits mined throughout the world. The Troodos ophiolite on Cyprus was mined extensively by the early Greeks for copper and tin. Mining tools dating from 2500 B.C. have been found in the underground mine workings. The ore provided some of the first bronze for the earliest Greek tools and sculptures. In effect, the Greeks were mining the ocean floor, which had been conveniently brought to the surface.

As the oceanic crust moves away from a spreading center, it eventually reaches the margin of the ocean basin. It is then either subducted into the mantle or collides head-on with another lithospheric plate and raises mountains. In the course of these events, fragments of oceanic crust are uplifted and exposed on land. These fragments of ancient oceanic lithosphere have been identified in various parts of the world. The ophiolites consist of an upper layer of marine sediments, a middle layer of pillow lava (basalts that have

Figure 121 *Composite cone of Mount Baker, Whatcom County, Washington.*

(Photo courtesy USGS)

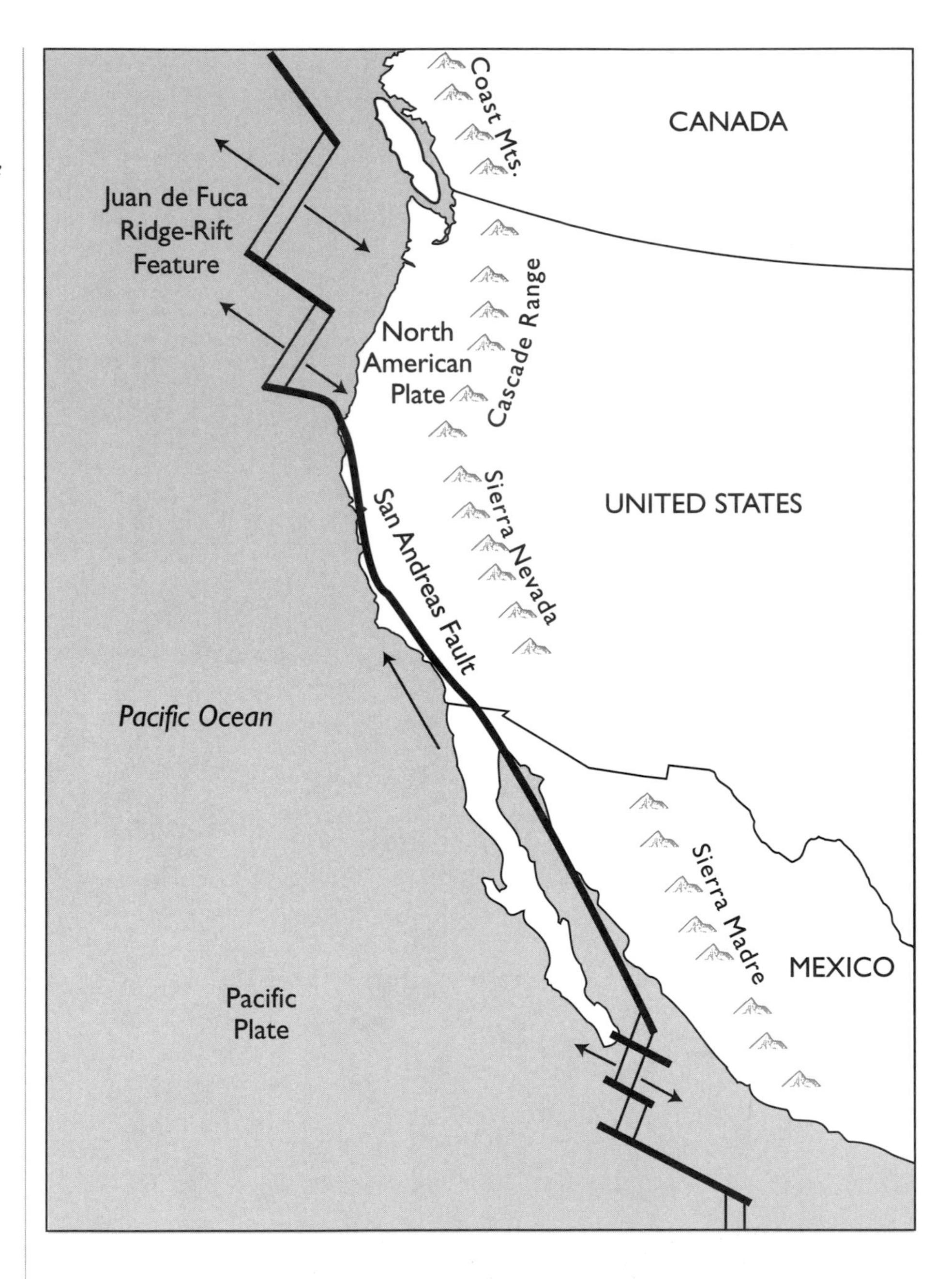

Figure 122 *The subduction of the Juan de Fuca plate into the Cascadia Subduction Zone is responsible for the volcanoes of the Cascade Range.*

erupted undersea), and a lower layer of dark, dense ultramafic rocks thought to be part of the upper mantle.

The metal ore deposits are at the base of the sedimentary layer just above the area where it contacts the basalt. Examples include ophiolite complexes exposed on the Apennines of northern Italy, the northern margins of the Himalayas in southern Tibet, the Ural Mountains in Russia, the eastern Mediterranean (including Cyprus), the Afar Desert of northeastern Africa, the Andes of South America, islands of the western Pacific such as the Philippines, uppermost Newfoundland, and Point Sol along the Big Sur coast of central California.

Another type of metal ore deposit formed at oceanic spreading centers is called a massive sulfide. These deposits contain sulfides of iron, copper, lead, and zinc. They occur in most ophiolite complexes (Fig. 123), such as the Apennine ophiolites, first mined by the ancient Romans. Massive sulfide deposits are mined extensively in other parts of the world for their rich ores of copper, lead, and zinc.

The theory of plate tectonics provides important clues about why certain ore bodies occupy their present locations. Hydrothermal activity, the movement of hot water in the crust, is a reflection of high heat flow, which is associated with plate boundaries. The Red Sea between Africa and Arabia and the Salton Sea of southern California both show evidence of recently transported metals at extensional plate boundaries, where plates are being pulled apart.

In the Salton Sea, the crust is thinning as blobs of magma rise toward the surface. This results in a hot, fractured crust whose salt-laden waters are tapped for their geothermal energy. The brine, which is eight times saltier than seawater, also dissolves metals in the sediments. The metals precipitate out of solution and are concentrated in fractures in the rock. The mineralization is

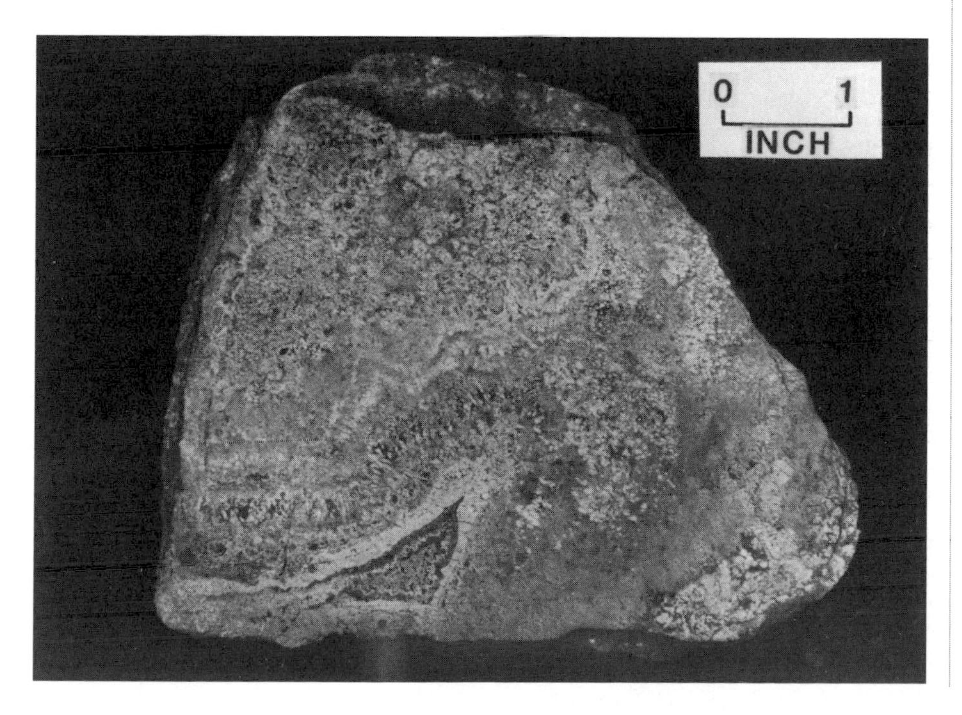

Figure 123 *Metal-rich massive sulfide vein deposit in ophiolite.*
(Photo courtesy USGS)

similar to that found in rifts that were opening more than 600 million years ago and are now mined for their copper, lead, and zinc content.

The Salton Sea area is still quite young, only 100,000 years old. Its metal deposits have not yet achieved ore grades. However, if given an additional half-million years, it should yield significant ores. Another area of high heat flow is associated with igneous activity at converging plate margins, where one plate is forced under another in subduction zones. Many hydrothermal ore deposits have been found in both new and ancient convergent margins, and doubtless many more future ore deposits will be found.

Plate tectonics has also aided in the exploration for oil and gas. The theory helps explain why petroleum reserves are located where they are and might suggest new sites to explore for petroleum. Oil and natural gas is commonly believed to have formed from the remains of abundant plant and animal life that lived in tropical regions tens of millions to hundreds of millions of years ago. This is true for coal as well. Plant fossils are actually found between coal layers, indicating their organic origin.

Deep burial and heat provided by the Earth's interior created a gigantic pressure cooker, which through geologic time has baked the organic compounds into hydrocarbons. The eventual drifting of the continents and erosion of the surface layers has brought the oil and gas deposits within easy reach of the driller's bit. As petroleum reserves become scarce, future reservoirs of oil and gas will have to be tapped in the deep-ocean basins.

GEOTHERMAL ENERGY

Much of the young mountain terrain in the western United States, as well as in Alaska and Hawaii, is of volcanic origin. It forms a well-locked treasure of geothermal energy used for generating electricity. The potential geothermal energy resource in the United States alone is estimated at twice the energy of the world's petroleum reserves. Just a single eruption of Kilauea on the main island of Hawaii could supply two-fifths the power requirements of the entire United States during the time of the eruption.

In areas lacking natural geysers, geothermal energy can be extracted from fractured hot dry rock with a method whereby water is injected into deep wells and steam is recovered. The dry hot rock resources are several thousand times greater than all petroleum reserves. Hot dry rocks lie beneath the surface in areas where the thermal gradients are two to three times greater than normal, about 100 degrees Celsius per mile of depth. The process of artificially making a geothermal reservoir within hot buried rocks is difficult and expensive. If successful, though, the potential is enormous.

Figure 124 *A geothermal generating plant at The Geysers near San Francisco, California.*

(Photo courtesy U.S. Department of Energy)

In a sense, the Earth's interior can be thought of as a natural nuclear power reactor because the heat is mainly derived by the decay of radioactive elements. Many steam and geyser areas around the world are generally associated with active volcanism located at plate margins. These are potential sites for tapping geothermal energy for steam heat and electrical power generation. Nations such as Iceland, Italy, Mexico, New Zealand, Russia, and the United States utilize underground supplies of superheated steam to drive turbine generators for electrical power production. Unfortunately, overproduction of steam fields such as The Geysers in California (Fig. 124), the largest geothermal electrical generating plant in the world, could rapidly deplete this valuable natural resource.

The geopressured energy deposits beneath the gulf coast off Texas and Louisiana are a hybrid of geothermal energy and fossil fuel in reservoirs of hot gas-charged seawater. The deposits formed millions of years ago when seawater was trapped in porous beds of sandstone between impermeable clay layers. Heat building up from below was captured in the seawater along with methane from decaying organic matter. As more sediment piled on, the hot gas-charged seawater became highly pressurized. Wells drilled into this formation not only tap geothermal energy but also natural gas, providing an energy potential equal to about one-third that of all coal deposits in the United States.

Geothermal energy could prove to be far more valuable in the long run than petroleum, coal, or even nuclear energy. Besides, it is nonpolluting. The Earth's internal heat will last for billions of years. Unlike limited resources of fossil fuels, geothermal energy, properly managed, has the potential of supplying our energy needs for millennia.

After seeing how plate tectonics has played a fundamental role in shaping the Earth, the next chapter will explain how plate tectonics has made our world a living planet.

8

THE ROCK CYCLE

THE BALANCE OF NATURE

This chapter examines the critical cycles important to the living Earth. The development of the theory of plate tectonics has led to a greater understanding of the geochemical carbon cycle, or simply rock cycle, that is extremely crucial for keeping our planet alive in the biologic sense. The recycling of carbon through the geosphere makes the Earth unique among planets. This is evidenced by the fact that the atmosphere contains large amounts of oxygen, which without the carbon cycle would have long since been buried in the geologic column. Fortunately, plants replenish oxygen by utilizing carbon dioxide, which plays a critical role as a primary source of carbon for photosynthesis and therefore provides the basis for all life.

Carbon dioxide along with methane and water vapor are important greenhouse gases that trap solar heat that would otherwise escape into space. In this respect, they operate similar to a thermostat that regulates the temperature of the planet. The circulation of carbon between the atmosphere and ocean maintains the balance between incoming and outgoing thermal energy, which determines the temperature of the planet. Therefore, any changes in the carbon cycle could have profound effects on the climate and, ultimately, on life.

THE ATMOSPHERE AND OCEAN

During the early formation of the Earth, volcanoes spewed out massive quantities of gases and steam during what is known as the "big burp." Volcanoes are responsible for a variety of products, including water vapor, nitrogen, carbon dioxide, methane, ammonia, sulfur dioxide, and other gases (Fig. 125). Water and carbon dioxide are especially abundant in magma, which help make it flow easily.

Early in the Earth's history, volcanoes erupted in much the same manner as they do today, only on a greater scale, with many going off all at once. They were also more violent due to the higher temperature of the Earth's interior and the larger amounts of volatiles in the magma. This made the eruptions highly explosive. The early volcanoes were gigantic by today's standards. They shot rock fragments and ash 100 miles or more into space. Since the Earth had no atmosphere to scatter the volcanic debris, it simply fell back around the volcanic vent, building up volcanoes to tremendous heights.

Some volcanoes were so massive they often could not support their own weight and came crashing down, forming enormous gaping craters called calderas. Fiery fountains of lava burst through cracks in the thin crust, paving over the Earth with thick layers of basalt. This formed vast basalt plains similar to those on the moon and Mars. The entire surface of the

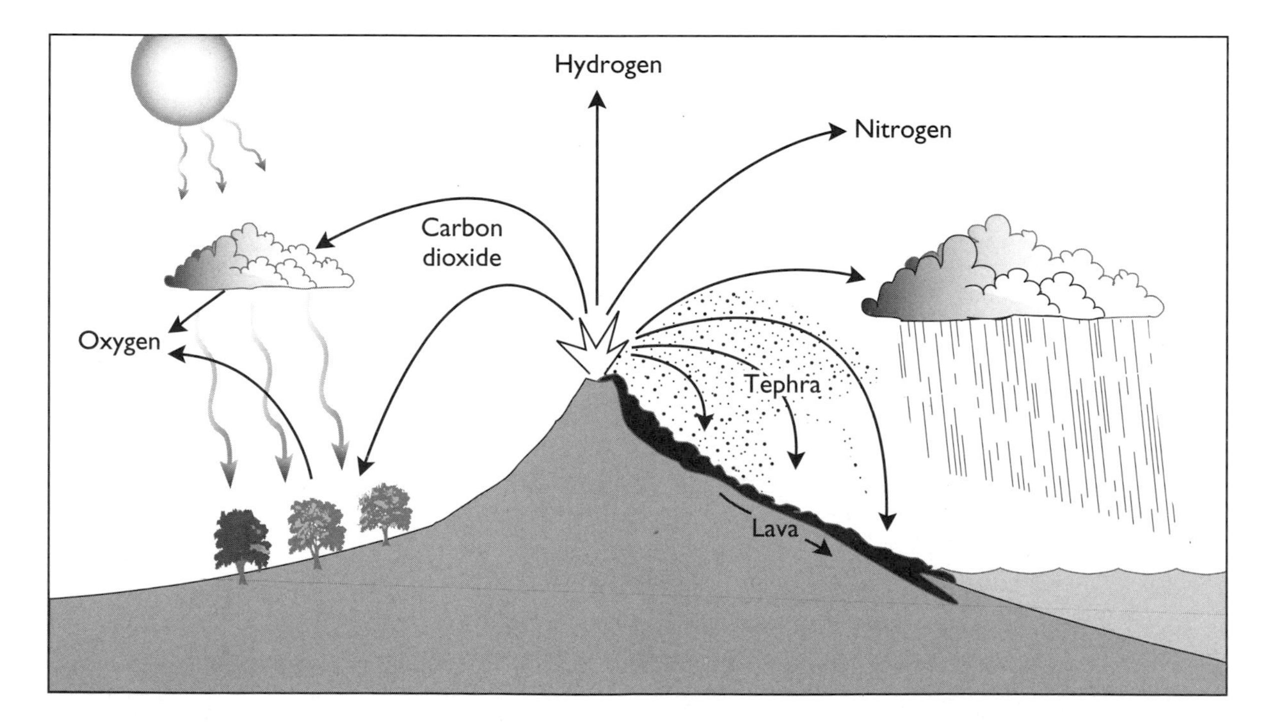

Figure 125 *The contribution of volcanoes includes lava, tephra, and gases.*

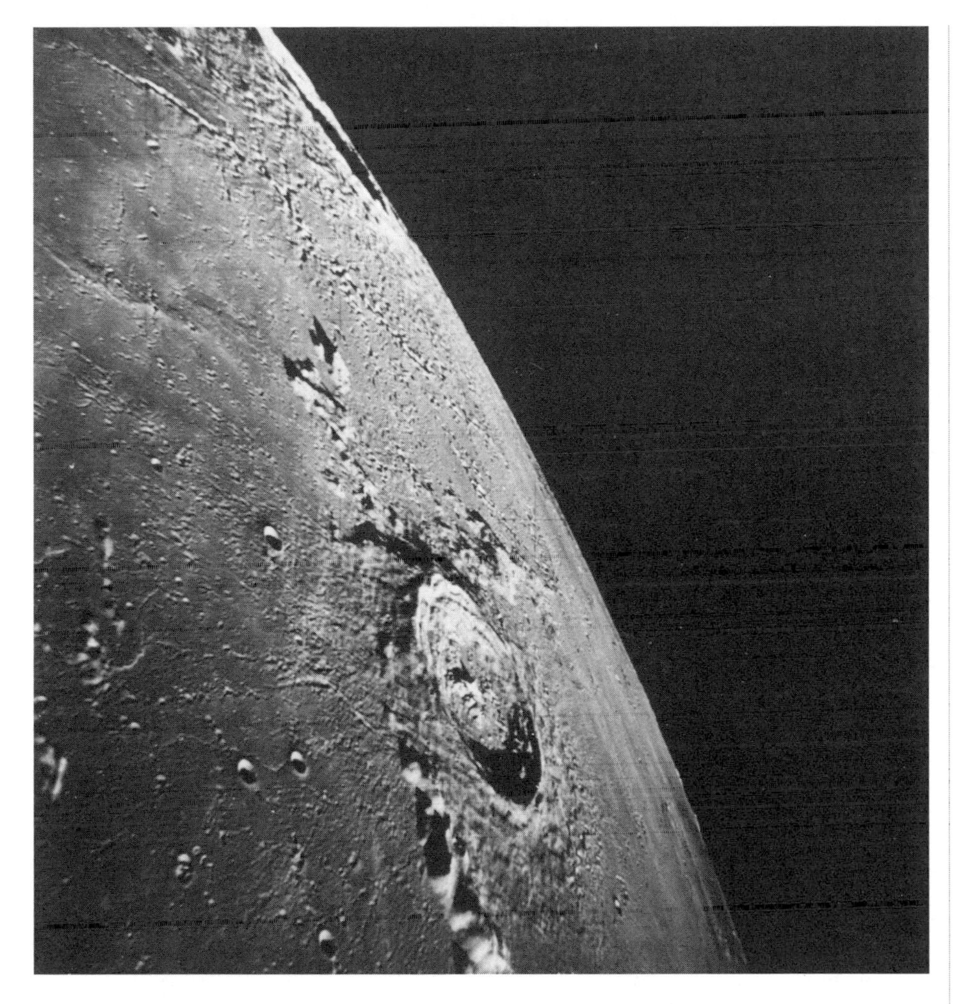

Figure 126 *A large meteorite crater on the lunar surface.*

(Photo courtesy NASA)

Earth was dotted with numerous active volcanoes erupting one after another in great profusion.

Icy visitors from outer space pounded the infant Earth and supplied substantial quantities of water vapor and gasses. The barrage of asteroids and comets began around 4.2 billion years ago and continued at a high rate of impact for another 300 million years. Other bodies in the solar system, including the Earth's own moon, show numerous craters from this massive bombardment (Fig. 126). Afterward, the rate of impact remained, for the most part, about what it is today. This was fortunate, for life would probably have had a poor chance of coming into existence if the Earth were continuously being bombarded by large meteorites.

Some meteorites that pounded the Earth were stony and composed of rock, many were metallic, and others were composed of water ice and frozen gas. Comets, which are essentially rocky material encased in ice, came from the outer reaches of the solar system. In addition, a great deal of carbon fell out of the sky from primitive meteorites known as carbonaceous chondrites, which are carbon-rich rocks left over from the formation of the solar system.

With all this heavy volcanic and meteoritic activity, the Earth quickly acquired a substantial atmosphere. Water vapor was so heavy that the atmospheric pressure was several times greater than it presently is. The surface was still very hot. Water vapor, carbon dioxide, methane, and ammonia, which broke down into nitrogen and hydrogen, produced a powerful greenhouse effect. These gases shrouded the Earth in a thick blanket of steam, making the planet look much like present-day Venus (Fig. 127).

The amount of carbon dioxide in the primordial atmosphere was about 1,000 times greater than at present. During the first billion years, the sun's output was about a third less than today. Carbon dioxide acted as a sort of thermal blanket, allowing the Earth to retain its heat. The greenhouse gasses kept the temperature of the early atmosphere well above the boiling point of water even though the sun shined only about as much as it does now on Mars. If the Earth had the current atmosphere at that time, it would be like Antarctica in the dead of winter. The ocean would be a solid block of ice.

The original oceanic crust was composed of basalt lava flows that erupted onto the surface long before the ocean basins began to fill with water. Then around 4 billion years ago, when the Earth finally cooled down, the rains fell in torrents, producing the greatest floods the planet has ever known. Deep meteorite craters and volcanic calderas rapidly filled, becoming huge bowls of water that spilled onto flat lava plains. Giant valleys were carved out as water rushed down the steep sides of tall volcanoes, which continued to spew steam and gases into the atmosphere. In addition, multitudes of icy comets continuously pounded the Earth, adding more water to the deluge.

When the rains ended and the skies finally cleared, the Earth was transformed into a glistening blue orb covered almost entirely with a deep global ocean. Volcanoes rising from the ocean floor dotted the seascape with a few scattered islands, but as yet no continents existed. The floor of the ocean was an alien world. Volcanoes continued to erupt undersea, and hydrothermal vents spewed out hot water containing sulfur and other chemicals (Fig. 128). In a short time, the sea turned from fresh to salty and contained all the ingredients necessary for the emergence of life. In effect, the ocean became a vast chemical factory, manufacturing all the substances needed to sustain living things.

THE HEAT BUDGET

The most significant influence on the climate is the atmosphere's ability to maintain living conditions by the greenhouse effect, which traps solar energy that would otherwise escape into space. Scientists recognized the mechanics of the greenhouse effect since before the turn of the 20th century. In 1896, the Swedish chemist Svante Arrhenius predicted the effects of atmospheric carbon dioxide on the climate. He estimated that a doubling of the concentration of carbon dioxide in the atmosphere would cause a global warming of about 5 degrees Celsius, surprisingly concurrent with present-day greenhouse models. Arrhenius also concluded that past glacial periods might have occurred largely because of a reduction of atmospheric carbon dioxide.

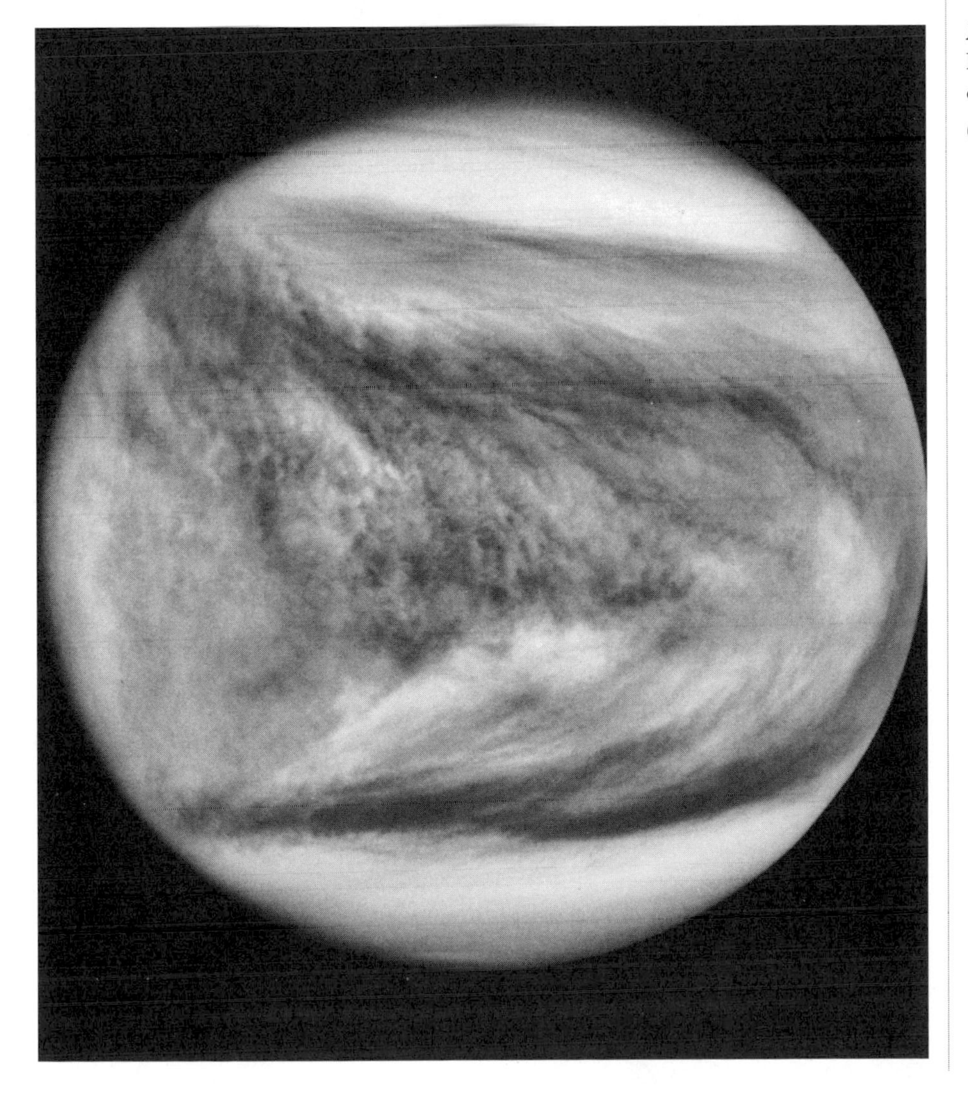

Figure 127 *Venus from* Pioneer Venus Orbiter *on December 26, 1980.*
(Photo courtesy NASA)

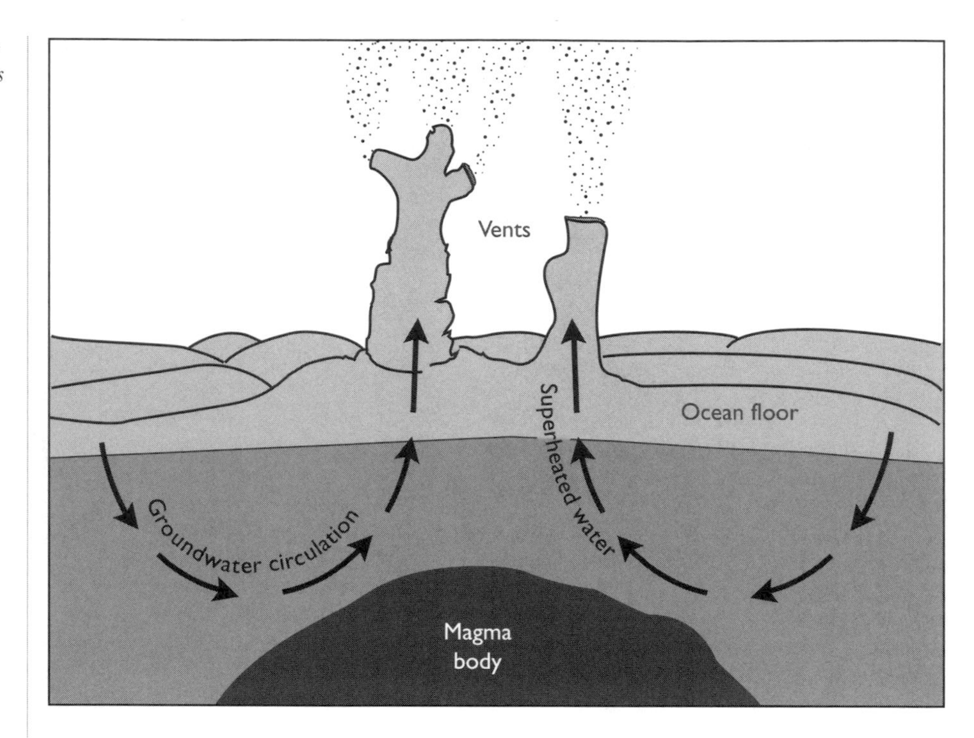

Figure 128 *The operation of hydrothermal vents on the ocean floor.*

The loss of carbon dioxide in the atmosphere by photosynthesis when single-celled plants first developed might have cooled the climate sufficiently to produce the first known ice age in geologic history beginning about 2.2 billion years ago. The burial of large amounts of carbon in the Earth's crust might have been the key to the onset of perhaps the greatest of all ice ages during the late Precambrian about 680 million years ago. The glaciations of the late Ordovician around 440 million years ago, the middle Carboniferous around 330 million years ago, and the Permo-Carboniferous around 290 million years ago might have been influenced by a reduction of atmospheric carbon dioxide to about one-quarter of its present value. Another glacial episode that occurred about 270 million years ago might have been triggered by the spread of forests across the land as plants adapted to living and reproducing out of the sea. The Earth began to cool as the forests removed atmospheric carbon dioxide, converting the carbon into organic matter that became coal, thereby burying substantial amounts of carbon in the crust.

Atmospheric scientists have acquired information on global geochemical cycles to understand what might have caused such a change in carbon dioxide concentration in the atmosphere. Data taken from deep-sea cores established that carbon dioxide variations preceded changes in the extent of the more recent glaciations. Possibly the earlier glacial epochs were similarly affected. The variations of carbon dioxide levels might not be the sole cause of glaciation. However, when combined with other processes, such as variations in the Earth's orbital motions, they could be a powerful influence. This might explain why the ice ages have turned on and off again throughout geologic history.

The strongest orbital cycles are the 23,000-year wobble, similar to the action of a spinning toy top, and the 41,000-year nodding, or altering the tilt angle, of the Earth's spin axis. They change the climate by redistributing sunlight across the planet. For example, the Earth's changing tilt might have triggered rapid drying of the Sahara some 4,000 years ago, changing a land of grasses and shrubbery into a brutal desert. Over the last 9,000 years, the planet's tilt has decreased from 24.15 degrees to 23.45 degrees, resulting in cooler summers in the Northern Hemisphere. The cooling reduced rainfall, which killed off vegetation and further reduced rainfall, causing widespread desertification in the Sahara.

Life appears to maintain oxygen and carbon dioxide in a perfect balance. Too much of one with respect to the other could have disastrous consequences. Life uses the atmosphere both as a source of raw materials, such as oxygen and nitrogen, and as a repository for waste products, such as carbon dioxide, an important greenhouse gas. In this manner, life has a direct input into the greenhouse effect. Living organisms can thus regulate the climate to their own benefit. Thus, without life, the Earth's climate system would be wildly out of control.

Therefore, life has owed its existence to the greenhouse effect since the very beginning. Large quantities of greenhouse gases in the early atmosphere maintained temperatures within tolerable limits for life to flourish, even though the Sun's output was lower than it is today. Fluctuations in the carbon dioxide content of the atmosphere have influenced major changes in the climate down through the ages. When the carbon cycle removed too much carbon dioxide from the atmosphere, temperatures plummeted and great ice sheets flowed across the land. When excessive volcanic activity added too much carbon dioxide to the atmosphere, temperatures soared and the Earth became a hothouse. Only when carbon dioxide levels remain fairly constant is the climate at its optimum for the benefit of all living things.

Life itself might have made major changes to maintain optimum living conditions. The biosphere, the portion of the Earth wherein life exists, appears to be able to control, to some extent, the environment by regulating the climate. This is similar to the way the human body regulates its temperature to optimize metabolic efficiency. For example, a certain species of plankton releases into the atmosphere a sulfur compound that aids in cloud formation. When the Earth warms, plankton growth is invigorated, releasing more cloud-forming sulfur compounds, which cool the planet and stabilize the temperature in an effective feedback mechanism.

The atmosphere plays a critical role in sustaining life by maintaining the balance of incoming solar radiation and outgoing infrared radiation. The Earth

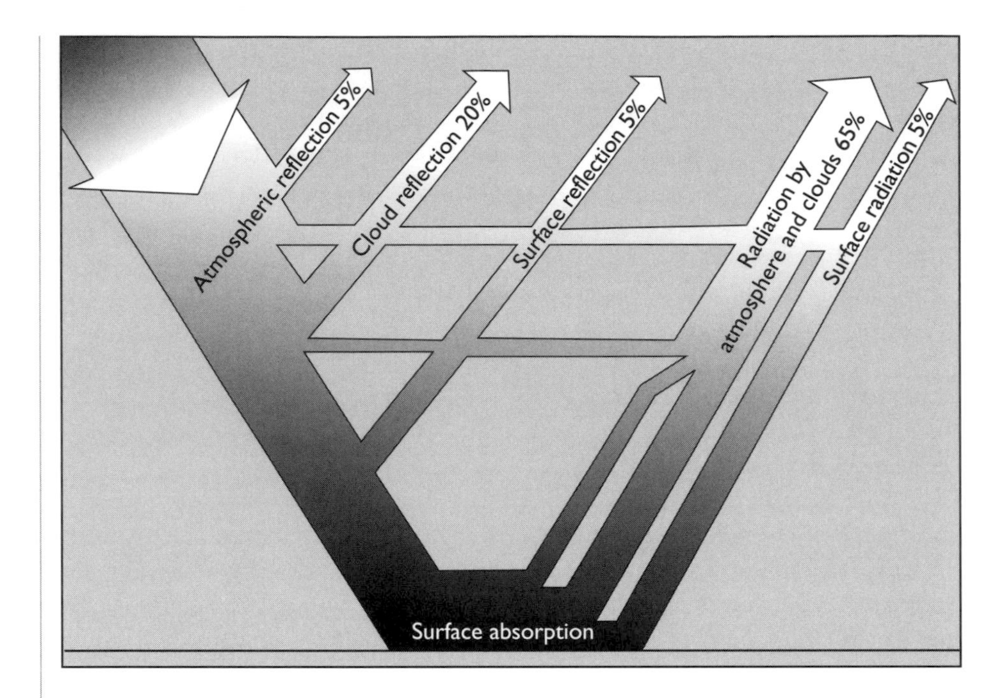

Figure 129 *The Earth's heat budget balances incoming and outgoing solar radiation.*

intercepts about one billionth of the Sun's rays. Only about half the solar energy reaches the surface, where 90 percent evaporates water to make the weather. The Earth must reradiate back into space the same amount of energy it receives from the Sun, or otherwise temperatures would become exceedingly hot. However, if the Earth emits too much infrared energy, temperatures would turn unbearably cold. This delicate balancing act is known as the Earth's energy or heat budget (Fig. 129). It is responsible for maintaining global temperatures within the narrow confines that make life possible.

When sunlight strikes the Earth's surface, it transforms into infrared energy, which is absorbed by the atmosphere and emitted to space. The angle that sunlight strikes the surface also determines the amount of solar energy being absorbed or reflected. In the tropics, the Sun's rays strike the Earth from directly overhead, and more solar radiation is absorbed on the surface than is reflected into space. In the polar regions, the Sun's rays strike the Earth at a low angle, and more solar radiation is reflected into space than is absorbed on the surface. If not for the distribution of heat by the atmosphere and ocean, the tropics would swelter in heat and the higher latitudes would shiver in cold.

The heat budget is also responsible for generating the weather. Warm air rises at the equator in narrow columns and travels aloft toward the poles. In the polar regions, the air liberates heat, cools, sinks, and returns to the equator, where it warms again in a continuous cycle. Currents in the ocean act in a similar manner, only more slowly, taking much longer to complete the journey. The middle latitudes, or temperate zones, become battlegrounds between warm, moist tropical air and cold, dry polar air. When these air masses clash, they create storms.

The oceans play a vital role in distributing solar energy. Solar radiation heats seawater. Thermal energy is transported by ocean currents; lost by conduction, radiation, and evaporation; and regained by precipitation (Fig. 130). Heat flow between the oceans and atmosphere is responsible for cloud formation. A tremendous amount of thermal energy is used to evaporate seawater into water vapor. When clouds move to other parts of the world, they liberate energy by precipitation, which circulates the ocean's heat around the world.

The heat budget mostly depends on the albedo effect, which is an object's ability to reflect sunlight (Table 11). Some things reflect solar energy better than others mainly due to their color. Light-colored objects, such as clouds, snow fields, or deserts, reflect more solar energy than they absorb. Dark-colored objects, such as oceans or forests, absorb more solar energy than they reflect. Most of the solar energy impinging on the ocean evaporates seawater. This energy is lost to space when water vapor condenses into rain.

One-third the solar energy is reflected back into space before it has a chance to heat the Earth. Most of this lost energy reflects off clouds. As a whole, however, clouds exert a net cooling influence on the planet, where the effect is much stronger at midlatitudes than in the tropics. High cirrus clouds retain the Earth's heat, whereas low stratus clouds block out the Sun and cool the surface. The solar energy that does manage to reach the Earth's surface

Figure 130 *Heat flow between the ocean and atmosphere is responsible for distributing the ocean's heat around the world.*

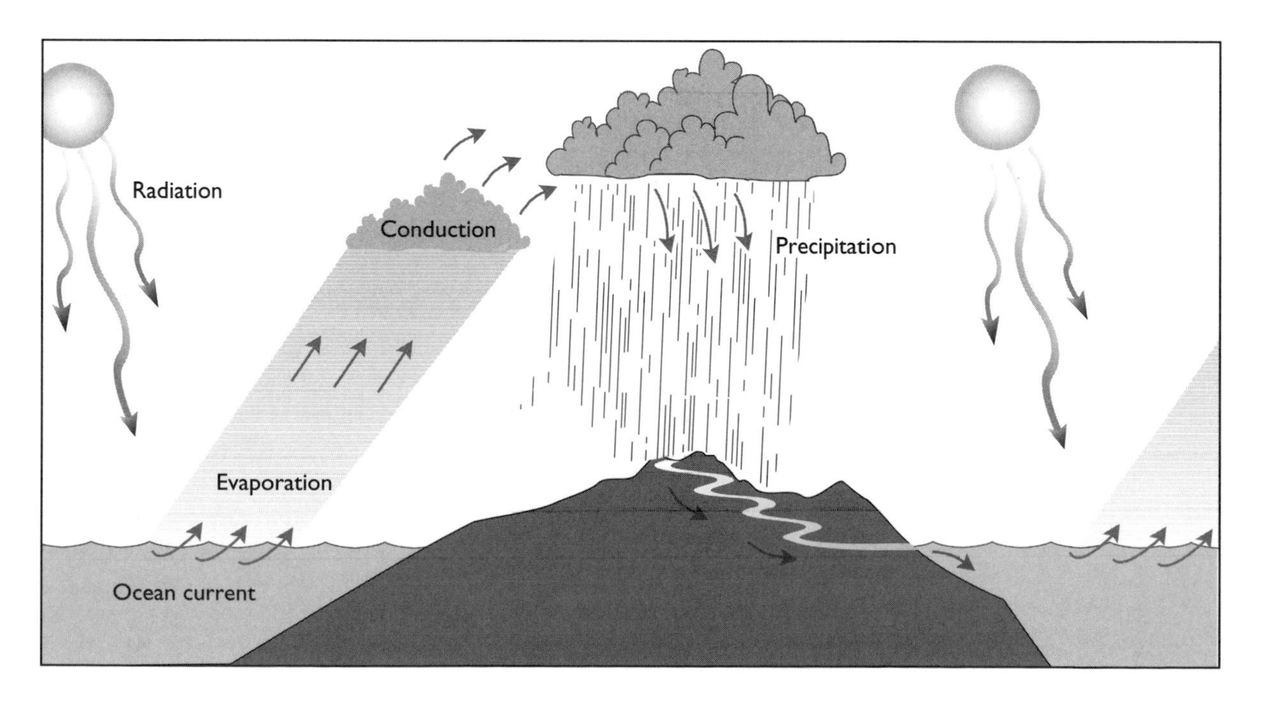

TABLE 11 ALBEDO OF VARIOUS SURFACES

Surface	Percent Reflected
Clouds, stratus	
< 500-feet thick	25–63
500–1,000-feet thick	45–75
1,000–2,000-feet thick	59–84
Average all types and thicknesses	50–55
Snow, fresh fallen	80–90
Snow, old	45–70
White sand	30–60
Light soil (or desert)	25–30
Concrete	17–27
Plowed field, moist	14–17
Crops, green	5–25
Meadows, green	5–10
Forests, green	5–10
Dark soil	5–15
Road, blacktop	5–10
Water, depending upon Sun angle	5–60

heats the ocean and the land and powers one of life's most important cycles—the hydrologic (water) cycle.

THE WATER CYCLE

The oceans cover about 70 percent of the Earth's surface with an average depth of about 2.5 miles, amounting to nearly 250 million cubic miles of water. Each day, some 1 trillion tons of water rain down onto the planet, most of which falls directly back into the sea. The movement of water on the Earth is one of nature's most important cycles. Without the transport of water over the land and back to the ocean, life as we know it would not exist.

The average journey water takes from the ocean to the atmosphere, across the land, and back to the sea again requires about 10 days. The journey is only a few hours long in the tropical coastal regions but might take as much as 10,000 years in the polar regions, returning to the sea as icebergs. This is what is known as the hydrologic cycle, or simply water cycle (Fig. 131).

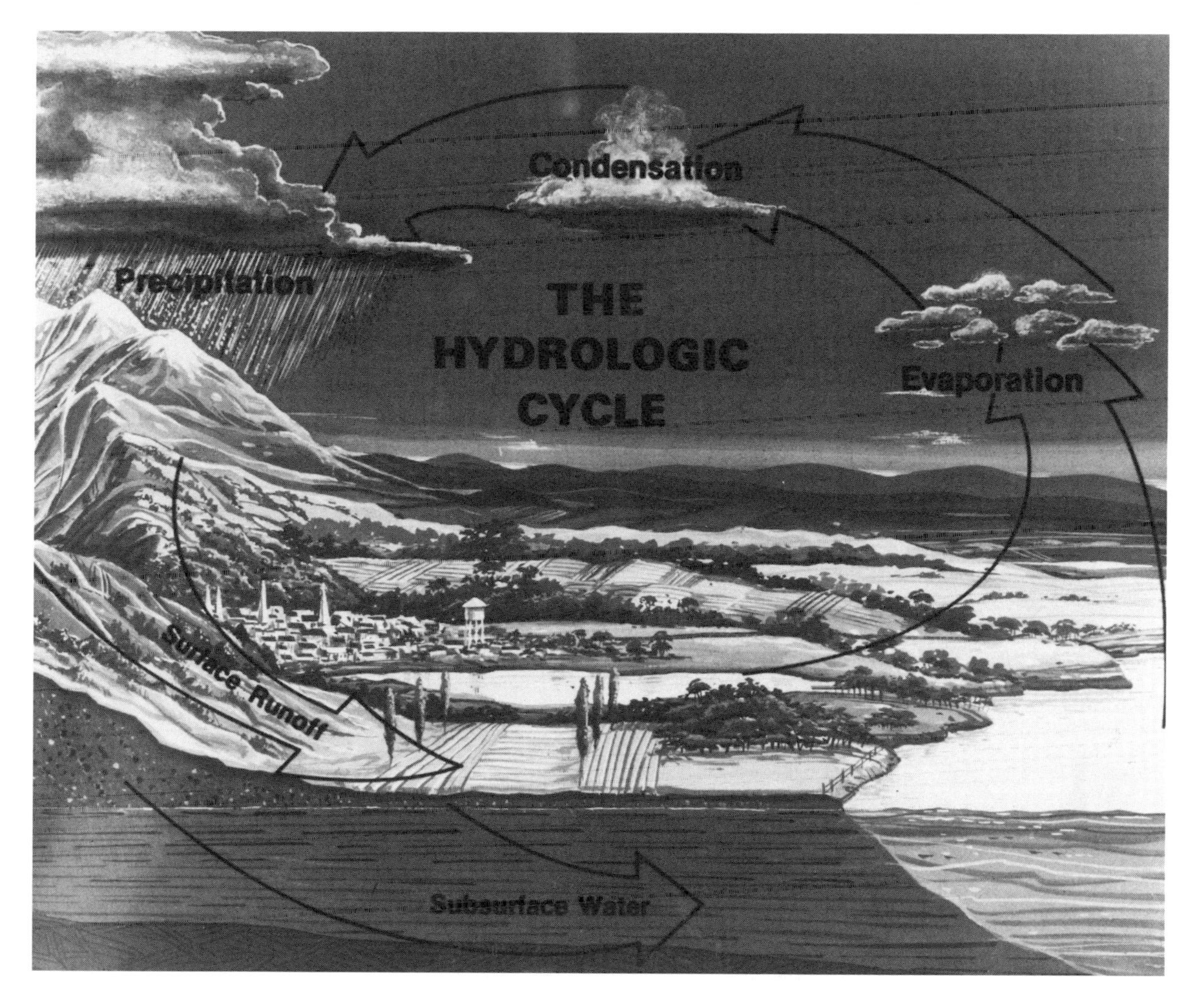

Figure 131 *The hydrologic cycle involves the flow of water from the ocean onto the land and back into the sea.*

(Photo courtesy USGS)

The fastest route water takes back to the ocean is by runoff in rivers and streams. This is perhaps the most apparent as well as the most important part of the water cycle. Surface runoff supplies minerals and nutrients to the ocean and cleanses the land. Acidic rainwater reacts chemically with metallic minerals on the surface, producing metallic salts carried in solution by rivers emptying into the sea. Rainwater also percolates into the ground, dissolves minerals from porous rocks, and transports these by groundwater.

Solid rock exposed on the surface is broken down chemically into clays and carbonates and mechanically into silts, sands, and gravels. Rivers carry the sediments to the shore. After reaching the ocean, the river's velocity falls off sharply and the sediment load drops out of suspension. Chemical solutions carried by the rivers are thoroughly mixed with seawater by currents and wave action. These substances are distributed evenly throughout the ocean with a mixing time of about 1,000 years.

The sediments reaching the seashore continually build the continental margins outward. Coarser sediments accumulate near shore, and progressively finer sediments settle out of suspension farther out to sea. As the shoreline advances seaward, the original fine sediments are overlain by coarser sediments. As the shoreline recedes due to rising sea levels, coarse sediments are overlain by fine sediments. This produces a sedimentary sequence of sandstones, siltstones, and shales. In addition, carbonates precipitate and accumulate in thick beds on the shallow ocean floor.

Most of the sedimentary deposits on the ocean floor are composed of detritus along with shells and skeletons of dead microscopic organisms that flourish in the sunlit waters of the mixed layer of the ocean, the topmost 250 feet. Detritus, which is generated by the weathering of surface rocks along with decaying vegetable matter, is carried by rivers to the edge of the continent and deposited out onto the continental shelf (Fig. 132). There, the material is picked up by marine currents.

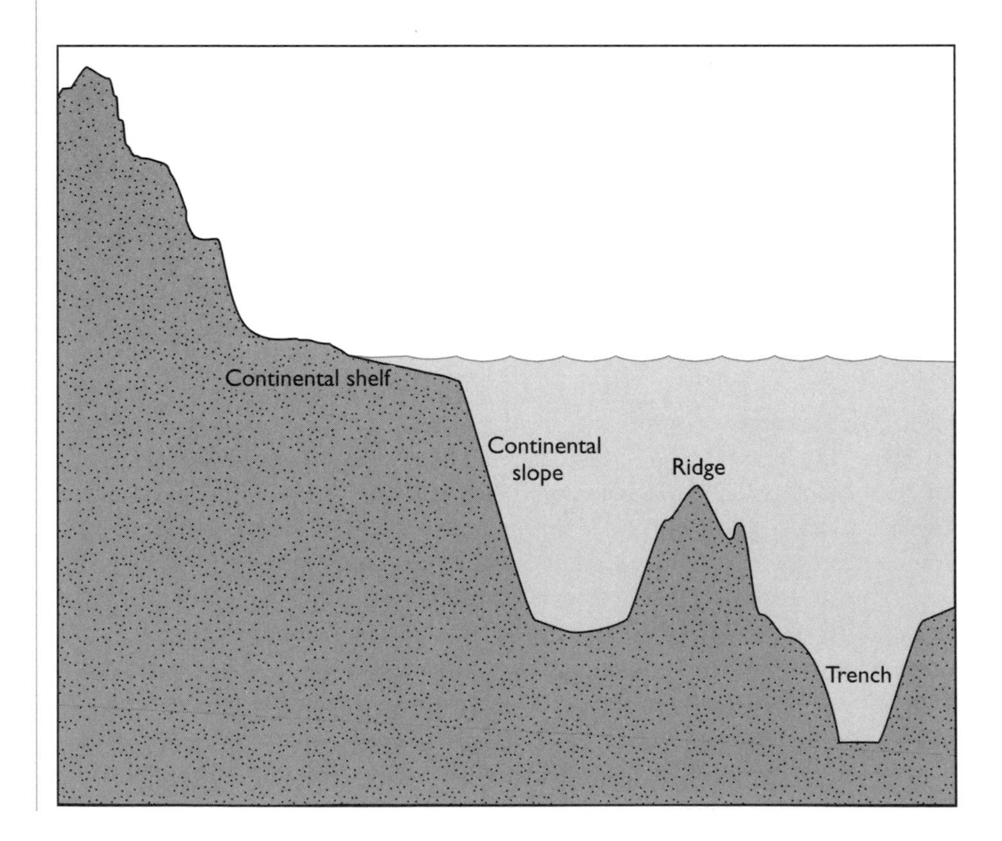

Figure 132 *Profile of the ocean floor.*

When the detritus reaches the edge of the continental shelf, it falls to the base of the continental slope under the pull of gravity. Also, a significant amount of terrestrial material is blown out to sea by dust storms. Approximately 15 billion tons of continental material reaches the outlets of rivers and streams annually. Most of the detritus is trapped near the outlets of rivers and on continental shelves. Only a few billion tons actually reach the deep sea.

Material from dead organisms contributes about 3 billion tons of sediment that accumulates on the ocean floor each year. The amount of accumulation is governed by the rates of biologic productivity, which are controlled in large part by ocean currents. Nutrient-rich water upwells from the ocean depths to the sunlit zone, where the nutrients are consumed by microorganisms. Areas of high productivity and high rates of accumulation are normally around major oceanic fronts and along the edges of major ocean currents.

The rate of marine life sedimentation is influenced by the ocean depth. The farther the shells have to descend, the less are their chances of reaching the bottom before dissolving in the cold waters of the abyss. Below the calcium carbonate compensation zone—the depth at which calcium carbonate readily dissolves, generally occurring about two miles deep—few shells are preserved as fossils. Preservation also depends on how quickly the shells are buried by sediments and protected from the corrosive action of seawater, thereby isolating them from the carbon cycle.

THE CARBON CYCLE

Not until the development of the theory of plate tectonics with its spreading ridges and subduction zones on the ocean floor was the mystery of the missing carbon dioxide finally solved. For without some means of removing excess atmospheric carbon dioxide, our planet could become nearly as hot as Venus, whose thick carbon dioxide atmosphere maintains surface temperatures sufficient to melt lead.

The entire volume of the world's oceans circulates through the crust at spreading ridges every 10 million years, similar to the annual flow of the Amazon, the world's largest river. This accounts for the unique chemistry of seawater and for the efficient thermal and chemical exchanges between the crust and the ocean. The magnitude of some of these chemical exchanges is comparable to the input of elements into the oceans by rivers carrying materials weathered from the continents. The most important of these chemical elements is carbon, which controls many life processes on the planet.

The geochemical carbon cycle (Fig. 133) is the transfer of carbon within the biosphere. It involves the interactions between the crust, ocean, atmos-

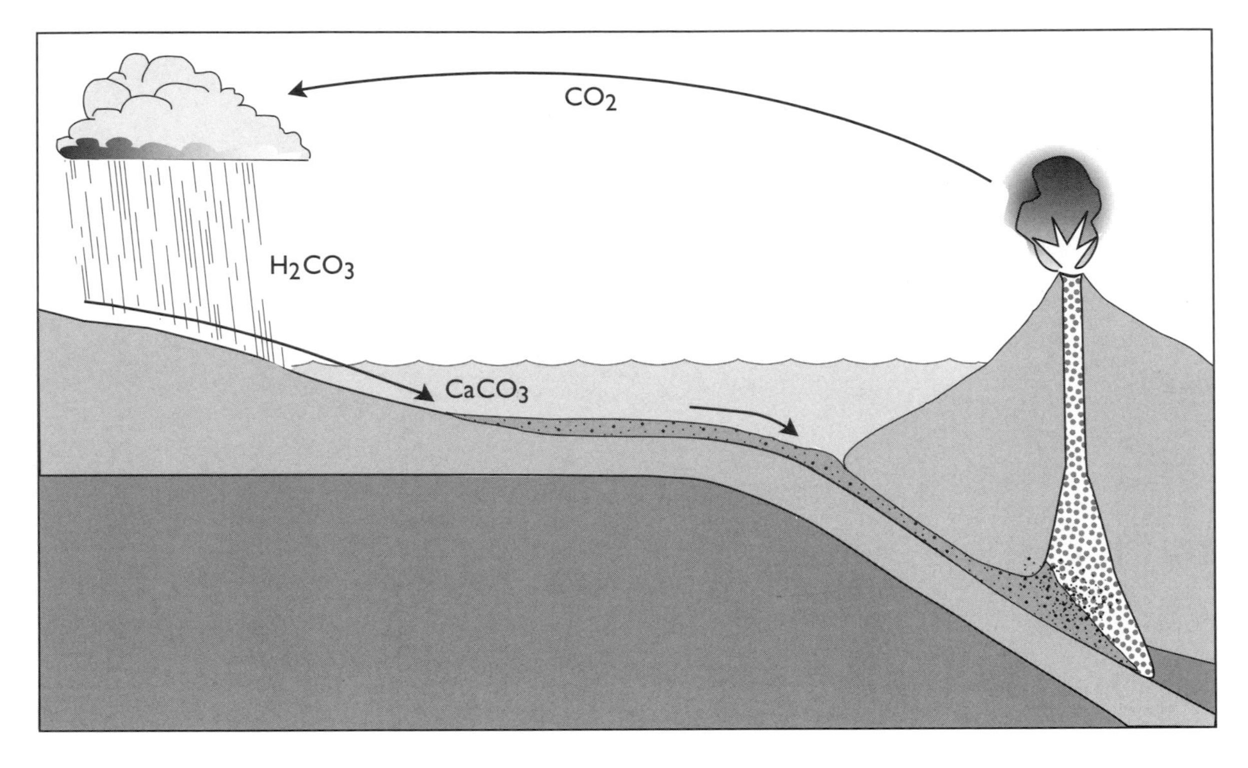

Figure 133 *The geochemical carbon cycle. Carbon dioxide in the form of bicarbonate is washed off the land and enters the ocean, where organisms convert it to carbonate sediments. These are thrust into the mantle, become part of the magma, and escape into the atmosphere from volcanoes.*

phere, and life. Many aspects of this important cycle were understood around the turn of 20th century, notably by the American geologist Thomas Chamberlain and chemist Harold Urey. However, only in the last few years has the geochemical carbon cycle been placed within the more comprehensive framework of plate tectonics.

The biologic carbon cycle is only a small component of this cycle. It is the transfer of carbon from the atmosphere to vegetation by photosynthesis, returning to the atmosphere when plants respire or decay. Only about a third of all chemical elements, mostly hydrogen, oxygen, carbon, and nitrogen, are recycled biologically. The vast majority of carbon is not stored in living tissue, however, but is locked up in sedimentary rocks. Even the amount of carbon contained in fossil fuels is meager by comparison.

Carbon dioxide presently comprises about 365 parts per million of the air molecules in the atmosphere, amounting to more than 800 billion tons of carbon. It is one of the most important greenhouse gases, which trap solar heat that would otherwise escape into space. Carbon dioxide thus plays a vital role in regulating the Earth's temperature. Major changes in the carbon cycle could have profound climatic effects. Carbon dioxide, therefore, operates somewhat like a thermostat to regulate the temperature of the planet. If temperatures began to fall, less water would evaporate from the ocean, chemical and biologic reactions would slow, and less carbon dioxide would be removed from the atmosphere even though the input from volcanoes and rift zones would remain somewhat constant. Conversely, if the carbon cycle generated too much carbon dioxide, the Earth would warm. Therefore, even slight changes in the carbon cycle could have considerable effects on the climate.

Atmospheric carbon dioxide reacts with rainwater to form a weak carbonic acid, which leaches minerals such as calcium and silica from surface rocks. Rivers transport dissolved calcium and bicarbonate to the sea, where they mix with seawater. The minerals are then taken up by marine organisms to make their shells. When the organisms die, their shells sink to the ocean bottom (Fig. 134). There, they slowly build up deposits of limestone if composed of calcium or of diatomite if composed of silica. If the scavenging of carbon dioxide from the atmosphere for the manufacture of carbonaceous sediments on the ocean floor continued unchecked, the atmosphere would soon be depleted of carbon dioxide.

The oceans play a major role in regulating the level of atmospheric carbon dioxide. In the upper layers of the ocean, which contain as much carbon dioxide as the entire atmosphere, the concentration of gases is in equilibrium with the atmosphere at all times (Fig. 135). The gas dissolves into the waters of the ocean mainly by the agitation of surface waves. If the ocean were lifeless without photosynthetic organisms to absorb dissolved carbon dioxide, much of its reservoir of this gas would escape into the atmosphere, more than tripling the present content.

Fortunately, the ocean is teeming with life. Marine organisms take up carbon dioxide as dissolved bicarbonates to build their carbonate skeletons and

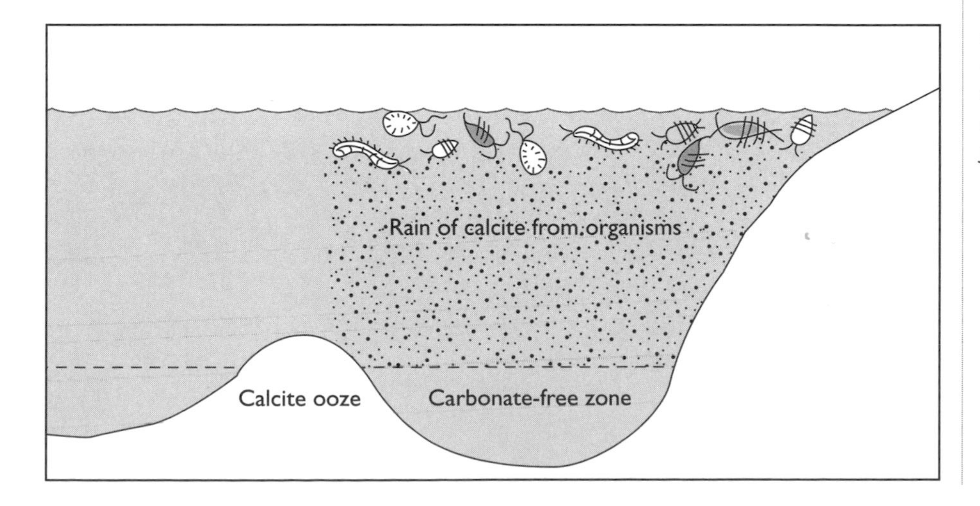

Figure 134 *The formation of limestone from carbonaceous sediments deposited onto the ocean floor.*

other supporting structures (Fig. 136). When the organisms die, their skeletons sink to the bottom of the ocean, where they dissolve in the deep waters of the abyss. Because of its large volume, the abyss holds the largest reservoir of carbon dioxide, containing 60 times more carbon than the atmosphere.

Sediments on the ocean floor and on the continents store most of the carbon. In shallow water, the carbonate skeletons build deposits of limestone, dolomite, and chalk. These bury carbon dioxide in the geologic column. The burial of carbonate in this manner is responsible for about 80 percent of the carbon deposited onto the ocean floor. The rest of the carbonate originates from the burial of dead organic matter washed off the continents.

The deep water, which represents about 90 percent of the ocean's volume, circulates very slowly and has a residence time on the order of about 1,000 years. It communicates directly with the atmosphere only in the polar regions, so its absorption of carbon dioxide is very limited there. The abyss receives most of its carbon dioxide in the form of shells of dead organisms and fecal matter, which sink to the bottom. Half the carbonate transforms back into carbon dioxide, which returns to the atmosphere. This occurs mostly by upwelling currents in the tropics. As a result, the concentration of atmospheric carbon dioxide is highest near the equator. If not for this process, in a mere 10,000 years, all carbon dioxide would be removed from the atmosphere. The loss of this important greenhouse gas would result in the cessation of photosynthesis and the extinction of life.

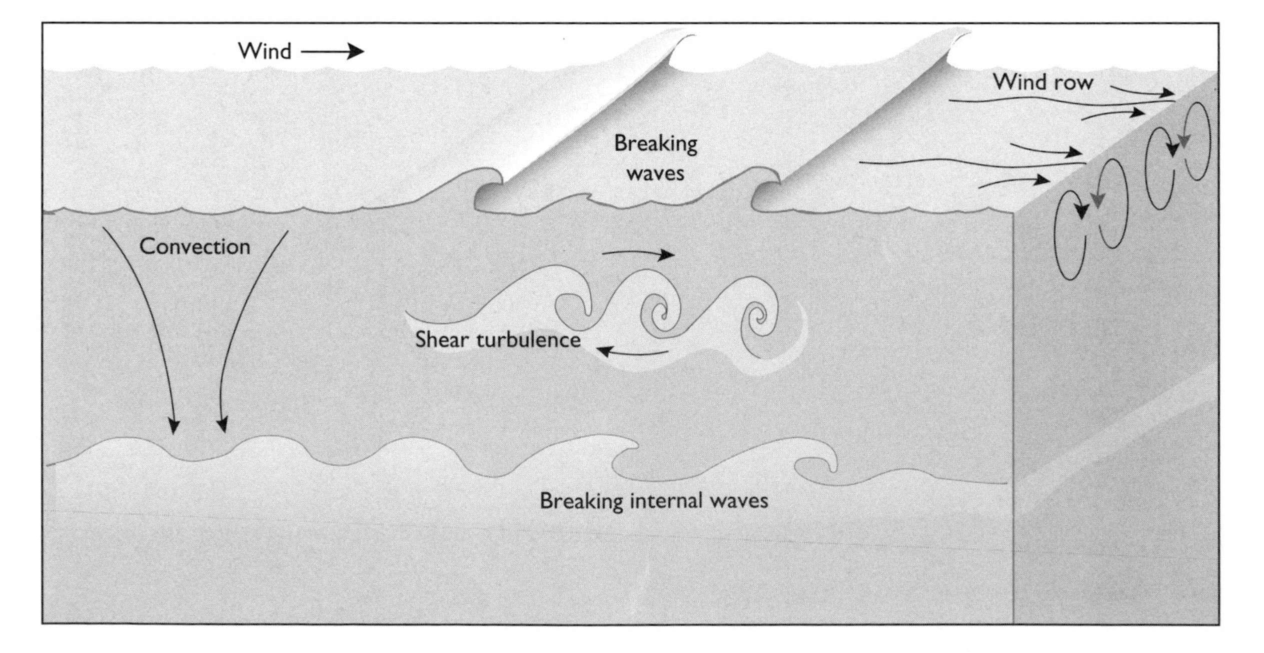

Figure 135 *Turbulence in the upper layers of the ocean induces the mixing of temperatures, nutrients, and gases.*

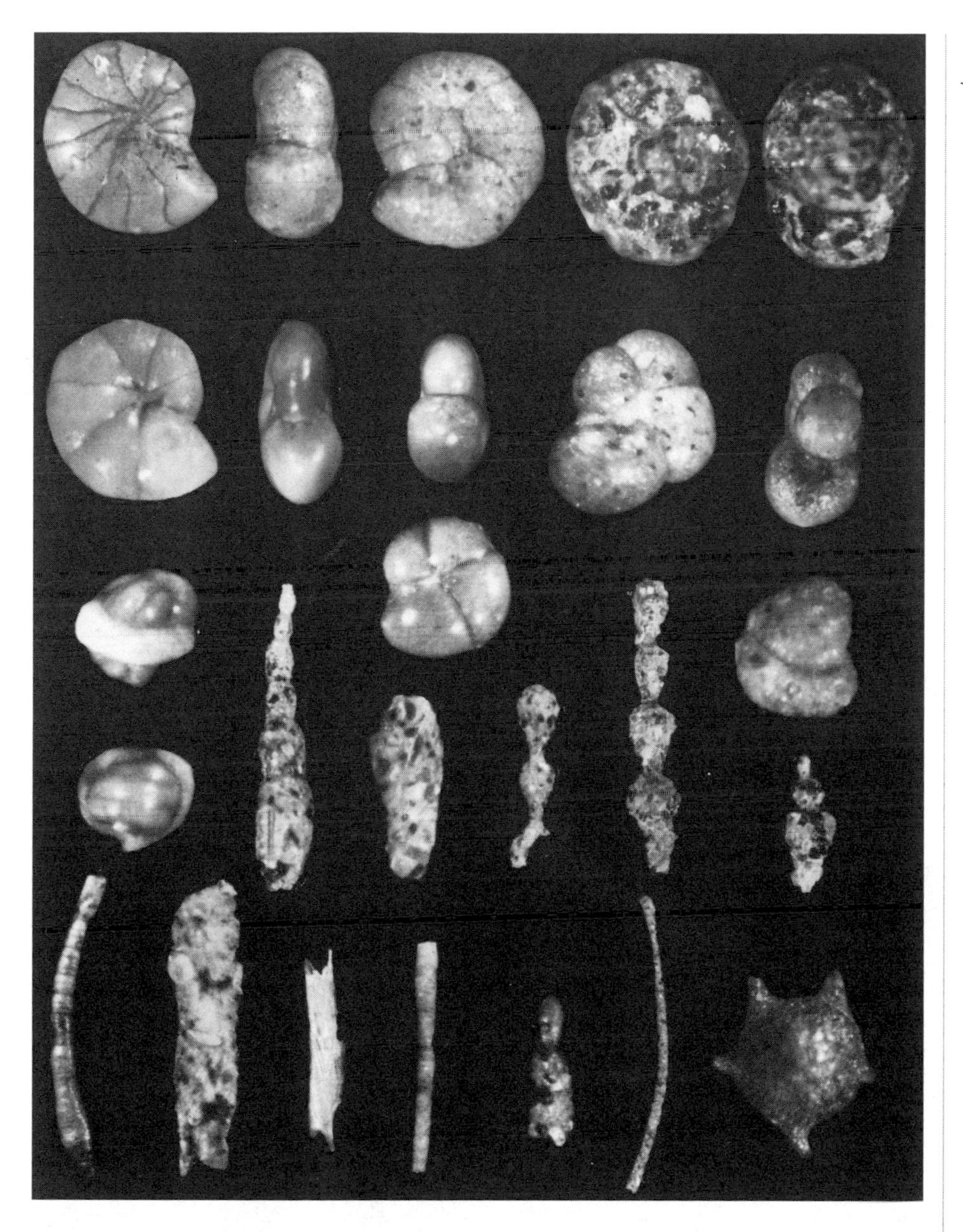

Figure 136 *Fossil foraminerfera of the North Pacific Ocean. Foraminifera were important limestone builders.*

(Photo by B. P. Smith, courtesy USGS)

THE NITROGEN CYCLE

Oxygen, hydrogen, carbon, and nitrogen comprise most of the elements of life. Nitrogen makes up 79 percent of the atmosphere and is one of the major constituents of living matter. Carbon, hydrogen, and nitrogen are the essential elements for manufacturing proteins and other biologic molecules. Nitrogen is practically an inert gas, however, and requires special chemical reactions to be used by nature. Therefore, much energy is required to make nitrogen combine with other substances.

Atmospheric nitrogen originated from early volcanic eruptions and the breakdown of ammonia, a molecule of one nitrogen atom and three hydrogen atoms. Ammonia was a large constituent of the primordial atmosphere and might have influenced the initiation of life. Unlike most other gases, which have been replaced or permanently stored in the crust, the Earth retains much of its original nitrogen. This is because life prevents all nitrogen from transforming into nitrate, which is easily dissolved in the ocean. Denitrifying bacteria in the ocean return the nitrate-nitrogen to its original gaseous state. Without this process, all the nitrogen in the atmosphere would have long ago disappeared, and the Earth would be left with only a fraction of its present atmospheric pressure.

The nitrogen cycle is a continuous exchange of elements between the atmosphere and biosphere by the action of organisms, such as nitrogen-fixing bacteria. All methods of nitrogen fixation need a source of abundant energy, mainly supplied by the Sun. The Earth also provides a source of energy used by animals living near hydrothermal vents on the deep ocean floor (Fig. 137).

Figure 137 *Cluster of tube worms and sulfide deposits around hydrothermal vents near the Juan de Fuca Ridge.*
(Photo courtesy USGS)

The decay of organisms after death releases nitrogen back into the atmosphere, thus forming a closed cycle.

Humans, however, have doubled the rate at which nitrogen gas in the atmosphere is chemically converted into compounds that can be used by plants and animals. The excess nitrogen has disrupted one of the planet's fundamental cycles. The increased nitrogen compounds in the atmosphere act as ozone destroyers, greenhouse gases, and pollutants. They are leaching soils of nutrients, increasing the acidification of surface waters, and clogging the seacoasts with nitrogen-hungry algae, which choke off other aquatic life.

Nitrogen oxides found in acid rain can be especially harmful to aquatic organisms. Nitrogen works like a nutrient, promoting the growth of algae. The algae block out sunlight and deplete water of its dissolved oxygen. This, in turn, suffocates other aquatic plants and animals. Widespread increases in nitrate levels along with higher concentrations of toxic metals, including arsenic, cadmium, and selenium, are occurring globally. The main factors contributing to this increase are fertilizer and pesticide runoff along with acid rain, which dissolves heavy metals in the soil. Acid rain also depletes the soil of some of its nutrients, including calcium, magnesium, and potassium. Some soils are so acidic they can no longer be cultivated.

Oxides of nitrogen produced in factory furnaces and by motor vehicles absorb solar radiation and initiate a chain of complex chemical reactions. In the presence of organic compounds, these reactions result in the formation of a number of undesirable secondary products that are very unstable, irritating, and highly toxic. High-temperature combustion yields nitrogen oxide along with gaseous nitric acid. Because of their minute size, these particles, called *aerosols,* scatter light that would normally heat the ground. Instead, the aerosols heat the atmosphere and cause a temperature imbalance between the atmosphere and the surface, creating abnormal weather patterns.

Nitrous oxides are produced by the combustion of fossil fuels especially under high temperatures and pressures created in coal-fired plants and internal-combustion engines. The tall chimneys of coal-fired plants send huge amounts of nitrous oxides high into the atmosphere. The gas eventually rises into the upper stratosphere, where it can break down the ozone layer that protects the Earth from ultraviolet radiation.

Deforestation might also threaten the erosion of the ozone layer by releasing nitrous oxide into the atmosphere. Clear-cutting of timber encourages soil bacteria to produce nitrous oxide, which is expelled into the atmosphere. The tremendous heat produced by the burning timber combines nitrogen and oxygen into nitrous oxide. Significant amounts of this gas escapes into the upper atmosphere. A continued depletion of the ozone layer with accompanying high ultraviolet exposures could reduce crop productivity and aquatic life, especially primary producers at the very bottom of the food chain.

Today's high-yield crops quickly deplete the soil of fixed nitrogen, which must be replenished by applying either organic fertilizers, the preferred method, or chemical fertilizers. These require large amounts of energy in their manufacture, which is usually supplied by fossil fuels. The natural supply of fixed nitrogen is limited, however, imposing a limit on world agriculture. Therefore, supplemental nitrogen must be supplied by chemical fertilizers if agriculture is to keep up with the ever-expanding human population.

CARBONATE ROCKS

Carbonate rocks, such as limestone, dolomite, and chalk, are formed by precipitation of carbonaceous minerals dissolved in seawater mostly by biologic activity as well as by direct chemical processes. Living organisms use these minerals to build their shells and skeletons composed of calcium carbonate. When the organisms die, their skeletons fall to the bottom of the ocean. Over time, the calcium carbonate comprising a calcite ooze builds up into thick limestone deposits (Fig. 138).

The most common precipitate rock is limestone. It is generally produced by biologic processes, as evidenced by an abundance of marine fossils in limestone beds. Some limestone is chemically precipitated directly from seawater. A minor amount precipitates in evaporite deposits from brines. Most limestones originated in the ocean, and some thin limestone beds were deposited in lakes and swamps. Many limestones form massive formations. They are recognized by their typically light gray or light brown color. Whole or partial fossils constitute many limestones, indicating their biologic origin.

Dolomite resembles limestone but is produced by the partial replacement of calcium in limestone with magnesium. The replacement can cause a reduction in volume, forming void spaces. For the last two centuries, geologists have been puzzled by the so-called dolomite problem. Ancient dolomite deposits were laid down in huge heaps such as those that created the Dolomite Alps in northern Italy. However today, little dolomite is being formed. Dolomite appears to have been made from the excrement of sulfate-consuming bacteria that apparently were far more prevalent in the past.

Chalk is a soft, porous carbonate rock that should not be confused with the chalk used on classroom blackboards, which is actually composed of calcium sulfate. Thick beds of chalk were deposited during the Cretaceous, which is how the period got its name—*creta* is Latin for "chalk." One of the largest chalk deposits are the chalk cliffs of Dorset, England, which, because they are so soft, severely erode during violent coastal storms.

The carbonate sediments were deposited in shallow seas, generally less than 100 feet deep. As calcareous sediments accumulate into thick deposits on

Figure 138 *Intensely folded limestone formation in Atacama Province, Chile.*

(Photo by K. Segerstrom, courtesy USGS)

the ocean floor, deep burial of the lower strata produces high pressures. This lithifies the beds into carbonate rock, consisting mostly of limestone or dolomite. If fine-grained calcareous sediments are not strongly lithified, they form deposits of soft, porous chalk. Limestones typically develop a secondary crystalline texture. This results from the growth of calcite crystals by solution and recrystallization following the formation of the original rock.

Some carbonate rocks were deposited in deep seas. However, the cold, high-pressure waters of the abyss, which contain the vast majority of free carbon dioxide, dissolves most calcium carbonate sinking to this level. The upwelling of deep ocean water, mainly in the tropics, returns to the atmosphere carbon dioxide lost by the carbon cycle, the circulation of carbon by geochemical processes.

Silica also dissolves in seawater in volcanically active areas on the seafloor such as at midocean spreading centers, from volcanic eruptions into the sea, and from weathering of siliceous rocks on the continents. Organisms such as diatoms

(Fig. 139) extract the dissolved silica directly from seawater to make their shells or skeletons. Accumulations of siliceous sediment on the ocean floor from dead organisms form diatomaceous earth, otherwise known as diatomite.

Evaporite deposits are produced in arid regions near shore where pools of brine, which are replenished with seawater during storms, evaporate, leaving salts behind. The salts precipitate out of solution in stages. The first mineral to precipitate is calcite, closely followed by dolomite. However, only

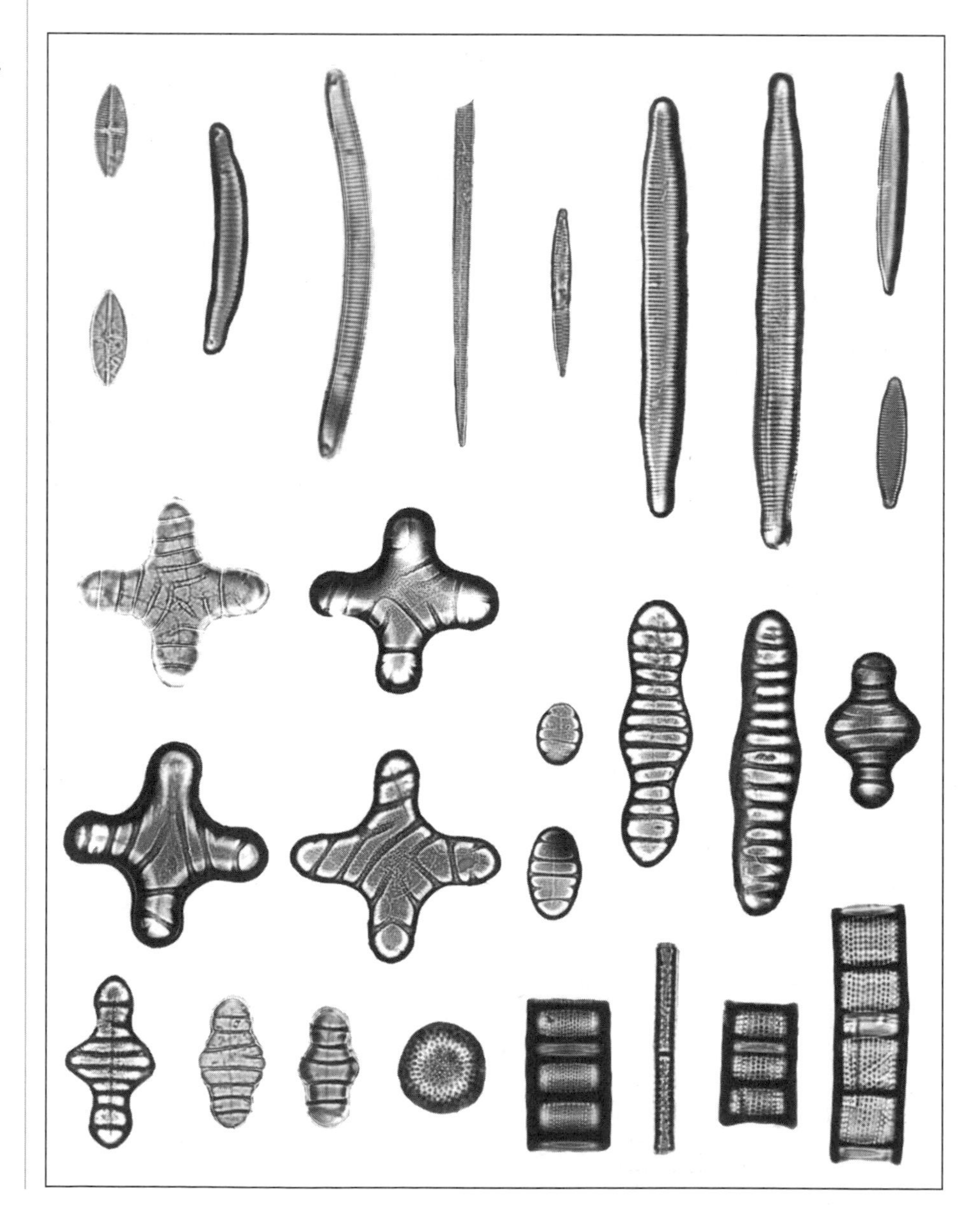

Figure 139 *Late Miocene diatoms from the Kilgore area, Cherry County, Nebraska.*

(Photo by G. W. Andrews, courtesy USGS)

minor amounts of these minerals are produced in this manner. After about two-thirds of the water has evaporated, gypsum precipitates. When nine-tenths of the water is removed, halite, or common salt, remains behind.

Much of the salt removed from the ocean is deposited in thick beds in nearly enclosed basins cut off from the general circulation of the sea. A large portion of the salt is also trapped in seawater between sediment grains on the ocean floor. When thrust deep inside the Earth at subduction zones, the water returns to the surface through volcanic eruptions. Also, seawater percolates through the oceanic crust and picks up minerals along the way. They are then pumped back to the surface by hydrothermal processes, which might help explain why the oceans remain salty.

VOLCANOES

Volcanic eruptions over subduction zones and lava flows at midocean ridges are the final stage of the rock cycle. The ocean floor is continuously being created at midocean ridges and destroyed in deep-sea trenches. When the seafloor is forced into the Earth's interior, carbon dioxide is driven out of carbonaceous sediments by the intense heat of the mantle. The molten magma along with its content of carbon dioxide rises to the surface to feed magma chambers beneath volcanoes and midocean ridges. The eruption of volcanoes and the flow of molten rock from midocean ridges resupplies the atmosphere with new carbon dioxide, making the Earth a great carbon-recycling plant.

Volcanoes are associated with crustal movements and occur on plate margins. They also play a direct role in regulating the Earth's climate. Large volcanic eruptions spew massive quantities of ash and aerosols into the atmosphere, which block out sunlight. Volcanic dust also absorbs solar radiation and indirectly heats the atmosphere, causing thermal disturbances and unusual weather.

Nearly 400 active volcanoes surrounding the Pacific Ocean are associated with subduction zones along the rim of the Pacific plate. Tectonic plates are subducted into the mantle by the collision of an oceanic plate with a continental plate or another oceanic plate. The lithosphere melts during its dive into the mantle. The lighter rock component, due to its greater buoyancy, works its way up into the crust to resupply magma chambers with new molten magma.

The composition of the magma also controls the type of eruption. Explosive eruptions occur when a viscous magma containing trapped volatiles and gases is kept from reaching the surface by a plug in the volcano's vent. As pressure increases, the obstruction is blown away along with most of the upper peak. Volcanoes associated with subduction zones, such as those in the western Pacific and in Indonesia (Fig. 140), are among the most explosive in the world.

In the Atlantic Ocean, volcanic activity is far less extensive and generally occurs at the Mid-Atlantic Ridge and in the West Indies. Many islands in the Atlantic are part of the spreading ridge system that extends above the sea (Fig. 141). Rift volcanoes account for about 15 percent of the world's known active volcanoes, and most are in Iceland and East Africa. Moreover, about 20 eruptions of deep submarine rift volcanoes are estimated to occur every year. Rift volcanoes on continents such as those in East Africa can be highly explosive.

The output of lava and pyroclastics for a single volcanic eruption varies from a few cubic yards to as much as five cubic miles. Rift volcanoes generate about 2.5 billion cubic yards per year of mainly submarine flows of basalt. Subduction zone volcanoes produce about 1 billion cubic yards of pyroclastic volcanic material per year. Volcanoes over hot spots produce about 500

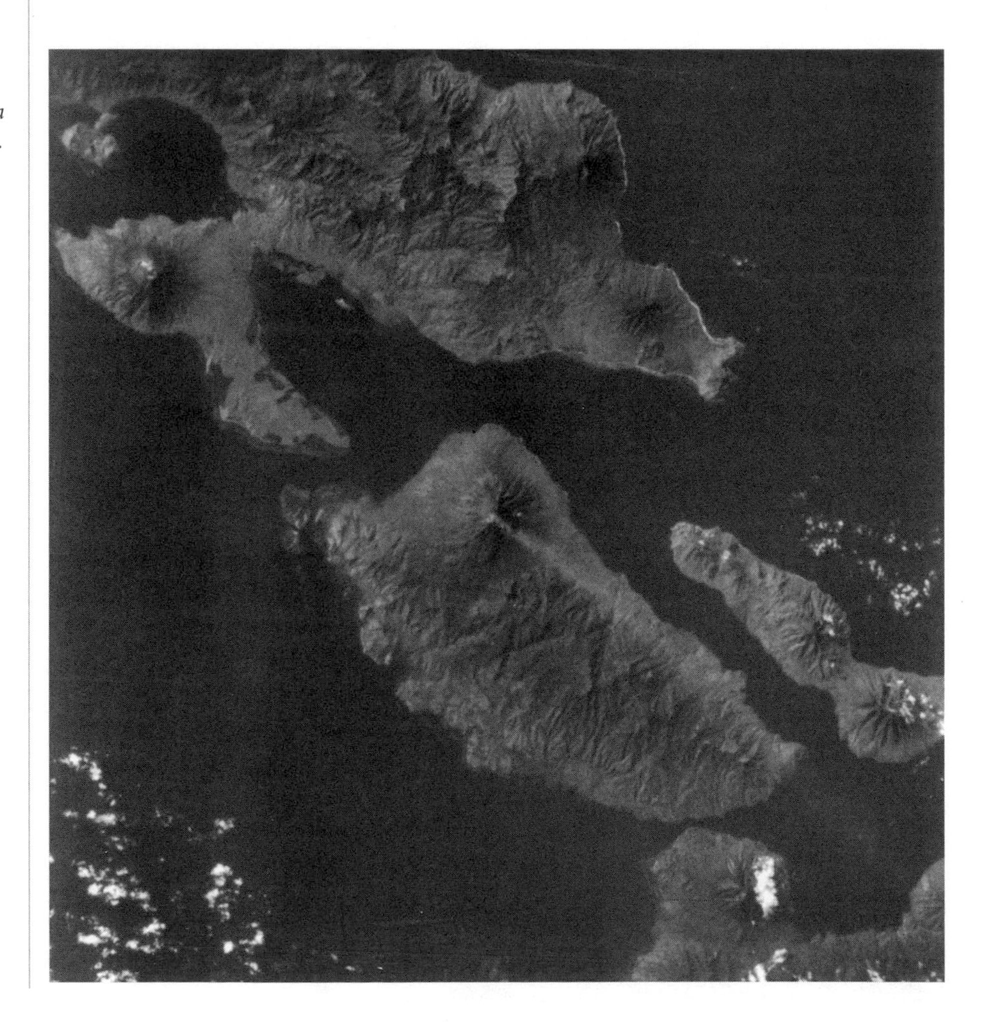

Figure 140 *An active volcano on Andonara Island, Indonesia, leaves a 30-mile-long train of ash.*
(Photo courtesy NASA)

Figure 141 *Birth of the new Icelandic island, Surtsey, in November 1963, located seven miles south of Iceland.*

(Photo courtesy U.S. Navy)

million cubic yards per year of mostly basalt flows in the oceans and of pyroclastics and lava flows on the continents.

Volcanic activity plays an all-important role in restoring the carbon dioxide content of the atmosphere. One of the most important volatiles in magma is carbon dioxide, which helps make it fluid. The carbon dioxide escapes from carbonaceous sediments when they melt after being forced into the mantle at subduction zones near the edges of crustal plates. The molten magma along with its content of carbon dioxide rises to the surface to feed volcanoes that lie on the edges of subduction zones and at midocean ridges. When the volcanoes erupt, carbon dioxide is released from the magma and returned to the atmosphere, and the cycle is complete.

After an examination of the Earth's life-sustaining processes, the next chapter will discuss how plate tectonics affects life itself.

9

TECTONICS AND LIFE

MAKING A LIVING PLANET

In 1977, while exploring the East Pacific Rise off the tip of Baha California, scientists in the research submarine *Alvin* discovered an oasis 1.5 miles below sea level. Species previously unknown to science were found living in total darkness among hydrothermal vents. Tube worms 10 feet tall swayed in the hydrothermal currents. Giant crabs scampered blindly across the volcanic terrain. Huge clams up to one foot long and clusters of mussels formed large communities around the vents (Fig. 142). The strange creatures discovered near these vents provide a dramatic example of how species develop in relation to their environments.

The East Pacific Rise is a 6,000-mile-long rift system along the eastern Pacific. It stretches from north of the Antarctic Circle to the Gulf of California and is the counterpart of the Mid-Atlantic Ridge. The undersea mountain range lies two miles beneath the surface of the sea. At the base of jagged basalt cliffs is evidence of active lava flows, including fields strewn with pillow lava. Exotic-looking chimneys called black smokers spew out hot water blackened with sulfide minerals. Others called white smokers eject hot water that is milky white.

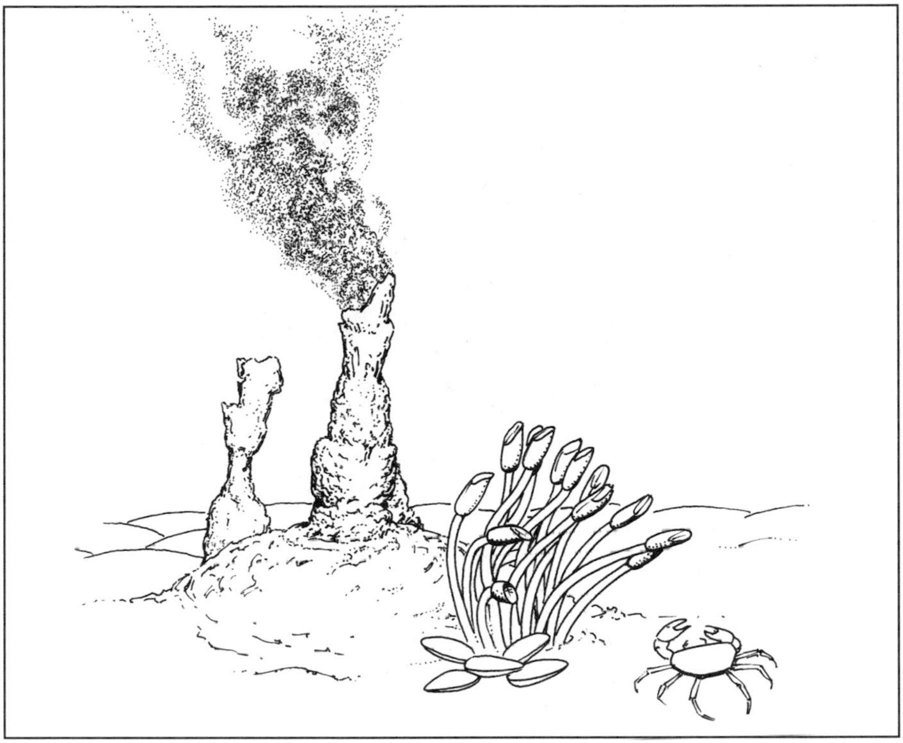

Figure 142 *Tall tube worms, giant clams, and large crabs cluster around hydrothermal vents on the deep ocean floor.*

The hot water originates from deep below the surface, where seawater percolating down through cracks in the ocean crust comes into contact with magma chambers below spreading centers. It then rises to the surface and is expelled through hydrothermal vents like undersea geysers (Fig. 143). In these volcanically active fields is a world that time forgot. The hydrothermal vents keep the bottom waters at comfortable temperatures, while the surrounding ocean hovers near freezing. The vents also provide valuable nutrients, making this the only environment on Earth completely independent of sunlight for its source of energy, which comes instead from the Earth's interior. This chapter examines how plate tectonics has played a major role in making our world a living planet.

THE BIOSPHERE

Like the Earth, many planets and their satellites possess a core, a mantle, a crust, and even an atmosphere or an icy hydrosphere. However, only the Earth contains a biosphere, which is the living component of the planet. The biosphere is more than just living beings. Life must also be integrated with the

Figure 143
Hydrothermal vents on the deep ocean floor provide nourishment and heat for bottom dwellers.
(Photo courtesy USGS)

geosphere, hydrosphere, and atmosphere to constitute a fully developed biosphere. Biologists have cataloged about 1.4 million species of plants and animals. Because many species have evaded detection, the total number could easily rise to 10 million or more, indicating the richness and diversity of life on this planet.

Because of plate tectonics, life was able to flourish. Possibly, active plate tectonics would not operate if Earth did not possess life as well. Lime-secreting organisms in the ocean remove carbon dioxide, an important greenhouse gas, from the atmosphere and store it in the bottom sediments. This keeps the Earth's surface temperature within the range needed for plate tectonics to operate, which in turn maintains living conditions on the planet.

Since life began, it has responded to a variety of chemical, climatological, and geographic changes in the Earth, forcing species either to adapt or to perish. Many dead-end streets along branches of the evolutionary tree are

TABLE 12 EVOLUTION OF THE BIOSPHERE

	Billions of Years Ago	Percent Oxygen	Biologic Effects	Event Results
Full oxygen conditions	0.4	100	Fish, land plants, and animals	Approach present biologic environs
Appearance of shell-covered animals	0.6	10	Cambrian fauna	Burrowing habitats
Metazoans appear	0.7	7	Ediacaran fauna	First metazoan fossils and tracks
Eukaryotic cells appear	1.4	>1	Larger cells with a nucleus	Redbeds, multi-cellular organisms
Blue-green algae	2.0	1	Algal filaments	Oxygen metabolism
Algal precursors	2.8	<1	Stromatolite mounds	Initial photo-synthesis
Origin of life	4.0	0	Light carbon	Evolution of the biosphere

found in the fossil record, itself only a fragmentary representation of all the species that have ever lived. Nearly every conceivable form and function has been tried, some more successful than others. Through this trial-and-error method of specialization, natural selection has chosen certain species to prosper while condemning others to extinction.

Few places on Earth are truly devoid of life. Species are found in the hottest deserts and coldest polar regions, including Antarctica's barren dry valleys (Fig. 144). They reside in the lowest canyons and tallest mountains. They also exist in the deepest oceans and the highest regions of the troposphere. Nor is life excluded from scalding hot springs or deep beneath the ground. Although species most frequently encountered on the Earth's surface seem to play the most dominant role in shaping the planet, the unseen microscopic creatures actually constitute the largest percentage of the total biomass and have the greatest influence.

About 80 percent of the Earth's breathable oxygen is generated by photosynthetic single-celled organisms that thrive in the ocean. Microorganisms such as bacteria play a critical role in breaking down the remains of plants and animals for recycling nutrients in the biosphere. Surface plants depend on bacteria in their root systems for nitrogen fixation. Bacteria live symbiotically in the gut of animals and aid in the digestion of food. Biologic processes are responsible for massive concentrations of minerals in the Earth's crust, including silicon, carbon, iron, manganese, copper, and sulfur. Simple organisms also comprise the bottom of the food chain, on which all life ultimately depends for its survival.

Figure 144 *The Wright Dry Valley, Taylor Glacier region, Victoria Land, Antarctica.*

(Photo by W. B. Hamilton, courtesy USGS)

Life on Earth constitutes a geologic force that is missing on all other bodies in the solar system. The evidence of biospheric processes in the Earth's history belongs to the field of biogeology. What might be the earliest fossilized remains of microorganisms dates back from nearly 4 billion years ago. The 3.8-billion-year-old carbonaceous sediments of the Isua formation in southwest Greenland show a depletion of carbon 13 with respect to carbon 12 determined by the environment. This indicates that biologic activity might have been in existence at a very early age. Therefore, life processes might have been operating for at least four-fifths of Earth history.

The oldest evidence of life is found in microfossils, which are the remains of ancient microorganisms along with stromatolites (Fig. 145), the layered structures formed by the accretion of fine sediment grains by colonies of primitive blue-green algae. The earliest stromatolites were found in sedimentary rocks of the Warrawoona group in Western Australia that are 3.5 billion years old. Associated with these rocks are cherts containing microfilaments, which are small, threadlike structures of possible bacterial origin. Similar cherts with microfossils of primitive bacteria were found in 3.3-billion-year-old rocks from eastern Transvaal in South Africa.

The abundance of chert in deposits older than 2.5 billion years indicates that most of the crust was deeply submerged during this time because the mineral precipitates in silica-rich seawater. The seas contained much more dissolved silica, which leached out of the volcanic rock pouring onto the ocean floor. Modern ocean water is deficient in silica because organisms such as sponges and diatoms (Fig. 146) extract it to build their skeletons. Massive deposits of diatomaceous earth in many parts of the world are a tribute to the great success of these organisms.

With this much time involved, life was able to bring about some dramatic and far-reaching changes. The first major alteration was the deposition of banded iron formations onto continental margins by iron-metabolizing bacteria. These formations are mined extensively for iron ore around the world. A second major change occurred with the transition of the gas content of the atmosphere and ocean from about one-quarter carbon dioxide to one-quarter oxygen by photosynthetic organisms. The oxygen in the upper atmosphere promoted the

Figure 145
Stromatolites in Helena Dolomite along the Dearborn River, Lewis and Clark County, Montana.
(Photo by M. R. Mudge, courtesy USGS)

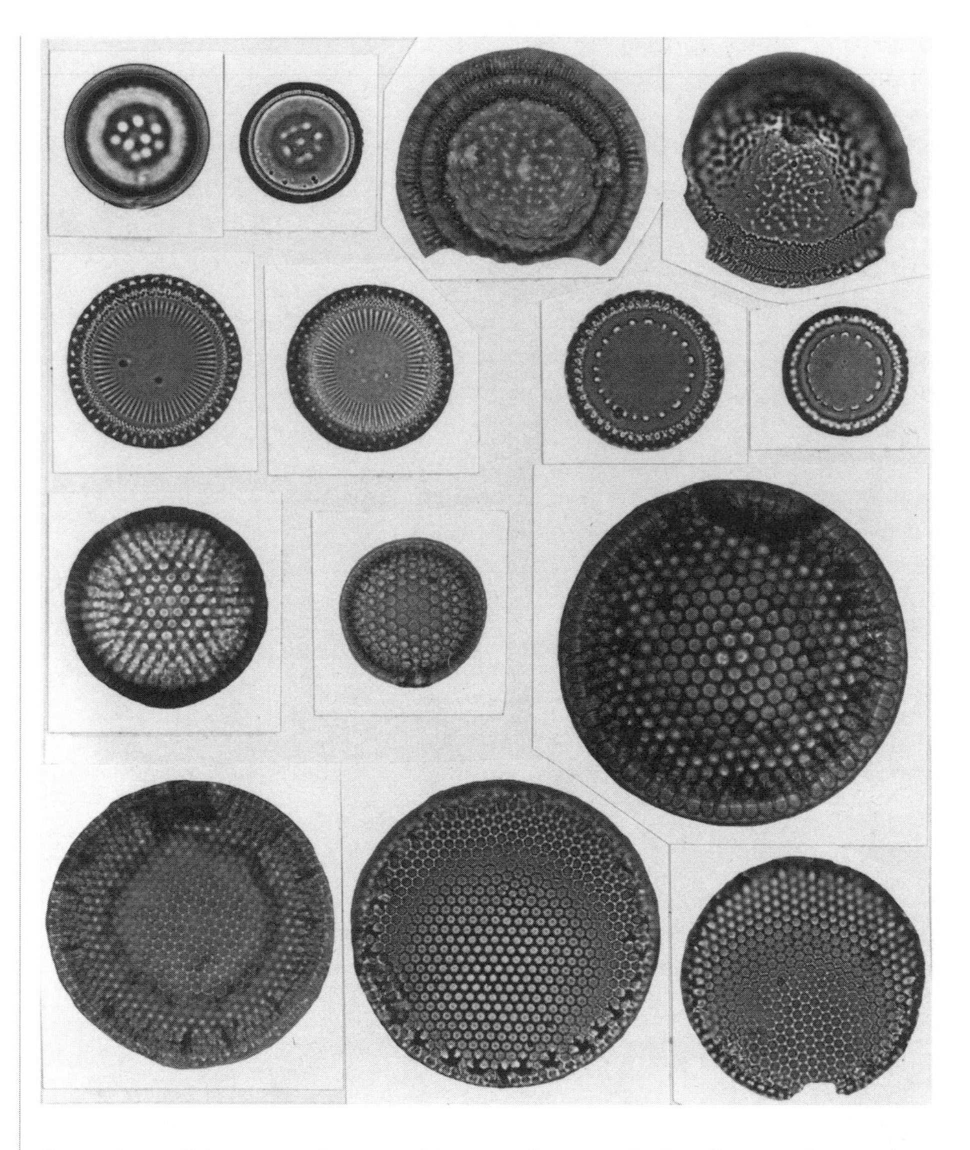

Figure 146 *Diatoms from the Choptank Formation, Calvert County, Maryland.*

(Photo by G. W. Andrews, courtesy USGS)

formation of the ozone layer, making conditions safe for plants and animals to conquer the land. This in itself constituted a major change on Earth.

THE GEOSPHERE

The geosphere interacts with the biosphere and directly influences it by changing the environment. The Earth has experienced 11 episodes of flood basalt volcanism over the past 250 million years (Fig. 147). These large eruptions created a series of separate overlapping lava flows that give many exposures worldwide a terracelike appearance, called traps, the Dutch word for "staircase." Many flood basalts are located near continental margins, where great rifts separated the present continents from Pangaea. Others such as the Columbia River basalts of the northwestern United States, which were responsible for a mass extinction 16 million years ago, are related to hot spot activity, with plumes of hot rocks from deep within the mantle rising to the surface.

The volcanic episodes were relatively short-lived events. Major phases lasted less than 3 million years. Furthermore, the episodes of volcanism appear to be somewhat periodic, occurring about every 32 million years. The timing of these major outbreaks correlates with the occurrence of worldwide mass extinctions of marine organisms. The largest mass extinction in the geologic record occurred at the end of the Paleozoic 250 million years ago and was responsible for the loss of more than 95 percent of all species. It also occurred simultaneously with the eruption of very large traps in Siberia, perhaps the largest outpouring of volcanic rock in Earth history.

During the eruption of a major basaltic lava flow, vigorous fire fountains inject large amounts of sulfur gases into the atmosphere (Fig. 148). Moreover, flood basalts release 10 times more sulfur than explosive eruptions. The gases are converted into acid, which has severe climatic and biologic consequences. A major alteration in the composition of the atmosphere also can affect the climate. Volcanoes spew massive quantities of ash and aerosols into the atmosphere, which block out sunlight. Volcanic dust also absorbs solar radiation, which heats the atmosphere, causing thermal imbalances and unstable climatic conditions. Heavy clouds of volcanic dust have a high albedo and reflect much of the solar radiation back into space

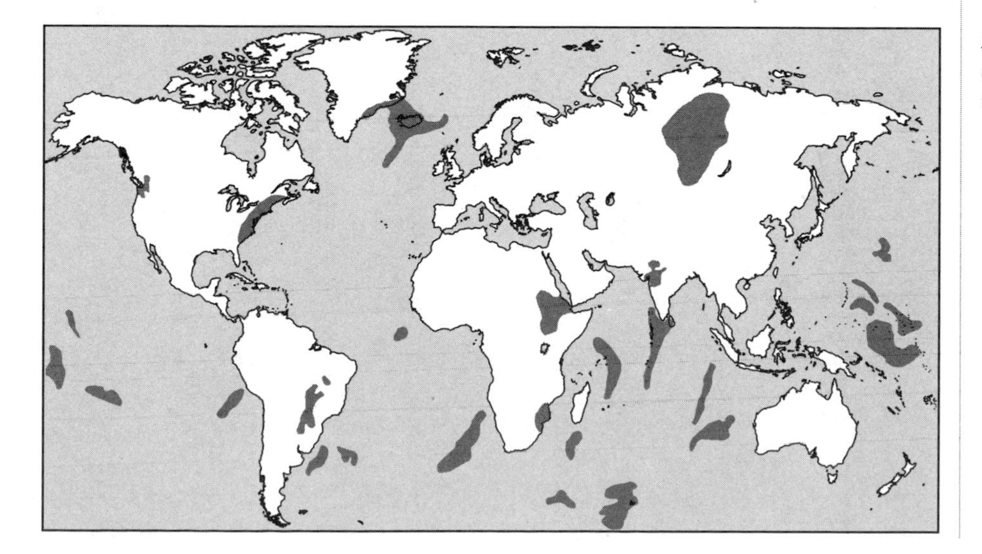

Figure 147 *Areas affected by flood basalt volcanism.*

(Fig. 149). This could shade the planet and lower global temperatures. The reduced sunlight might also cause mass extinctions of plants and animals by lowering the rate of photosynthesis.

A reduction of insolation, or the amount of solar radiation reaching the surface, by 5 percent might result in a drop in global temperatures by as much as 5 degrees Celsius. If maintained long enough, this temperature drop would be sufficient to initiate an ice age. The long-term cooling would allow glaciers to expand and lower the sea level, which would limit marine habitat area. The lowered temperature could also adversely affect the geographic distribution of species, confining warmth-loving organisms to the tropics.

Extensive volcanic activity 100 times more intense than at present could produce strong acid rain showers. These, in turn, could cause widespread destruction of terrestrial and marine species by defoliating plants and altering the acid/alkali balance of the ocean. Acid gases spewed into the atmosphere might also deplete the ozone layer, allowing deadly solar ultraviolet radiation to bathe the planet, thereby eliminating life on the surface.

Massive volcanic eruptions might have had a hand in the extinction of the dinosaurs. At the end of the Cretaceous, 65 million years ago, a giant rift opened on the west side of India, and huge volumes of molten lava poured onto the surface. Nearly 500,000 square miles of lava covering an area about

Figure 148 *Lava fountain and lake at Kilauea Volcano during 1959 and 1960 eruptions.*

(Photo by D. H. Richter, courtesy USGS)

the size of France erupted over a period of 500,000 years. The eruptions consisted of about 100 lava flows. They blanketed much of west-central India, known as the Deccan Traps (Fig. 150), in layers of basalt hundreds of feet thick. It was the largest volcanic catastrophe in the last 200 million years.

Figure 149 *The volcanic ash cloud from the 1980 eruption of Mount St. Helens.*

(Photo courtesy NOAA)

When the eruptions began, India was drifting toward southern Asia. The rift separated India from the Seychelles Bank, which was left behind as the subcontinent continued its journey northward. In the Amirante Basin on the southern edge of the Seychelles Bank, about 300 miles northeast of Madagascar, lies a remarkably intact circular depression about 200 miles wide that appears to be a large impact structure. A massive meteorite impact in the Amirante Basin might have triggered the great lava flows that created the

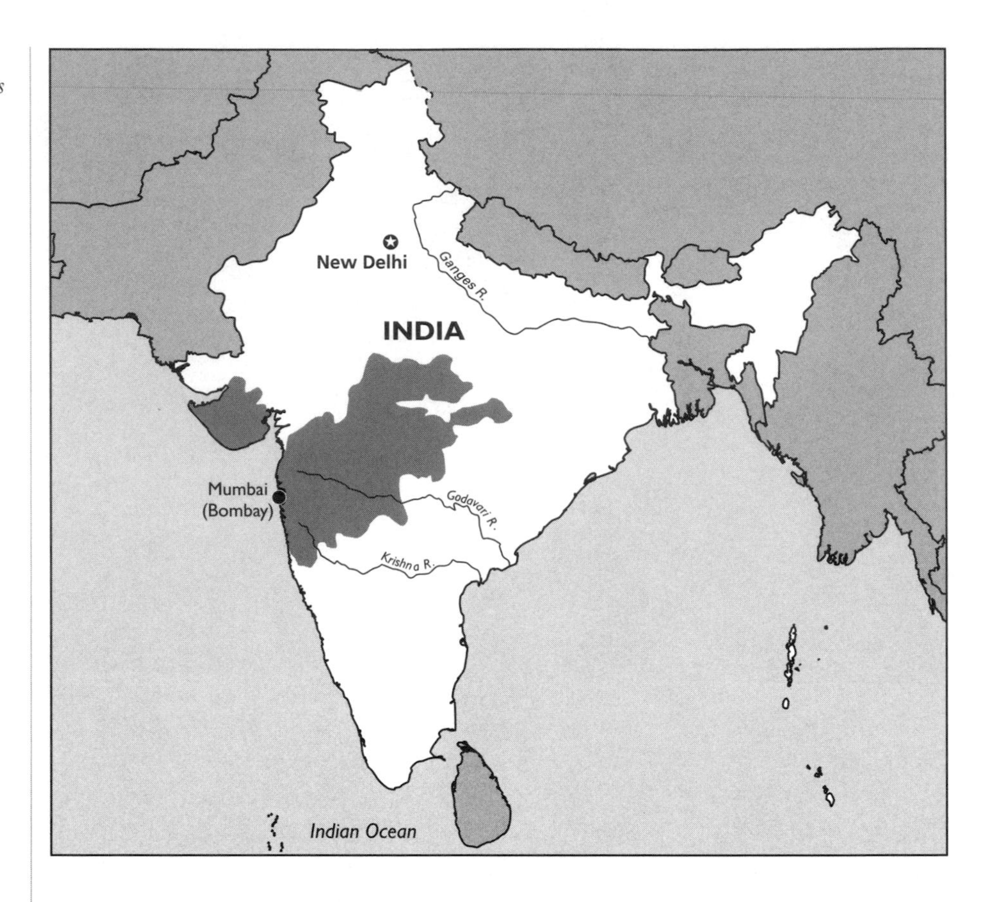

Figure 150 The Deccan Traps flood basalts in India.

TABLE 13 FLOOD BASALT VOLCANISM AND MASS EXTINCTIONS

Volcanic Episode	Million Years Ago	Extinction Event	Million Years Ago
Columbia River, USA	17	Low-mid Miocene	14
Ethiopian	35	Upper Eocene	36
Deccan, India	65	Maastrichtian	65
		Cenomanian	91
Rajmahal, India	110	Aptian	110
South-West African	135	Tithonian	137
Antarctica	170	Bajocian	173
South African	190	Pliensbachian	191
E. North American	200	Rhaectian/Norian	211
Siberian	250	Guadalupian	249

Deccan Traps and the Seychelles Islands. Quartz grains shocked by the high pressures generated by the impact were found lying just beneath the immense lava flows, suggesting that they might be linked to the impact. Perhaps such large meteorite impacts create so much disturbance in the Earth's thin outer crust they can induce massive volcanic eruptions.

Shocked quartz and iridium, an isotope of platinum that is extremely rare in the crust but relatively abundant on asteroids and comets, is found in the Cretaceous-Tertiary (K-T) boundary clay throughout the world (Fig. 151). The K-T sediments are thought to have originated from the fallout of a giant meteorite impact. They might also have resulted from massive volcanic eruptions taking place over a period of hundreds of thousands of years. Explosive volcanic eruptions can produce similar shocked quartz grains. Volcanoes whose magma source lies deep within the mantle, such as Kilauea on Hawaii, can produce sig-

Figure 151 *Geologists point out the Cretaceous-Tertiary boundary at Browie Butte outcrop, Garfield County, Montana.*

(Photo by B. F. Bohor, courtesy USGS)

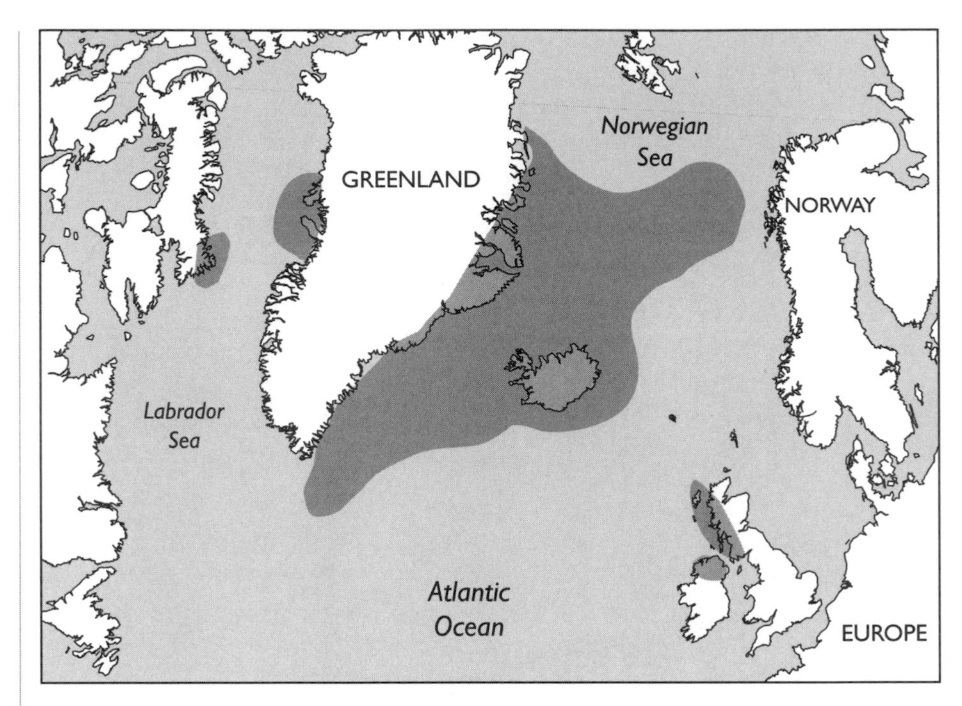

Figure 152 *Extensive volcanic activity during the opening of the North Atlantic 57 million years ago led to the extinction of species.*

nificant quantities of iridium. The Deccan eruptions could have emitted large amounts of iridium, which would account for the anomalously high concentrations in the K-T boundary layer. Moreover, the microspherules found at the K-T contact believed to be the product of impact melt could also have been produced by large volcanic eruptions.

An enormous plateau of basalts along the coasts of Greenland and Scotland might have erupted at the same time the dinosaurs and 70 percent of all other species abruptly disappeared. Evidence of substantial explosive volcanism was found in an extensive region from the South Atlantic to Antarctica. The eruptions might have dealt a major blow to the climatic and ecological stability of the planet, forcing species to become extinct. Another significant change in the global environment followed 10 million years later during volcanic eruptions along the continental margins of the North Atlantic (Fig. 152) when many deep-sea foraminifers and land mammals became extinct. Ocean temperatures were also warmer than at any other time during the past 70 million years.

TECTONICS AND EVOLUTION

Plate tectonics and continental drift have played a prominent role in the history of life practically since the very beginning. Changes in the relative configuration of the continents and the oceans had a far-ranging influence on the environment, climate conditions, and the composition of species. The changes in continental shapes significantly affected global temperatures, ocean currents, productivity, and many other factors of fundamental importance to life. Plate tectonics is therefore among the strongest forces influencing evolutionary changes.

The motions of the continents had a major impact on the distribution, isolation, and evolution of species. Many different environments result in a wide variety of species. The continual variations in world ecology had a substantial influence on the course of evolution and accordingly on the diversity of living organisms. Therefore, evolutionary trends varied throughout geologic time. They occurred in response to major environmental changes as natural selection adapted organisms to the new conditions forced on them by environmental factors affected in large part by continental drift.

Crustal plates in motion are continuously rearranging continents and ocean basins. When continents rift apart, they override ocean basins. This makes seas less confined, which raises global sea levels. The rising seas inundate low-lying areas inland of the continents, dramatically increasing the shoreline and shallow-water marine habitat area. The expansion of the inhabitable area can thus support a larger number of species.

When all the continents were welded into the supercontinent Pangaea near the end of the Paleozoic around 250 million years ago, a great diversity of plant and animal life evolved in the ocean as well as on land (Fig. 153). The formation of Pangaea marked a major turning point in the evolution of life, during which the reptiles emerged as the dominant species. Most of the Pangaean climate was equable and fairly warm throughout the year. Much of the interior of Pangaea, however, was a desert whose temperatures fluctuated wildly between seasons, with scorching summers and freezing winters. The temperature extremes might have contributed to the widespread extinction of land-based species during the late Paleozoic. These changes also explain why the reptiles, which adapt readily to hot, dry climates, replaced the amphibians as the dominant land species.

The vast majority of marine species live on continental shelves or shallow-water portions of islands and subsurface rises at shallow depths generally less than 600 feet. The richest shallow-water faunas live in the tropics, which contain large numbers of highly specialized species. When progressing to higher latitudes, diversity gradually falls off until in the polar regions, less than 10 percent as many species exist as in the tropics. Moreover, twice the species diversity occurs in the Arctic Ocean, which is surrounded by continents, than in the Antarctic Sea, which surrounds a continent.

Biologic diversity mostly depends on the stability of the food supply. This is largely affected by the shape of the continents, the width of shallow

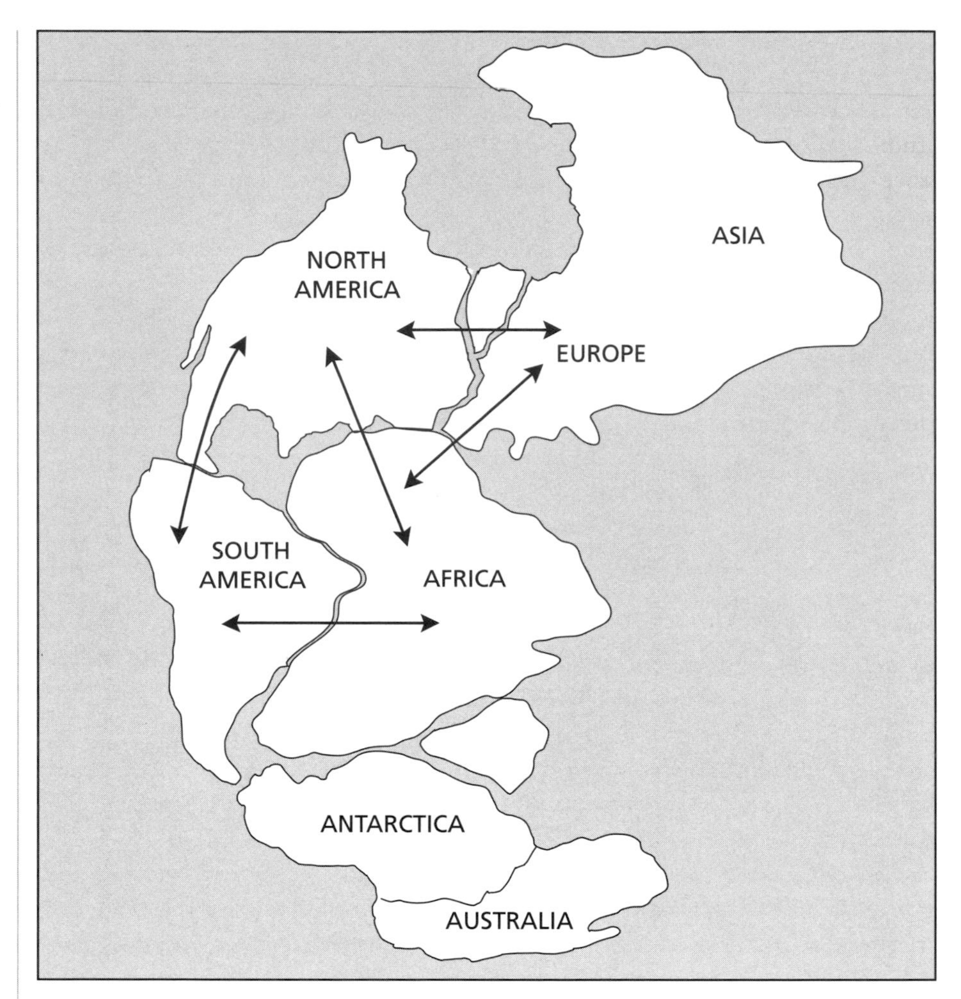

Figure 153 *The fitting together of the continents showing possible migration routes of plants and animals.*

continental margins, the extent of inland seas, and the presence of coastal mountains. Erosion from mountains pump nutrients into the sea, fueling booms of marine plankton and increasing the food supply for animals higher up the food chain. Organisms with more food are more likely to thrive and diversify into different species. Mountains that arise from the seafloor to form islands increase the likelihood of isolation of individual animals, which increases the chances of forming a new species.

The greatest diversity is off the shores of small islands or small continents in large oceans. There, fluctuations in nutrient supplies are least affected by the seasonal influences of landmasses. Diversity is lowest off large continents, particularly when they face small oceans, where shallow-water seasonal variations are the greatest. Diversity also increases with distance from large continents.

Species living in different oceans or on opposite sides of the same ocean evolve separately from their overseas relatives. Even along a continuous coastline, major changes in species generally correspond to the climate because latitudinal and climatic changes create barriers to shallow-water organisms. The deep-sea floor, generated at oceanic spreading ridges, provides another formidable barrier to the dispersal of shallow-water organisms. These barriers partition marine faunas into more than 30 geographic provinces, with only a few common species living in each province. The shallow-water marine faunas represent more than 10 times as many species than are found in a world with only a single province. For example, around 200 million years ago, when the single large continent Pangaea was surrounded by a global ocean, species diversity was at an all-time low.

Diversity is greatly influenced by seasonal changes. These include variations in surface and upwelling ocean currents that affect the nutrient supply, which in turn cause large fluctuations in productivity. Therefore, the greatest diversity among species is off the shores of small islands or small continents facing large oceans. There fluctuations in the nutrient supply are least affected by seasonal effects of landmasses, which often sponsor episodes of glaciation or other environmental swings.

Species diversity is also affected by continental motions. When the continents were assembled into Pangaea, a continuous shallow-water margin ran around the entire perimeter of the supercontinent, with no major physical barriers to the dispersal of marine life. Furthermore, the seas were largely confined to the ocean basins, leaving the continental shelves mostly exposed. Consequently, habitat area for shallow-water marine organisms was very limited, accounting for the low species diversity. As a result, marine biotas were more widespread but contained comparatively fewer species.

Similar conditions might have occurred during the late Precambrian around 680 million years ago, when another supercontinent named Rodinia was in existence. During the Cambrian period, it rifted into perhaps four or five large continents. Their extended shoreline might have caused the explosion of new species during that time. Twice as many phyla (organisms that share the same general body plan) were living during the Cambrian than before or since. Never were so many experimental organisms in existence (Fig. 154), none of which have any counterparts today. When Pangaea broke up and the resulting continents migrated to their current positions, diversity again increased to unprecedented heights, providing our present-day world with a rich variety of species.

The movement of continents changes the shapes of ocean basins. This affects the flow of ocean currents, the width of continental margins, and the abundance of marine habitats. When a supercontinent breaks up, more continental margins are created, the land lowers, and the ocean rises, providing a larger habitat area for marine organisms. During times of highly active continental movements, volcanic activity increases. This is especially true at midocean spreading centers, where crustal plates are pulled apart by upwelling magma from the upper mantle. It is also true at subduction zones, where crustal plates are forced into the Earth's interior and remelted to provide the raw materials for new crust. The amount of volcanism could affect the composition of the atmosphere, the rate of mountain building, and the climate, all of which invariably have an effect on life.

The drifting of the continents during the Cenozoic isolated many groups of mammals, and they evolved along independent lines. Around 40 million years ago, Australia drifted away from Antarctica, which acted as a bridge between it and South America. The separation isolated the continent from the rest of the world. It is now inhabited by strange egg-laying mammals called monotremes. These include the spiny anteater and platypus, which should rightfully be classified as surviving mammal-like reptiles. When the platypus was first discovered, it was thought to be a missing link between mammals and their ancient ancestors.

Marsupials are believed to have originated in North America around 100 million years ago. They migrated to South America, crossed over Antarctica when the two continents were still attached to each other, and landed in Australia prior to it breaking away from Antarctica. The Australian group consists of kangaroos, wombats, and bandicoots, while opossums and related species occupy other parts of the world.

Figure 154 *Early Cambrian marine fauna.*

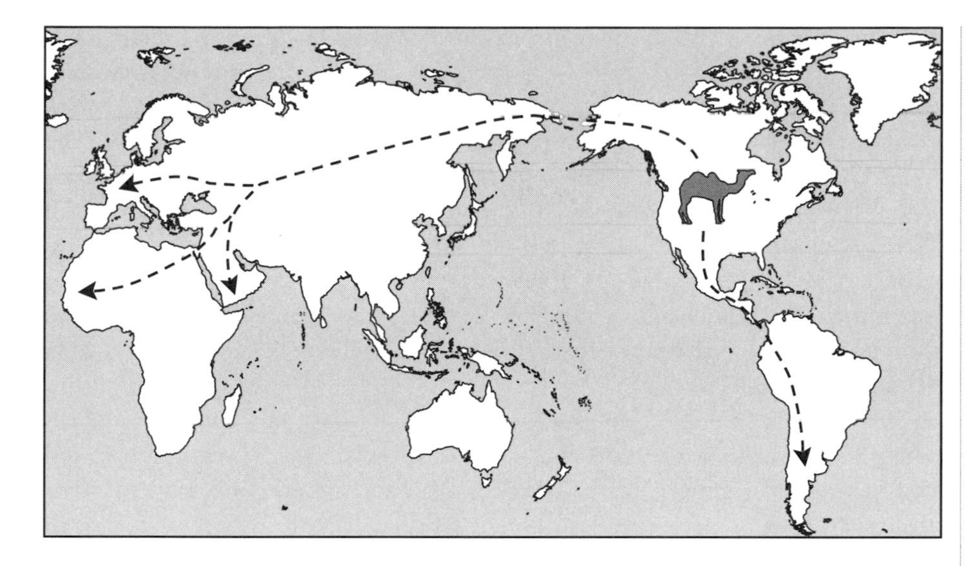

Figure 155 The migration of camels from North America to other parts of the world.

Camels originated in North America about 25 million years ago and migrated to other parts of the world by crossing over connecting land bridges (Fig. 155). Horses, which evolved at about the same time, took a similar route out of North America. Madagascar, which broke away from Africa about 125 million years ago, has none of the large mammals that occupy the mainland except the hippopotamus, which somehow managed to cross over to the island after it had already drifted some distance away from the African continent.

TECTONICS AND EXTINCTION

When considering all the great upheavals in the Earth throughout geologic history, how life managed to survive to the present remains a puzzle. Episodes of mass extinctions of species correlate with cycles of terrestrial phenomena. The most pervasive of these is a 300-million-year cycle of convection currents in the Earth's mantle. Convection within the mantle is the very driving force behind plate tectonics and all geologic activity taking place on the Earth's surface.

During periods of rapid mantle convection, supercontinents tend to break up. This leads to the compression of ocean basins, which causes a rise in sea level and a transgression of the seas onto the land. It also increases volcanism, which in turn increases the carbon dioxide content of the atmosphere, resulting in a strong greenhouse effect and rising global temperatures. During times of low mantle convection, the entire landmass is assembled into a supercontinent. This causes a widening of ocean basins with a consequent drop in global sea levels and a regression of the seas from the land. The fluctuation in sea levels is directly related to vertical tectonic movements on the continents. As the continents rise, the ocean lowers. Moreover, a reduction of atmospheric carbon dioxide occurs due to low levels of volcanism, resulting in lower global temperatures.

Throughout most of the Earth's history, continents and ocean basins have been continuously reshaped and rearranged by crustal plates in motion. When continents break up, they override ocean basins. This makes the seas less confined and raises global sea levels several hundred feet. Low-lying areas inland of the continents are inundated by the sea, dramatically increasing the shoreline and the shallow-water marine habitat area, which can support a larger number of species.

Mountain building associated with the movement of crustal plates alters patterns of river drainages and climate. In turn, this affects terrestrial habitats. Raising land to higher elevations, where the air is thinner and colder, allows glaciers to form, especially in the higher latitudes. Continents scattered in all parts of the world can also interfere with ocean currents, which affect global heat distribution.

When continents assemble into a supercontinent, land no longer impedes the flow of ocean currents. They can therefore distribute heat more evenly around the planet and maintain more uniform global temperatures. The ocean basins also widen, causing sea levels to drop. A substantial and rapid fall in sea level could directly and indirectly influence the biologic world. It would increase the seasonal extremes of temperature on the continents, thereby increasing environmental stress on terrestrial species.

A drop in sea level also forces inland seas to retreat. This results in continuous, narrow continental margins around the continents. These, in turn, reduce the shoreline, radically limiting the marine habitat area. Moreover, unstable near-shore conditions result in an unreliable food supply. Extinctions tend to increase toward the shore because intertidal environments fluctuate much more than those farther out to sea. Many species cannot cope with the limited living space and food supply, and they die out in tragically large numbers. This occurred at the end of the Paleozoic, when the vast majority of marine species became extinct.

During the final stages of the Cretaceous, when the seas were departing from the land and the level of the ocean began to drop, the temperatures in a broad tropical ocean belt known as the Tethys Sea began to fall. This might explain why the Tethyan species that were the most temperature sensitive suffered the heaviest extinction at the end of the period. Species that were amazingly successful in the warm waters of the Tethys were totally decimated when temperatures dropped. Afterward, marine species took on a more modern appearance as ocean bottom temperatures continued to plummet.

TECTONICS AND CLIMATE

The positions of the continents with respect to each other and to the equator determined climatic conditions. When most of the land huddled near the equatorial regions (Fig. 156), the climate was warm. However, when lands wandered into the polar regions, the climate turned cold, spawning episodes of glaciation. Over the last 100 million years, changes in climate have been largely controlled by plate tectonics. Furthermore, during times of highly active continental movements, a greater amount of volcanic activity occurs, especially at midocean ridges and subduction zones. Volcanism could affect the composition of the atmosphere and the rate of mountain building, which ultimately affect the climate.

When all the lands were welded into the supercontinent Pangaea around 250 million years ago, much of the land was located near the tropics, where more of the Sun's heat could be absorbed. This contributed to higher global temperatures. Also, oceans in the high latitudes were less reflective than land. Therefore, they absorbed more heat, which further moderated the climate. Moreover, without land in the polar regions to interfere with the movement of warm ocean currents, both poles remained ice free year-round. This resulted in little variation in temperature between the high latitudes and the tropics.

When Pangaea began to break up, the climate of the Cretaceous was extremely warm, with average global temperatures 5 to 10 degrees Celsius warmer than they are today. When the continents drifted toward the poles by the end of the period, however, they disrupted the transport of poleward oceanic heat and replaced heat-retaining water with heat-losing land. As the cooling progressed, the land accumulated snow and ice, creating a greater reflective surface. This further lowered global temperatures and sea levels. The

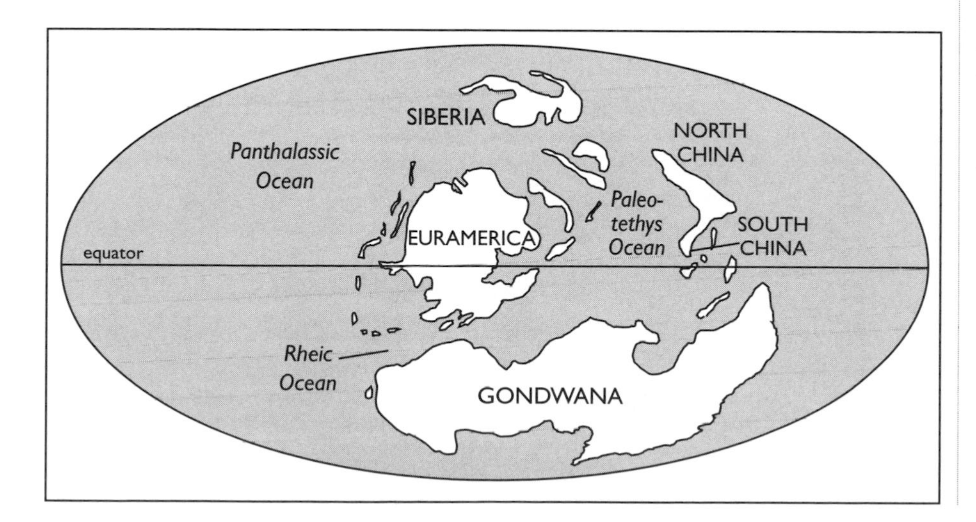

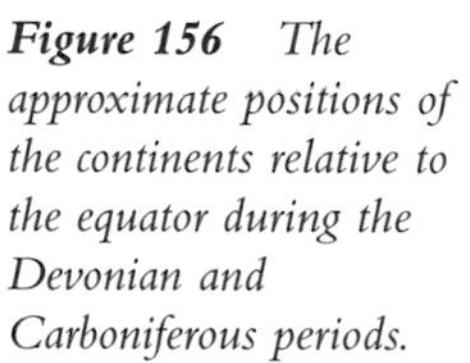
Figure 156 *The approximate positions of the continents relative to the equator during the Devonian and Carboniferous periods.*

most important factor controlling the geographic distribution of marine species is ocean temperature. Therefore, climatic cooling is primarily responsible for most crises among seafaring creatures.

The location of land in the polar regions is often the cause of extended periods of glaciation because high-latitude land has a higher albedo and a lower heat capacity than the surrounding seas. This condition encourages the accumulation of snow and ice. The more land area in the higher latitudes the colder and more persistent is the ice, especially when much of the land is at higher elevations, where glaciers can grow easily. Furthermore, replacing land in the tropics with oceans has a net cooling effect because land located in the lower latitudes absorbs more of the sun's heat, while oceans reflect solar rays back into space. Also, by increasing the land area in the high latitudes, where snow falls steadily with little melting, a permanent polar glacial climate is established.

After the glaciers are securely in place, the high reflectivity of snow and ice tends to perpetuate them and sustain glaciation. This is true even if the once high land were to sink to the level of the sea due to a decrease in crustal buoyancy by the weight of the ice. Increasing the weight on the land can also interfere with the flow of mantle currents, possibly causing an increase in volcanism. The greater volcanic activity increases the amount of ash injected into the atmosphere, which shades the planet and lowers global temperatures, maintaining glacial conditions.

The ocean bottom influences how much heat is carried by ocean currents from the tropics to the poles. When Antarctica separated from South America and Australia and moved over the South Pole some 40 million years ago, a circumpolar Antarctic ocean current was established. This current isolated the frozen continent, preventing it from receiving warm poleward-flowing waters from the tropics, causing it to become a frozen wasteland. During this time, warm salty water filled the ocean depths, while cooler water covered the upper layers.

In the Tethys, a large shallow sea that separated Eurasia from Africa, warm water became top-heavy with salt due to high evaporation rates and little rainfall, causing it to sink to the bottom. Meanwhile, ancient Antarctica, which was much warmer than at present, generated cool water that filled the upper layers of the ocean, causing the entire ocean circulation system to run backward. Then around 28 million years ago, Africa slammed into Eurasia, turning off the tap of warm water flowing to the poles and placing Antarctica into a deep freeze. The cold air and ice cooled the surface waters and made them heavy enough to sink and flow toward the equator, providing the Earth with the ocean circulation system it has today (Fig. 157).

Changes in the deep-ocean circulation also coincided with the late Eocene extinctions, in which all the archaic mammals abruptly disappeared 37 million years ago. The extinctions also eliminated many European species of marine life when the continent was flooded by shallow seas. The separation of Greenland from Europe during this time might have caused frigid Arctic

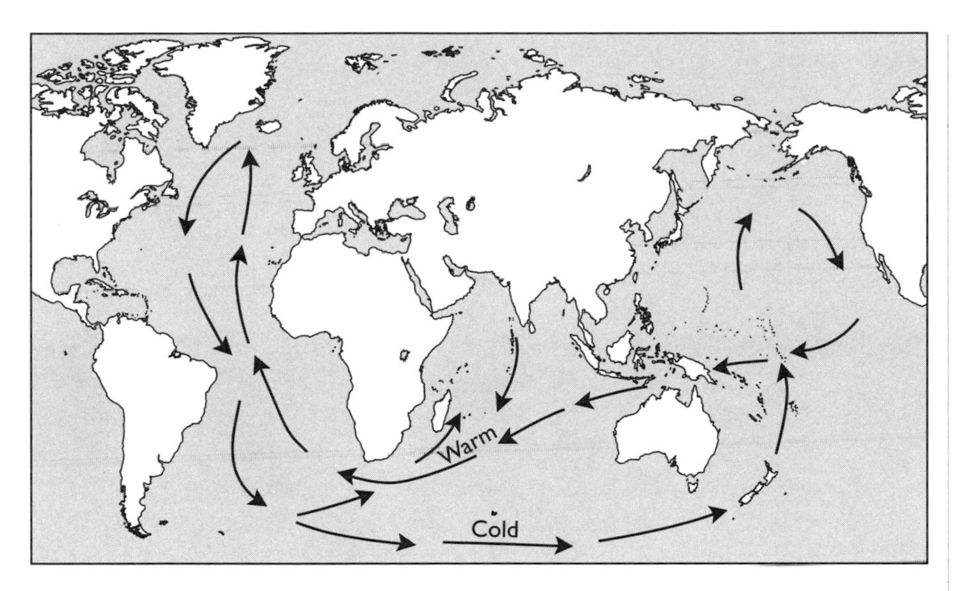

Figure 157 *The ocean conveyor belt, transporting warm and cold water over the Earth.*

waters to drain into the Atlantic, significantly lowering its temperature and causing the disappearance of most types of foraminiferans. In the ensuing Oligocene epoch, the seas were drained from the land as the ocean withdrew to perhaps its lowest level since the last several hundred million years. The seas remained depressed for the next 5 million years.

These cooling events removed the most vulnerable of species so that those living today are more robust and capable of withstanding the extreme environmental swings that occurred during the last 2 million years, when glaciers spanned much of the Northern Hemisphere (Fig. 158). The ice ages

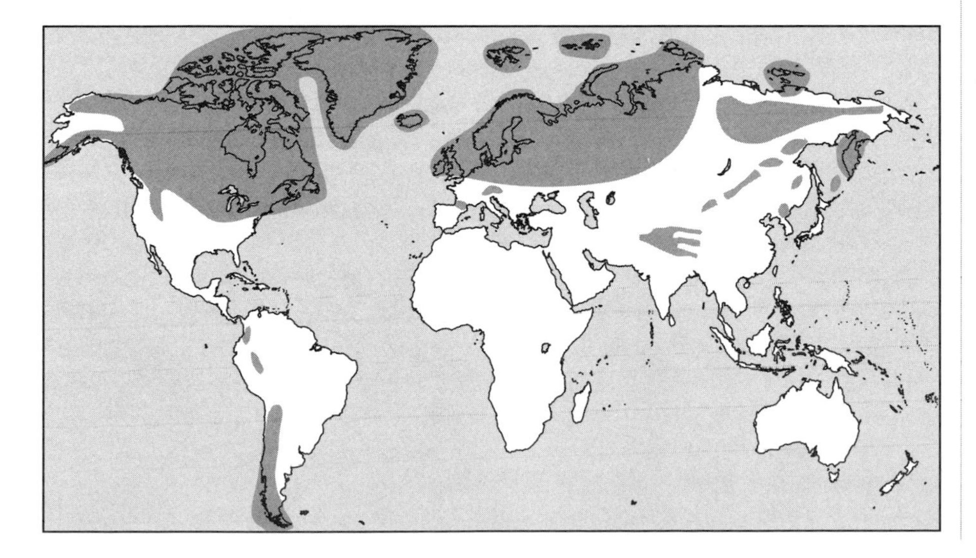

Figure 158 *Extent of glaciation during the last ice age.*

came and went almost like clockwork about every 100,000 years or so. They will most likely continue for perhaps several more million years until plate tectonics again gathers all the continents together at the equator. This will alter the face of the planet and the composition of its life, just as it has done for most of the Earth's history.

TECTONICS OF THE FUTURE

The Earth and its life forms are constantly evolving. In the past, our planet was much different from today. It will continue to change. In the future, the world will hardly be recognizable. In just a few million years from now, major alterations in the Earth's surface will take place. For example, California westward of the San Andreas Fault will continue to drift northward, finally ending just below Alaska as the plate upon which it rides plunges down the Aleutian Trench. At the same time, Baja California, which rifted apart from mainland Mexico 6 million years ago, will scoot along the west coast of the United States, finally coming to rest just below present-day Canada. Other scraps of land will continue to crash into the North American continent, continually enlarging its landmass.

The East African Rift Valley will eventually widen and flood with seawater. It will form a new subcontinent similar to Madagascar, which broke away from Africa about 125 million years ago and became an isolated landmass with independently evolving species. The crustal fragment will then drift toward India, possibly colliding with the subcontinent. India itself will continue to press against South Asia, distorting the Asian continent and causing tremendous earthquakes to rumble across the countryside. The crumpling of the crust will continue to raise the Himalayas and Tibetan Plateau, while huge rivers draining the region will add new land to the Gulf of Bengal. Africa and Arabia will continue diverging from each other, ever widening the Red Sea and the Gulf of Aden, finally opening a passage into the Mediterranean Sea.

The Atlantic Ocean will continue to expand at the expense of the Pacific as new ocean floor is added at the Mid-Atlantic Ridge. The Pacific plate will continue to shrink as the ocean floor is swallowed up by subduction zones in the western Pacific. This jostling of crustal plates will shift the positions of the continents (Fig. 159). The Hawaiian Islands in the central Pacific will gradually disappear beneath the ocean waves. A new island name Loihi will rise some 11,000 feet from the ocean floor and emerge in a huge fiery outburst. Volcanoes throughout the rest of the world will erupt one after another as the Earth vents its excess heat to the surface during a time of rapid continental movements, producing an extraordinarily warm climate.

The Americas will separate at Panama, with North America drifting westward and South America drifting into the South Pacific. The opening of

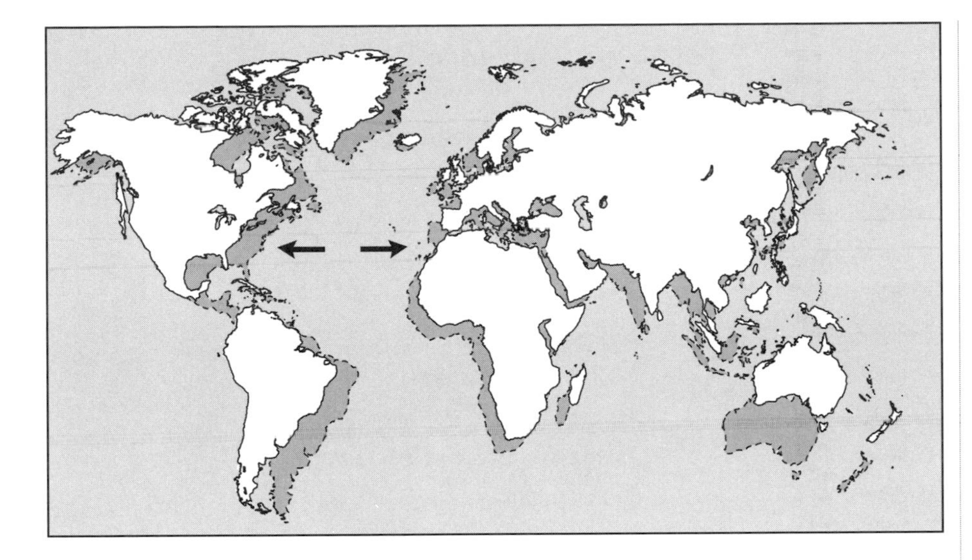

Figure 159 Drift of the continents 50 million years from now. Dashed lines indicate present positions of continents.

the Panama Isthmus will halt the migration of plants and animals between the two continents. It will also allow westward-flowing currents from the Atlantic into the Pacific, resulting in a lively exchange of marine species between the two oceans. Furthermore, the strong current flow will shut off the warm Gulf Stream circum-Atlantic current, placing Europe into a deep freeze. The loss of this great heat-transfer mechanism will initiate a new ice age, whose massive glaciers will spread out from the polar regions and plow up land far southward. The drop in sea level several hundred feet will advance the shoreline in places hundreds of miles seaward.

The retreating glaciers at the end of the ice age will leave behind a huge desert wasteland, with long, sinuous eskers, thousands of hillocks of glacial till called drumlins, and expansive glacial lakes that dwarf the present Great Lakes. The outwash of glacial meltwater gorged with sedimentary debris will flood over the continents, carving out new river channels while clogging old streams with mounds of sand and gravel. Plants and animals forced to retreat southward to avoid the harsh ice age conditions will slowly return to recolonize land once claimed by the glaciers.

Africa will continue to press against Eurasia. The Mediterranean Sea caught in the middle will be squeezed dry. New mountain ranges will rise out of thick sediments that have been accumulating in the Mediterranean Basin for millions of years. Australia will drift northward, possibly colliding with Southeast Asia as the plate upon which it rides dives into the gaping Java Trench. At the same time, Japan and Taiwan will crash into eastern China, adding more land to the Asian continent.

TABLE 14 THE MAJOR ICE AGES

Time (years)	Event
10,000–present	Present interglacial
15,000–10,000	Melting of ice sheets
20,000–18,000	Last glacial maximum
100,000	Most recent glacial episode
1 million	First major interglacial
3 million	First glacial episode in Northern Hemisphere
4 million	Ice covers Greenland and the Arctic Ocean
15 million	Second major glacial episode in Antarctica
30 million	First major glacial episode in Antarctica
65 million	Climate deteriorates, poles become much colder
250–65 million	Interval of warm and relatively uniform climate
250 million	The great Permian ice age
700 million	The great Precambrian ice age
2.4 billion	First major ice age

Antarctica will drift away from the South Polar region, where it had remained stationary for millions of years. The thawing of the ice continent will raise global sea levels several hundred feet and flood coastal areas in places hundreds of miles inland, greatly reducing the land area. Antarctica will then become a lush green continent. It will have new species of plants and animals and an entirely independent ecosystem separate from the rest of the world just like Australia had when the southern contents separated from Gondwana.

North and South America will continue to slide across the Pacific and then reverse direction and head back toward Eurasia as new subduction zones open outside the continental margins surrounding the Atlantic. The Mid-Atlantic Ridge will become an extinct rift ultimately becoming consumed by the subduction of the seafloor into the mantle. Eventually, all continents will reunite into a single large supercontinent called Neopangaea. Then, the process of continental breakup and drift will begin again, forming new continents and oceans that will bear no resemblance to those of today. The Earth will become a strange new world.

As the Earth ages, the lithosphere will continue to thicken, and the asthenosphere will become more viscous due to the cooling of the mantle after most of the radioactive heat sources have decayed into stable elements.

As the plates continue to thicken and activity in the asthenosphere slows, the movement of the plates becomes sluggish. Eventually, they will come to a complete halt. When this happens, perhaps some 2 billion years from now, volcanoes will no longer erupt, earthquakes will cease, and the Earth will become just another dead planet.

In the next chapter, we will leave our earthly home and go out into space in search of plate tectonics on other bodies in the solar system.

10

TECTONICS IN SPACE

EXPLORING THE SOLAR SYSTEM

The geologic activity on Earth, driven by heat flowing from the interior, constantly reshapes and rearranges the surface as crustal plates collide, as volcanoes expel hot rocks and gases through cracks in the crust, and as earthquakes wrench the crust at plate boundaries. The Earth is not the only body in the solar system to possess this activity. The inner terrestrial planets, Mercury, Venus, and Mars, had similar origins. Except for the Earth and possibly Venus, they are now tectonically dead. Their volcanoes no longer erupt, and their faults no longer quake.

However, several moons of the outer planets, Jupiter, Saturn, Uranus, and Neptune, appear to remain active. Volcanic eruptions have actually been observed on Jupiter's moon Io. Indeed, that moon is so active it rivals even the Earth in its fusillade of fiery blasts. This chapter takes a journey through the solar system, looking for evidence of plate tectonic activity on other worlds.

LUNAR TECTONICS

Our Earth and its Moon appear to have formed at roughly the same time some 4.6 billion years ago. The moon's crust is 30 to 60 miles thick, over twice as thick as the crust of the Earth. The Moon has a brittle mantle about 700 miles thick. Below the mantle is a relatively dense core with a radius of about 300 miles that might be partly molten. Early in its history, the Moon's surface was melted by a massive meteorite bombardment, which was responsible for most of its present-day terrain features (Fig. 160).

Beginning about 4.2 billion years ago and continuing for several hundred million years, huge basaltic lava flows welled up through the weakened crater floors and flooded great stretches of the Moon's surface, filling and submerging many meteorite craters. The dark basalts covered some 17 percent of the lunar surface with lava that hardened into smooth plains called maria, named so because when viewed from Earth, they resemble seas. The composition of the basalts indicates they came from a deep-seated source, much deeper than magmas from the Earth's interior. This resulted in massive outpourings of molten rock onto the surface, producing oceans of basaltic lava.

Numerous narrow, sinuous depressions, or rills, slice through the lava flows. Many of these emanate from craters that appear to be volcanic in origin. In some areas, wrinkles break the surface of the lava flows that might have been caused by moonquakes, which explains the extensive seismic activity sensed by lunar probes. Also, areas of previous volcanic mountain building span the Moon's surface. Ridges reach several hundred feet high and extend for hundreds of miles. The last of the basalt lava flows hardened about 2 billion

Figure 160 *The Moon from* Apollo 11 *on July 1969.*

(Photo courtesy NASA)

years ago. Except for several fresh meteorite impacts, the Moon looks much the same today as it did then.

Lunar craters as much as 250 miles and more across, larger than any found on Earth, destroyed most of the Moon's original crustal rocks. A mammoth depression on the Moon's south pole called the Aitken Basin is 7.5 miles deep and 1,500 miles across, or about a quarter of the Moon's circumference. It appears to have been initially created by a gigantic asteroid or comet that penetrated deep into the mantle. Water ice up to several hundred feet thick detected inside the massive crater might some day serve as a source of water for future lunar bases.

Moon rocks bought back during the Apollo missions (Fig. 161) vary in age from 3.2 to 4.5 billion years old. The oldest rocks are primitive and have not changed significantly since they first originated from molten magma.

Figure 161 *Moon rock brought back by* Apollo 15.

(Photo courtesy NASA)

These rocks formed the original lunar crust and are called Genesis Rock. They are composed of a coarse-grained, feldspar granite that formed deep in the Moon's interior. The youngest rocks are volcanic in origin and were melted and reconstructed by giant meteorite impacts. When the basalt flows ended, new Moon rock formation appears to have ceased, and no rocks are known to be younger than 3.2 billion years.

All lunar rocks are igneous, meaning they were derived from molten magma. Together they form the regolith, which is composed of loose rock material on the surface. The rocks include coarse-grained gabbro basalt, meteoritic impact breccia, pyroxine peridotite, glass beads known as chondrules derived from meteorite impacts, and dust-size soil material. Samples of lunar rocks brought back by Apollo astronauts contained mostly impact breccias, which are debris from meteorite impacts. Apparently, only a portion of the Moon is of volcanic origin as originally thought. The regolith is generally about 10 feet thick but is thought to be thicker in the highlands. Because of its dark basalts, the Moon's surface is a poor reflector of sunlight. Only about 7 percent of the light is reflected, which makes the Moon one of the darkest bodies in the solar system.

MERCURIAN TECTONICS

If Mercury and the Moon were placed side by side, they would display many striking resemblances. Images of Mercury taken during the *Mariner 10* encounter in March 1974 (Fig. 162) could easily be mistaken for the far side of the Moon. About the only major difference is that Mercury does not have jumbled mountainous regions and wide lava plains, or maria. Instead, it has long, low-winding cliffs that resemble fault lines several hundred miles long. A network of faults indicated by a series of curved scarps, or cliffs, crisscrossing Mercury's surface apparently formed as the planet cooled and shrank.

Like the Moon, Mercury is heavily scarred with meteorite craters from a massive bombardment some 4 billion years ago. The huge meteorites broke through the fragile crust and released torrents of lava that paved over the surface. Several large craters are surrounded by multiple concentric rings of hills and valleys. They possibly originated when a meteorite impact caused shock waves to ripple outward, similar to the waves formed when a stone is tossed into a pond. The largest of these craters, named Caloris, is 800 miles in diameter and 3.6 billion years old. Objects striking Mercury probably had higher velocities than those hitting the other planets due to the sun's stronger gravitational attraction at this range.

The compositions of Mercury and the Moon are very similar as well and comparable to the interior of the Earth. Mercury has an appreciable magnetic

Figure 162 *A simulated encounter with Mercury by* Mariner 10 *in March 1974.*

(Photo courtesy NASA)

field, indicating the presence of a large metallic core, which also accounts for the planet's high density. The core is twice as massive relative to Mercury's size as that of any other rocky planet. Above the core lies a silicate mantle, comprising only a quarter of the radius. In contrast, the Earth's mantle takes up nearly half the radius and almost 70 percent of the planet's volume. Mercury might have formed an iron core of conventional proportions while much of its outer rocky mantle was blasted away by large meteorite impacts during the formation of the solar system. Indeed, giant impacts early in their histories might have sent the planets down vastly different evolutionary pathways.

Mercury has the widest temperature extremes of any planet. During the day, temperatures soar to 300 degrees Celsius, hot enough to melt lead. At night, temperatures plummet to –150 degrees. This large difference is mainly due to the planet's closeness to the Sun, a highly elliptical orbit that ranges between 29 and 43 million miles from the Sun, a slow rotation rate of 1.5 times for every revolution around the Sun (which keeps Mercury's dark side away from the Sun for long periods), and a lack of an appreciable atmosphere to spread the heat around the planet.

Any gases and water vapor formed by volcanic outgassing or degassing from meteorite impacts quickly boiled off due to the excessive heat and low escape velocity needed to break away from the low gravity. The tenuous atmosphere that remains contains small amounts of hydrogen, helium, and oxygen. They probably originated from the direct infall of cometary material and outgassing of the few remaining volatiles in the interior. Traces of water ice might still exist in permanently shaded regions near the poles.

Mercury revolves around the Sun every 88 Earth days and rotates on its axis every 59 Earth days. This unusual orbital feature requires Mercury to go twice around the Sun before completing a single day, which from dawn to dusk equals to 176 Earth days. By comparison, Earth and Mars might have acquired their high rotation rates by glancing blows from large asteroids during their formation. Because of its small size, which allowed all the planet's internal heat to escape into space early in its life, Mercury is now a tectonically dead planet.

VENUSIAN TECTONICS

Venus is the Earth's sister planet in many ways. It is nearly the same size, mass, and composition and contains a substantial atmosphere composed almost entirely of the greenhouse gas carbon dioxide. This produces surface temperatures of about 460 degrees Celsius, three times hotter than without an atmosphere. The surface of Venus is relatively young, with an average age of less than 1.5 billion years. This is comparable to the Earth 3 billion years ago before plate tectonics began to change its identity. Therefore, Venus is the best place in the solar system to view Earth as it existed in the early years.

As on Earth, the geology on Venus still appears to be highly active. The jagged surface on Venus seems to have been shaped by deep-seated tectonic forces and volcanic activity eons ago, providing a landscape totally alien compared with the Earth. Mapping the surface of Venus by the *Magellan* spacecraft in orbit around the planet has revealed a crumpled and torn surface typical of jostling segments of crust comprising movable tectonic plates. The horizontal motion of the crust has compressed it into ridges and troughs comparable only to those found on Earth.

Folded and broken crust, formed by the collision of tectonic plates, built mountains. Some of these bear a considerable resemblance to the Appalachians, which were formed by the collision of North America and Africa. The 36,000-foot-high Maxwell Montes dwarfs Mount Everest by over a mile. Faults resembling San Andreas shoot through the surface, displacing large chunks of crust. Although Venus possesses continental highlands in the northern hemisphere, its ocean basins lack one essential ingredient—water.

Venus also appears to be very volcanically active, much more so than Earth. Venus might resemble our planet during its early stages of development. Large volcanic structures on the surface of Venus suggest the Venusian volcanoes are much more massive than those on Earth. To hold up these gigantic volcanoes, the lithosphere beneath would have to be 20 to 40 miles thick. Radar maps made by data supplied by orbiters along with surface elevations from Earth-based radar have revealed large volcanic structures on the surface of Venus. Very low frequency radio waves called whistlers picked up by the orbiters are believed to originate when lightning arcs across volcanic dust clouds.

A region known as Beta Regio (Fig. 163) appears to possess numerous large volcanoes, some of which are up to three miles high. A huge shield volcano known as Theia Mons has a diameter of more than 400 miles, much larger than any volcano on Earth. Large anomalies in the gravity field seem to indicate that some highlands on Venus are buoyed up by rising plumes of mantle rock, which are thought to feed active volcanoes. An abundance of sulfur gases in the atmosphere suggests that Venus has recently undergone massive volcanic eruptions.

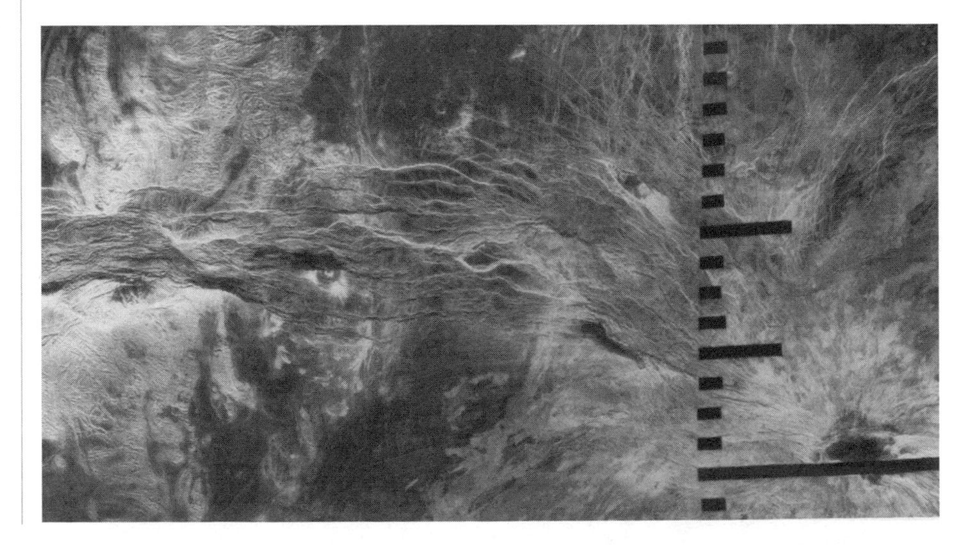

Figure 163 *The central part of Beta Regio from the Venus probe* Magellan.

(Photo courtesy NASA)

Figure 164 *An artist's rendition of the Venus rift valley, which at 3 miles deep, 175 miles wide, and 900 miles long is largest canyon in the solar system.*

(Photo courtesy NASA)

Radar mapping has revealed large, elevated, plateau-like tracts rising 3 to 6 miles above the surrounding terrain. Elongated ridges and circular depressions more than 50 miles in diameter might have resulted from large meteorite impacts. Most of the impact craters on Venus appear to have been erased 500 to 700 million years ago when widespread volcanic eruptions paved over a large portion of the planet. Many craters have been altered by later lava flows, implying that Venus has been volcanically active for much of its life. A great rift valley measuring 8 miles deep, 175 miles wide, and 900 miles long might well be the grandest canyon in the solar system (Fig. 164). The jagged surface on Venus appears to have been shaped by deep-seated tectonic forces and volcanic activity, producing a landscape totally alien to anything found on Earth.

The surface on Venus appears to be remarkably flat (Fig. 165), much more so than on Earth or Mars. Two-thirds of the planet has a relief of less than 3,000 feet. Scattered rocks on the surface observed by Soviet landing craft were angular in some places but flat and rounded at other locations. This suggests strong wind erosion. The surface rocks have a density identical to terrestrial granites. The soil has a composition similar to basalts on Earth and its moon. On the rugged equatorial highlands of Aphrodite Terra, the jumbled debris of a massive landslide matches the largest avalanches on Earth.

One curious aspect about Venus is the lack of a moon even though the planet's origins are thought to be very similar to the Earth. The Venusian moon might have crashed into its mother planet or escaped into orbit around

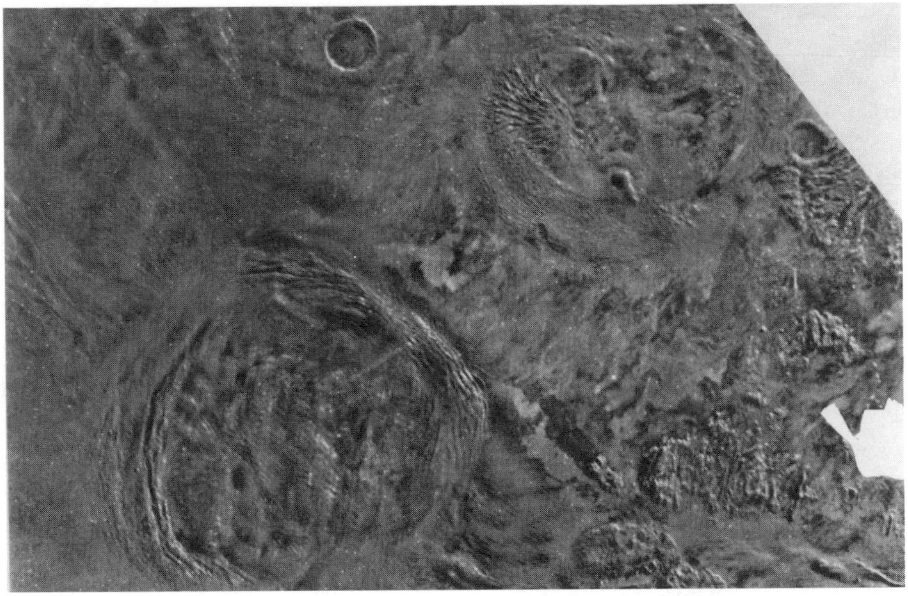

Figure 165 *Radar image of Venus' northern latitudes by* Venera 15.
(Photo courtesy NASA)

the Sun. Perhaps Mercury, which is comparable in size to the Earth's moon, was once a moon of Venus. This might indicate that the creation of our Moon was a unique event in the evolution of the solar system.

The internal structure of Venus is thought to be much like the inner workings of the Earth. Venus possesses a liquid metallic core surrounded by a mantle and a rocky crust. The core constitutes about one-quarter of the planet's mass and half its radius. An almost complete absence of a magnetic field, however, might be due to the planet's slow rotation and lack of an Earthlike inner core, which prevents the generation of the electrical currents needed to produce a magnetic field.

The density of the Venusian mantle ranges from 5.6 times the density of water near the top to 9.5 near the core. The density of the crust averages 2.9, similar to the Earth's crust. The presence of the three radioactive elements uranium, thorium, and potassium in the crust are in amounts comparable with those found on Earth. Therefore, the amount of radiogenic heat in the interior of Venus is assumed to be similar to that of the Earth's interior. As with the Earth, the best escape for excess internal heat is volcanism.

Upwelling of magma below the volcanic region might be responsible for horizontal movements in the crust in much the same manner that global tectonics operate on Earth. This has created a major linear rift system flanked by volcanic structures. Isolated peaks have also been identified near the equator, and these structures are thought to be individual volcanoes.

Orbiting spacecraft have observed large circular features as wide as several hundred miles and relatively low in elevation. These structures are attributed to gigantic volcanic domes that have collapsed, leaving huge, gaping holes called calderas surrounded by folds of crust as though massive bubbles of magma had burst on the surface. Much of the volcanic activity is concentrated in a huge cluster of volcanoes and lava flows 6,000 miles across and centered on the equator. The volcanic centers, numbering in the thousands, might have originally formed outside the equator. However, increased mass deposited onto the planet's surface might have tipped Venus over until the cluster was positioned at the equator, the most stable place for it to be.

Venus was thought to have counterparts to the rifts and collision zones created by active plate tectonics such as those on Earth. However, more detailed images from *Magellan* suggest that the Venusian surface is a single shell largely devoid of global plate tectonics. Only a small percentage of the heat flow from Venus is by rifting. On Earth, in contrast, 70 percent of its interior heat is lost by seafloor spreading. Venus therefore appears to be a dry, fiery planet whose surface is locked in an immobile shell and unable to shift its crust like the Earth does.

MARTIAN TECTONICS

Like the Moon and Mercury, the surface on Mars is highly cratered. However, Mars also has many terrain features produced by wind and water (Fig. 166) as well as by tectonic activity, including massive volcanic eruptions. The southern hemisphere is rough, heavily cratered, and traversed by huge channel-like depressions, where massive floodwaters possibly flowed some 700 million years ago. The southern highlands are a highly cratered region resembling the Moon but lack volcanoes and other geologic scars that mark the planet's northern lands.

A deeply cut, sinuous channel in Mars' Nanedi Vallis is perhaps the strongest evidence that water existed on the planet's surface for prolonged periods. The giant Argyle Planitia impact basin in the southern highlands is about 750 miles wide and over a mile deep. It shows evidence that liquid water once flowed on Mars. It contains layers of material that appears to be sediment from a huge body of water held by the basin millions of years ago. Three networks of channels lead into the basin from the south, whereas other channels slope northward out of the basin. Like Earth, Mars has two polar ice caps. The southern ice cap might have once grown large enough for water to flow into the basin, forming a huge, icy lake that spilled over the northern side of the basin and carved out deep channels. The northern ice cap holds a significant amount of the planet's water and, if melted, would cover the planet to a depth of about 40 feet.

The northern hemisphere is smooth, only lightly cratered, and dotted with numerous extinct volcanoes. The *Mars Global Surveyor* spacecraft showed that

Figure 166 *The landscape of Mars from* Pathfinder *lander showing boulders surrounded by windblown sediment.*
(Photo courtesy NASA)

much of the planet's northern hemisphere is a low-lying plain thousands of miles wide and roughly centered on the north pole, while the rest of the planet is ancient highlands. Perhaps a massive body slammed into the northern hemisphere and altered the surface, which would also explain the smaller number of impact craters. Another idea is that Earthlike tectonic forces and perhaps even an ancient ocean have shaped the northern lowlands. The relief of the land is remarkably low, making the region the flattest surface in the entire solar system.

Mars is only about half the size of the Earth. Like the Earth's moon, it has been tectonically dead for what appears to be well more than 2 billion years. However, the darkness of some pyroclastics suggests that they are no older than a few million years. Therefore, Mars could have been a volcanically active planet throughout most of its history. Data gathered from the *Mars Global Surveyor* spacecraft detected magnetic stripes in the highly cratered highlands similar to those found on Earth at oceanic rifts. They suggest that in the past, Mars had Earthlike plate tectonics. The Martian stripes are 10 times wider than those on Earth however, implying either faster spreading or slower magnetic field reversals. They are also far less regular in shape and spacing and not nearly so symmetrical on either side of the spreading center, which became the clinching evidence for plate tectonics on Earth.

The climate system on Mars is highly active. The polar regions are capped by a several-miles-thick mixture of frozen carbon dioxide, water ice, and windblown dust. The entire planet displays evidence of wind erosion and sedimentation. Violent seasonal dust storms whip up winds of 170 miles per hour for weeks at a time and are responsible for the planet's red glow. Winds play an important role by stirring the surface sediments and scouring out ridges and grooves in the surface. Deep layers of windblown sediment have accumulated in the Martian polar regions, and dune fields larger than any on Earth lie in areas surrounding the north pole.

Most of the Martian landscape is a cold, dry, desolate wasteland etched by deep valleys and crisscrossing channels, bearing the unmistakable imprint of flowing water sometime in the planet's long history. Channels dug out of the surface seem to imply that the climate in the past was significantly different from today. Perhaps geothermal heat from the planet's interior kept Mars warm enough to sustain liquid water on its surface. Flowing streams might have fanned out of the highlands and disappeared into the deserts as do flash floods on Earth.

The site of the robotic *Pathfinder* Mission sits at the mouth of an ancient flood channel called Ares Vallis (Fig. 167), which is 1 mile deep and 60 miles wide. Apparently, the equivalent of all the water in the Great Lakes once poured through the valley in only a few weeks, draining a large section of the Martian surface in a catastrophic flood. The great flood transported a variety of interesting rocks and other material from a long distance away. Rounded pebbles, cobbles, and conglomerates tumbled by running water are a sure sign that floods rushed through the region.

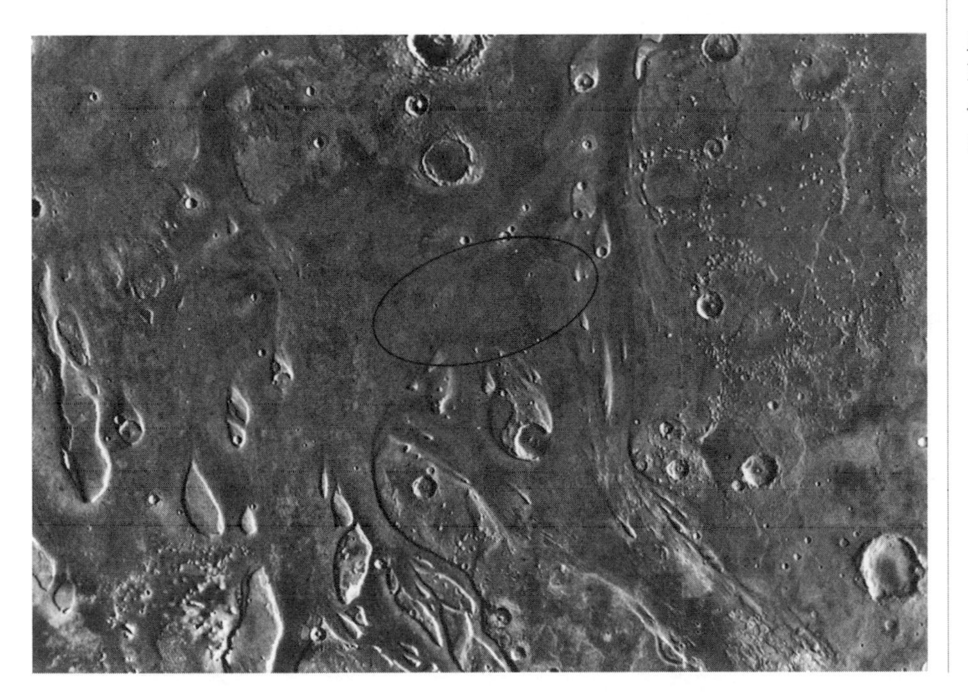

Figure 167 *The Mars* Pathfinder *landing site in Ares Vallis.*

(Photo courtesy NASA)

Branching tributaries running across the Martian surface look similar to dry river beds. Apparently, heat generated either by volcanic activity or by meteorite bombardments melted subterranean ice, resulting in great floods of water and flowing mud gushing forth and carving out gigantic ditches that rival anything found on Earth. The largest canyon, Valles Marineris (Fig. 168), measures 3,000 miles long, 100 miles wide, and 4 miles deep. It could hide several Grand Canyons with room to spare. The canyon is thought to have formed by slippage of the crust along giant faults similar to those of the East African Rift Valley. This activity is thought to have been accompanied by volcanism just like that on Earth.

Images returned by orbiting spacecraft revealed a Martian northern hemisphere with numerous volcanoes. The largest of these is Olympus Mons (Fig. 169), which covers an area about the size of Ohio. It has an elevation of 75,000 feet, over twice as high as Mauna Kea, Hawaii, the tallest volcano on Earth. The Martian volcanoes closely resemble shield volcanoes similar to those that built the main island of Hawaii. Their extreme size might have resulted from the absence of plate movements. Rather than forming a chain of relatively small volcanoes as though they were assembled on a conveyor belt when a plate moved across a volcanic hot spot, a single very large cone developed due to the crust remaining stationary over a magma plume for long periods.

Impact craters are notably less abundant in the regions where volcanoes are most numerous. This indicates that much of the volcanic topography on Mars formed after the great meteorite bombardment some 4 billion years ago. The southern hemisphere has a more highly cratered surface that is comparable in age to the lunar highlands, which are about 3.5 to 4.0 billion years old. Mars has many odd, elongated craters possibly created when minimoons in orbit around the planet plowed into the Martian surface at a low angle. The discovery of several highly cratered and weathered volcanoes indicates that volcanic activity began early and had a long history. However, even the fresh-appearing volcanoes and lava planes of the northern hemisphere are probably very ancient.

Although Mars is in proximity to the asteroid belt and the number of meteorite impacts were expected to be high, organic matter from infalling carbonaceous chondrites should have been abundant. Yet not a single trace of organic compounds was uncovered by Mars landers. Apparently, because the thin Martian atmosphere has only about 1 percent of the atmospheric pressure of Earth, organic compounds are actively destroyed by the Sun's strong ultraviolet radiation.

Seismographs on Mars landers detected no seismic activity, indicating that the planet might be tectonically dead. Apparently, horizontal crustal movement of individual lithospheric plates does not occur. The crust is probably much colder and more rigid than the Earth's crust. This explains why

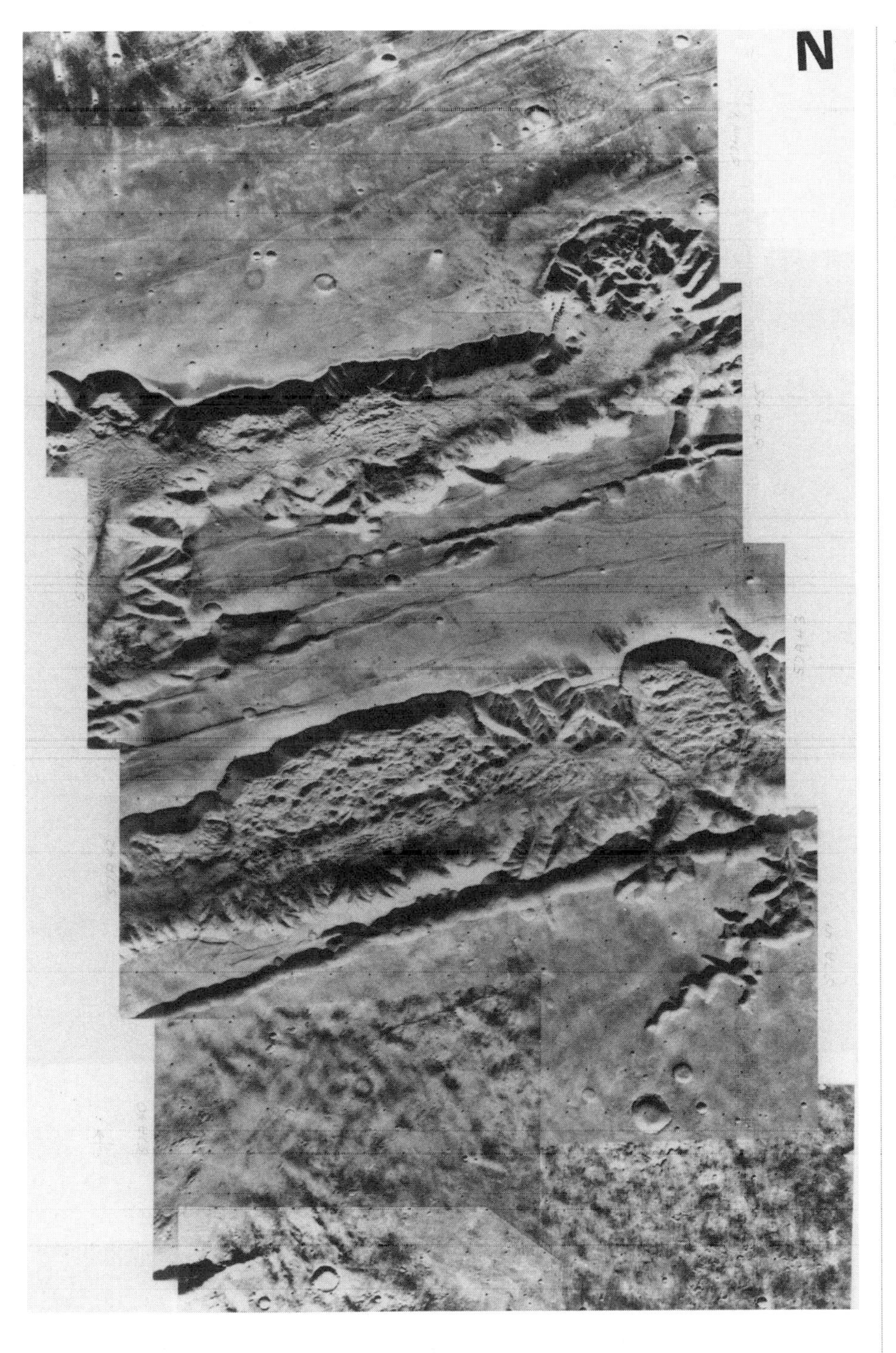

Figure 168 *A mosaic of the Mars surface at the west end of the Valles Marineris canyon system. These two canyons are over 30 miles wide and nearly 1 mile deep.*

(Photo courtesy NASA)

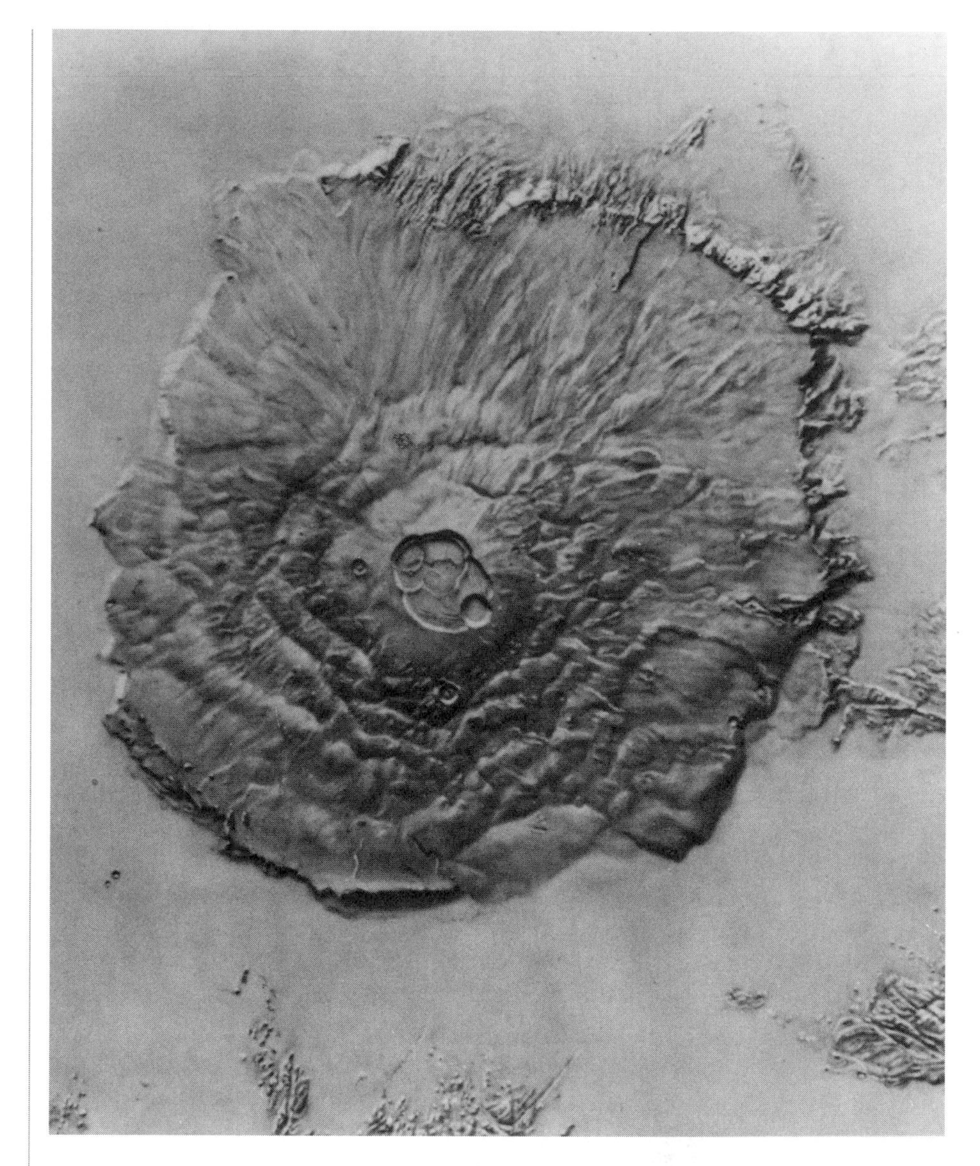

Figure 169 *The 350-mile diameter Olympus Mons, the largest volcano on Mars.*

(Photo by M. H. Carr, courtesy USGS)

folded mountain ranges do not exist on Mars. Because it has far less mass than the Earth, Mars could not have generated and stored enough heat to keep its interior in a semimolten state. Furthermore, Mars has no appreciable global magnetic field because it either has no metallic core or the core is not fluid due to a low internal temperature. It does have many small magnetic fields with different orientations scattered all over the surface.

Because no evidence of plate tectonics was found, heat could not be generated by friction between plates. Therefore, in order for Mars to exhibit

TABLE 15 MAJOR VOLCANOES OF MARS

Volcano	Height (Miles)	Width (Miles)	Age (Million Years)
Olympus Mons	16	300	200
Ascraeus Mons	12	250	400
Pavonis Mons	12	250	400
Arsia Mons	12	250	800
Elysium Mons	9	150	1,000–2,000
Hecates Tholus	4.5	125	1,000–2,000
Alba Patera	4	1,000	1,000–2,000
Apollinaris Patera	2.5	125	2,000–3,500
Hadriaca Patera	1	400	3,500–4,000
Amphitrites Patera	1	400	3,500–4,000

recent volcanism under these conditions, pockets of heat would have to exist just below the surface. Evidence for polar wandering on Mars indicates the entire crust might have shifted as a single plate due to instabilities in the interior resulting from the planet's rotation. This movement could have been caused by convection currents in the mantle, forcing molten rock to the surface, where great outpourings of magma produced maria similar to the vast lava plains on the Earth's Moon.

Mars' two tiny moons, Phobos (Fig. 170) and Deimos, are an enigma. The Martian satellites are oblong chunks of rock in nearly circular orbits. Their small sizes, blocky shapes, and low densities suggest they were captured bodies from the nearby main asteroid belt. However, such a capture seems a highly unlikely event. For this to happen twice would be nearly impossible. Instead, the moons might be the last surviving remnants of a ring of debris blasted into orbit when a huge asteroid more than 1,000 miles wide collided with Mars some 4 billion years ago. Phobos, which is not much more than an irregular rock 14 miles across, is also in a decaying orbit and will fall to Mars in about 30 million years.

JOVIAN TECTONICS

The large outer bodies of the solar system are referred to as the giant gaseous planets. Because of their low densities, gases are believed to make up the bulk of their masses. This alone indicates that these bodies are completely

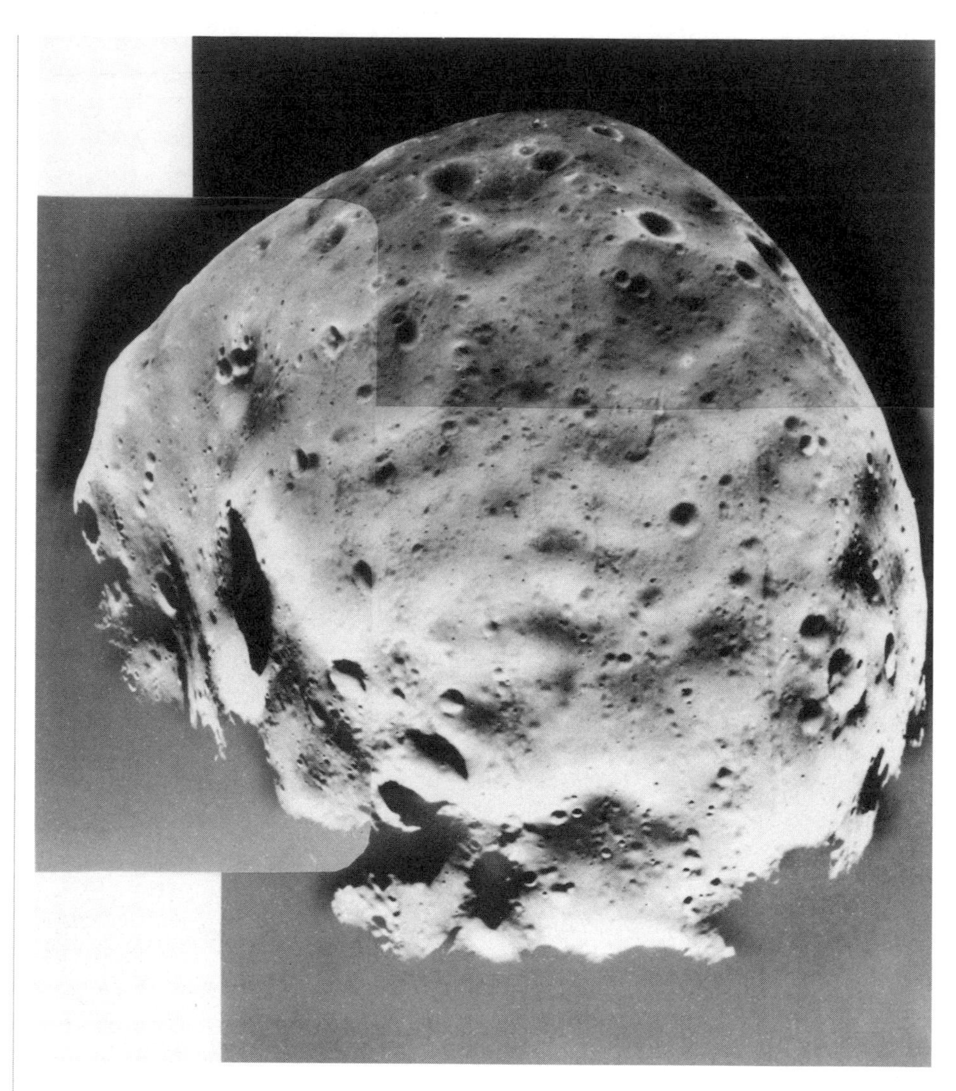

Figure 170 *Mars' inner moon Phobos, measuring 13 miles across, is thought to be a captured asteroid.*
(Photo courtesy NASA)

different from the inner, dense, terrestrial planets. Although the large planets are spectacular in their own right, they are not as geologically impressive as their moons.

Jupiter and its satellites resemble a miniature solar system (Fig. 171) comprising 15 moons. The four largest moons, which were first discovered by Galileo in 1610, travel in nearly circular orbits with periods ranging from 2 to 17 days. The largest of the Galilean moons, Callisto and Ganymede, are about the size of Mercury. The two smaller Galilean moons, Europa and Io (pronounced EYE-oh), are about the size of the Earth's moon. The surface of Callisto, the outermost of the Galilean satellites, is densely cratered. It resembles the far side of the Earth's moon, except the meteorite impacts were made on a frozen crust of dirty ice. A prominent bull's-eye structure appears to be a large impact basin.

Ganymede is the largest moon in the solar system. Its surface is less cratered than Callisto and resembles the near side of the Earth's moon. It has densely cratered regions and smooth areas, where young icy lavas covered the scars of older craters. Some smooth areas are heavily cratered and splintered by fractures. Numerous grooves, faults, and fractures suggest tectonic activity in the not-too-

Figure 171 *Jupiter and its four planet-size moons from* Voyager 1 *in March 1979.*

(Photo courtesy NASA)

Figure 172 *Ganymede is Jupiter's largest moon and shows ridges and grooves that probably resulted in deformation of the icy crust.*

(Photo courtesy NASA)

distant past (Fig. 172). A complex intersecting network of branching bright bands consisting of furrows and ridges crisscross the moon's surface. However, the lack of significant craters suggests that the surface formed fairly recently.

Ganymede is the only moon known to have a magnetic field. The satellite keeps the same side always facing Jupiter. On the side facing away from the planet is a large circular region of dark cratered terrain, with closely spaced parallel ridges and troughs from 3 to 10 miles across. They formed when a huge meteorite smashed into the soft crust and fractured it, producing an enormous pattern of concentric rings. Later, water filled the fractures and froze, forming one of the most impressive sights the solar system has to offer.

Europa has a smooth icy surface, devoid of major impact craters, and is crisscrossed by numerous linear features that are thousands of miles long and 100 miles wide (Fig. 173). These might be fractures in the icy crust that were filled with material erupted from below. Pictures from the *Galileo* spacecraft reveal a complex network of ridges and fractures, some of which resemble features formed by plate tectonics on Earth. In places, the surface of Europa looks like a giant jigsaw puzzle that has been pulled apart. Through the cracks flowed huge outpourings of slushy ice that buried all of Europa's terrain features, including its impact craters.

The moon probably formed after the great meteorite bombardment period, 4 billion years ago, when impacts were much more prevalent than they are today. An intricate tangle of crisscrossing ridges on Europa suggests icy volcanic eruptions reminiscent of Earth's midocean ridges, with ice rising to the surface at the central ridge and spreading away to form new crust. Europa appears to have a core about as dense as that of Earth and an icy crust broken into giant ice floes at least six miles thick. Below it might be liquid water heated by undersea volcanoes. This would be the only other known water ocean in the entire solar system.

The most intriguing of the Jovian moons is Io. It is the innermost moon. Its size, mass, and density are nearly identical to the Earth's moon. Due to a gravitational tug-of-war between Jupiter and Ganymede, with Io caught in

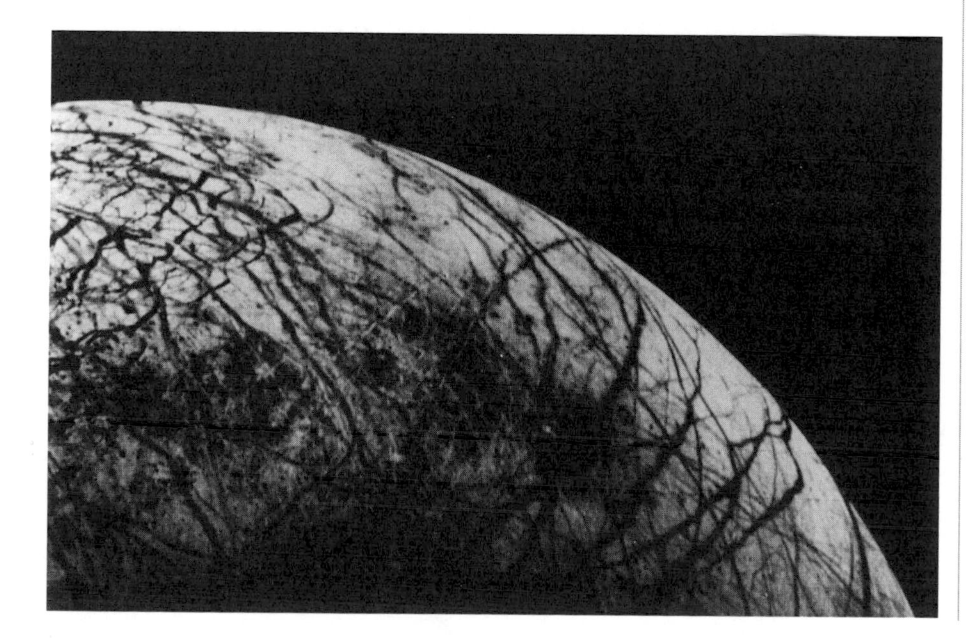

Figure 173 *The Jovian moon Europa showing a complex array of streaks that indicate the crust has been fractured and filled with materials from the interior.*

(Photo courtesy NASA)

Figure 174 *Io, the most volcanic body in the solar system, from the* Galileo *spacecraft.*

(Photo courtesy NASA)

the middle, frictional heat has kept Io's interior in a constant molten state with temperatures above 370 degrees Celsius, with hot spots exceeding 1,000 degrees. So much heat is pouring out of Io, it could have melted thousands of times over. This produces widespread volcanism over the entire moon's surface, which has more than 100 active volcanoes (Fig. 174). This makes Io possibly the most volcanically active body in the solar system. Io's volcanism also produces a surface that is highly colorful due to the multicolored sulfurous rocks, which paint the moon in hues of yellow, orange, and red.

The almost total lack of impact craters indicates that the surface of Io has recently, within the last million years, been paved over with large amounts of basaltic lava enriched in iron and magnesium. Several major volcanoes are erupting at any given moment on Io, whereas on Earth fewer major volcanic eruptions occur in a century. The largest volcanoes such as Pele, named for the Hawaiian volcano goddess, eject volcanic material in huge umbrella-shaped plumes that rise to an altitude of 150 miles or more and spread ejecta over areas as wide as 400 miles. A fire fountain flowing from a caldera near Io's north pole shot a curtain of lava some 5,000 feet high.

Io's volcanoes are not only more numerous, but they are also more energetic than those on Earth. Material is projected as though shot from a high-powered rifle at speeds of more than 2,000 miles per hour. This makes Io's volcanoes immensely more powerful than the most explosive volcanoes on Earth. Most of the large volcanoes, whose life spans are relatively short—only a few days or weeks—congregate in one hemisphere. The smaller, more numerous volcanoes gather in the opposite hemisphere.

The large volcanoes spew out enormous quantities of lava that quickly obliterate any craters that form. A typical area on Io receives up to several inches of fresh lava each year. The larger volcanoes pour out hot, sulfurous lava. The smaller ones eject sulfur dioxide, which quickly freezes, producing white, snow-capped volcanic terrain. The tallest volcanoes are as high as Mount Everest. Since sulfur is too weak to support such huge structures, they are probably made of silicate rock, produced by volcanic eruptions similar to those on Earth.

SATURNIAN TECTONICS

Saturn, which is similar in size and composition to Jupiter but with only a third of its mass, possesses the strangest set of moons in the solar system (Fig. 175). Its 17 moons range in size from an asteroid to larger than Mercury. All but the outer

Figure 175 *Saturn and its moons from* Voyager 1 *in November 1980.*

(Photo courtesy NASA)

two moons have nearly circular orbits, lie in the equatorial plane (the same plane as the rings), and keep the same side facing their mother planet, as does the Earth's moon.

Saturn's moons have densities less than twice the density of water, indicating they are composed mostly of ice. For most of Saturn's moons, the composition is 30 to 40 percent rock and 60 to 70 percent ice by weight. A surface reflectance, or albedo, between 60 and 90 percent for most moons suggests that they are coated with ice, making them highly reflective. With an albedo of nearly 100 percent, Enceladus, the second major moon outward from Saturn, is the most reflective body in the solar system. Its icy surface heavily dotted with meteorite craters and long rills appears to have been resurfaced by volcanic activity.

Hyperion appears to consist of fragments that recombined after collision with another large object had shattered it. Iapetus, the second outermost moon, is half black and half white, which makes it disappear east of Saturn and reappear west of the planet. The dark side might have been formed by volcanic extrusions composed not of hot lava but instead a slurry of ammonia, ice, and a dark material of possible organic origin.

Rhea, the second largest of Saturn's moons, has a densely cratered surface similar to the highlands on Mercury and the Earth's moon. Although sharing much of the same terrain features as Rhea, Dione is also Saturn's second densest moon. Tethys has a branching canyon, 600 miles long, 60 miles wide, and several miles deep, spanning the distance between the north and south polar regions. It also has a giant impact crater that is more than two-fifths the diameter of the moon itself.

Titan, which is larger than Mercury, is the only moon in the solar system to possess a substantial atmosphere. Moreover, its atmosphere is even denser than the Earth's. It is composed of compounds of nitrogen, carbon, and hydrogen and is believed to resemble the Earth's atmosphere during its infancy. This makes Titan possibly the best place in the solar system to look for the precursors of life. It is also the only known body in the solar system besides the Earth whose surface is partially covered by a liquid. Admittedly, its oceans are composed of liquid methane at temperatures of −175 degrees Celsius, and its continents are made of ice.

URANIAN TECTONICS

Most of the moons of Jupiter, Saturn, and Uranus look as though they were resurfaced, cracked, and modified by flows of solid ice. Yet the moons seem to lack the internal energy sources needed to make these extensive changes. Oberon and Titania, the largest and outermost of the Uranian moons, are both a little less than half the size of the Earth's moon. Oberon has a few features that resemble faults, but it shows no evidence of geologic activity. Its surface is fairly uniform gray and rich in water ice. Bright rays, assumed to be clean, buried ice brought to the surface, shoot out from around several meteorite craters. The surface is saturated with large craters, some more than 60 miles across. On the floor of several larger craters, volcanic activity has spewed out a mixture of ice and carbonaceous rock from the moon's interior.

The surface of Titania bears strong evidence of tectonic activity, including a complex set of rift valleys bounded by extensional faults where the crust is being pulled apart (Fig. 176). Titania's surface was also heavily cratered in its

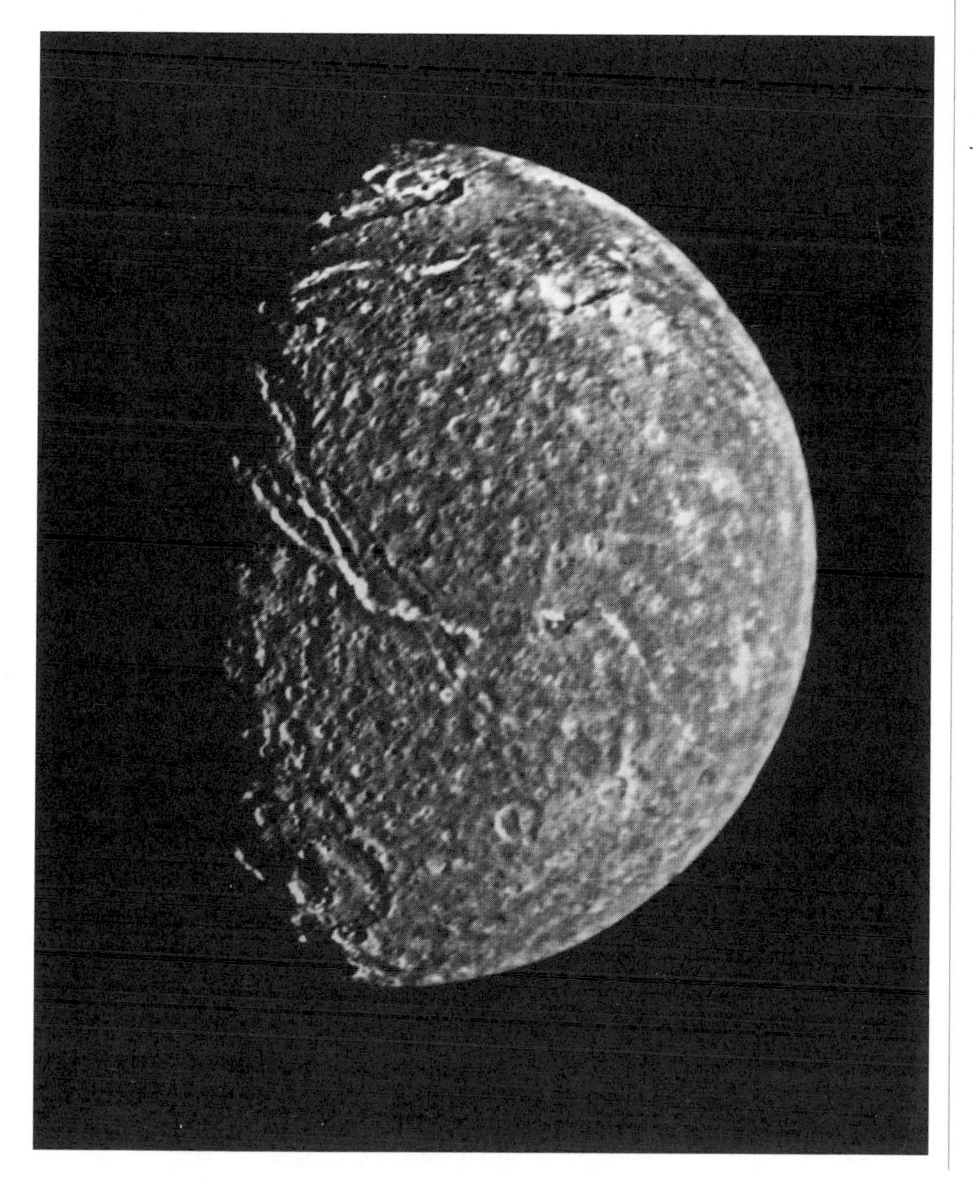

Figure 176 *Titania, the largest Uranian moon, shows a large trenchlike feature near the day-night boundary, which suggests tectonic activity.*

(Photo courtesy NASA)

early history. However, many larger craters were erased when the moon was resurfaced by volcanism, which spewed water onto the crust. Some large craters have disappeared because they were flooded with water or because the soft icy crust collapsed. When the water in the interior began to freeze and expand, the entire surface stretched. Meanwhile, the crust ripped open, enormous blocks of ice dropped down along the faults, and water upwelled through the cracks to form smooth plains.

Umbriel is about a third as large as the Earth's moon. It is heavily cratered but lacks rays, which gives it a nearly bland appearance. This indicates the moon might be covered with a thick blanket of uniform, dark material composed of ice and rock. This also makes Umbriel the darkest of the Uranian moons, whereas Ariel is the brightest.

Images of Ariel and Miranda reveal evidence of solid-ice volcanism never before observed in the solar system. A canyonlike feature on Ariel called Brownie Chasma resembles a graben, which is a crack in the crust formed when the surface is pulled apart, causing large blocks to drop downward. The walls of the chasm are about 50 miles across, and the chasm floor bulges up into a round-topped ridge about a mile high.

Ariel's density suggests that its composition is mostly ice. However, the material rising up from the chasm floor cannot be liquid water because this would form a flat surface before it froze. Furthermore, the moon appears to lack any energy sources to maintain water in a liquid state. The material oozing up through the cracks would therefore have to be a plastic form of ice that piles up into a ridge instead of simply running out over the surface. Ariel is the least cratered of the major moons, indicating its surface was remade repeatedly. The moon was resurfaced by volcanic extrusion of a viscous mixture of water and rock that flowed glacierlike from deep cracks similar to the flowage of lava from extensional faults on Earth.

With a surface temperature of –200 degrees Celsius, the viscous flows do not travel very far and flatten out before they freeze. The presence of substances such as ammonia, methane, and carbon dioxide in ice buried beneath the surface could make it more buoyant than the surrounding rock-ice mixture. This would make the ice rise up through the cracks and erupt out onto the surface, producing a landscape seen nowhere else in the solar system.

Miranda is perhaps the strangest world yet encountered. It might have acquired its bizarre terrain when the moon was shattered by comet impact and the pieces reassembled in a patchwork fashion (Fig. 177). It is the smallest of the major moons of Uranus, with a diameter of only 300 miles. For its modest size, though, it packs more terrain features than any moon. The surface is covered with densely cratered rolling plains, up to 200-mile-wide parallel belts of ridges and groves shaped like racetracks, and a series of chevron-shaped scarps. The entire landscape is cut by huge fracture zones

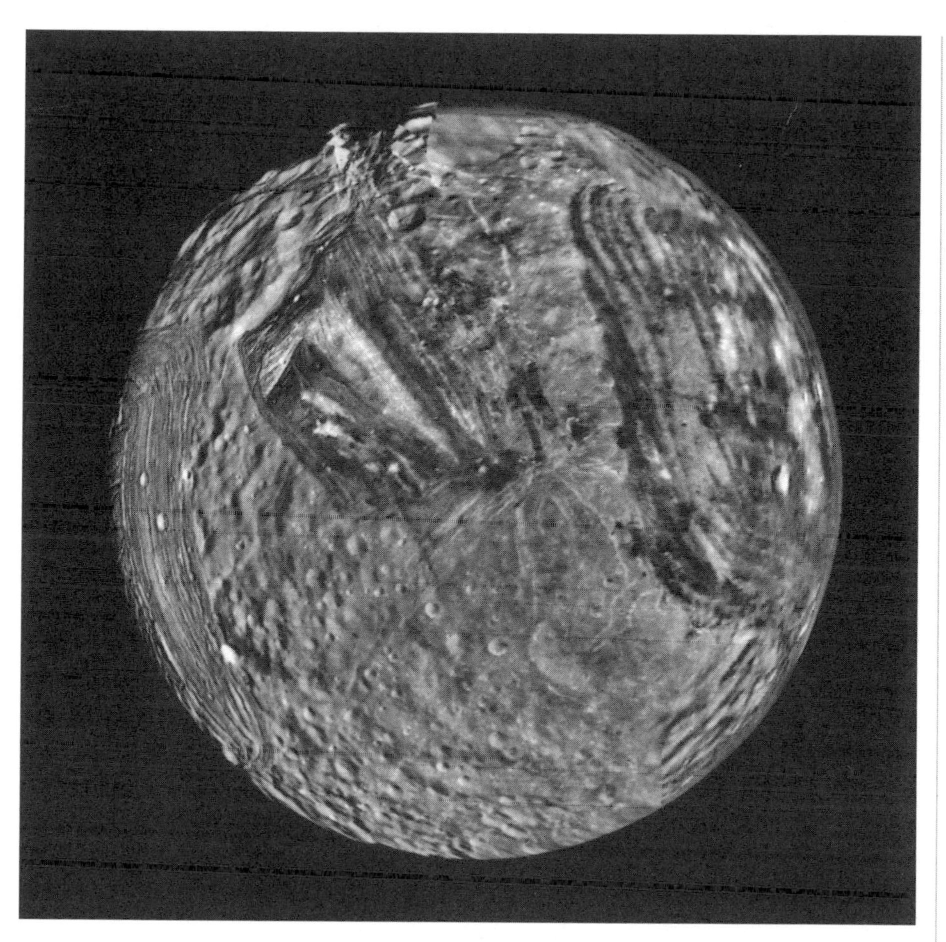

Figure 177 *Uranus's moon Miranda from* Voyager 2 *in January 1986 showing the strangest terrain features of any moon.*

(Photo courtesy NASA)

that encircle the moon, creating fault valleys whose steep, terraced cliffs reach a height of 12 miles.

TRITON TECTONICS

Triton, the largest moon of Neptune, appears to have been captured because its orbit is tilted 21 degrees to Neptune's equator and is in a retrograde direction, the only large moon in the solar system known to do so. Triton, Pluto, and other ice-covered residents of the outer solar system are thought to have been formed by an agglomeration of icy planetesimals, or comets. Triton also turns out to be the second most volcanically active body in the solar system behind Jupiter's moon Io. Geysers of carbon-rich material shoot as high as five miles above the moon's surface, similar to the outbursts of comets as they near the Sun.

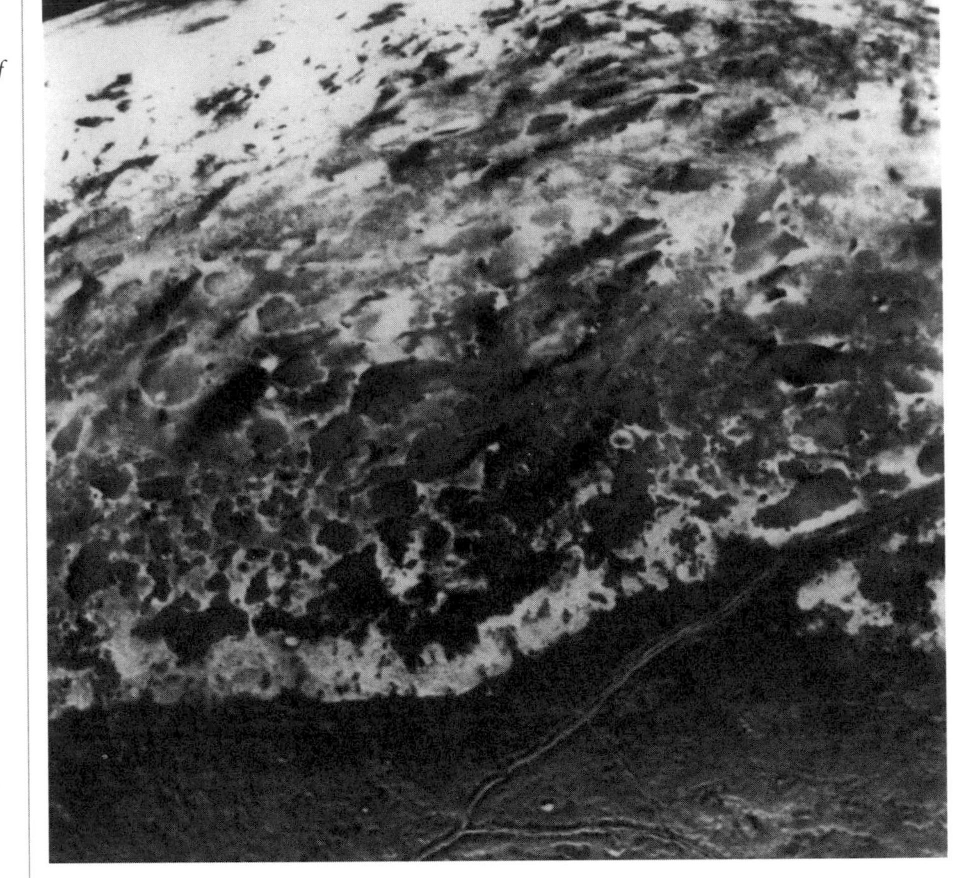

Figure 178 *Neptune's moon Triton from* Voyager 2 *in August 1989, showing evidence of extensive gas eruptions.*
(Photo courtesy NASA)

Triton's internal heat has long since vanished. Therefore, the source of energy needed to power its volcanic eruptions remains a mystery. What is left of Triton's meager heat from lingering radioactive decay is apparently enough to melt its interior of exotic ices to produce lavas or reshape and renew the surface. The surface appears to have been recently repaved as evident by the small numbers of meteorite impacts on it. An old surface would have many craters.

Images sent back by *Voyager* 2 in August 1989 showed dark plumes that seemed to indicate highly active volcanism on a body thought to have been tectonically dead for 4 billion years. Apparently, gigantic nitrogen-driven geysers spew out fountains of particles high into the thin, cold atmosphere that are strewn across the surface (Fig. 178). As on Ariel, icy slush oozes out of huge fissures and ice lava forms vast frozen lakes, providing one of the most bewildering landscapes in the solar system.

CONCLUSION

Plate tectonics is a relatively new field of science that has only recently (within the last couple of decades) gained worldwide acceptance. Many holdouts of the old scientific dogma believe in the geologic theory of uniformitarianism, whereby the Earth has operated in the past the same as it does today. An old saw among geologists is "I would not have seen it if I had not believed it." In other words, preconceived notions affect one's perspective, something that is hard to change. Alfred Wegener ran into this difficulty when trying to convince his colleagues about his theory of continental drift. Over the years, scientists debated the theory, most treating it with scorn, many with indifference, and only a few with respect.

Even today, many questions pertaining to plate tectonics remain unanswered. Many exceptions to rigid rules have been found throughout the world. Although classical plate tectonic theory works well for oceanic crust, it has a difficult time explaining the motion of the continents. Much controversy still exists, prompting heated arguments among earth scientists, especially in areas such as plate margins and continental interiors. In these areas, plate tectonic theory is not fully adequate and requires additional work. Nevertheless, the theory of plate tectonics has been highly successful and a powerful tool for exploring for petroleum and minerals our society desperately needs to maintain its economic health.

GLOSSARY

aa lava (AH-ah) a lava that forms large jagged, irregular blocks
abrasion erosion by friction, generally caused by rock particles carried by running water, ice, and wind
abyss (ah-BIS) the deep ocean, generally over a mile in depth
accretion the accumulation of celestial dust by gravitational attraction into a planetesimal, asteroid, moon, or planet
aerosol a mass of minute solid or liquid particles dispersed in the air
aftershock a smaller quake following the main earthquake
age a geological time interval smaller than an epoch
albedo the amount of sunlight reflected from an object and dependent on the object's color and texture
alluvium (ah-LUE-vee-um) stream-deposited sediment
alpine glacier a mountain glacier or a glacier in a mountain valley
andesite (an-di-zite) an intermediate type of volcanic rock between basalt and rhyolite
anticline folded sediments that slope downward away from a central axis
Apollo asteroids asteroids that come from the main belt between Mars and Jupiter and cross the Earth's orbit
aquifer (AH-kwe-fer) a subterranean bed of sediments through which groundwater flows
ash fall the fallout of small, solid particles from a volcanic eruption cloud
asperite (AS-per-ite) the point where a fault hangs up and eventually slips, causing earthquakes

asteroid a rocky or metallic body whose impact on the Earth creates a large meteorite crater

asthenosphere (as-THE-nah-sfir) a layer of the upper mantle from about 60 to 200 miles below the surface that is more plastic than the rock above and below and might be in convective motion

astrobleme (as-TRA-bleem) eroded remains on the Earth's surface of an ancient impact structure produced by a large cosmic body

back-arc basin a seafloor-spreading system of volcanoes caused by extension behind an island arc that is above a subduction zone

basalt (bah-SALT) a dark volcanic rock that is usually quite fluid in the molten state

basement rock subterranean igneous, metamorphic, granitized, or highly deformed rock underlying younger sediments

batholith (BA-tha-lith) the largest of intrusive igneous bodies and more than 40 square miles on its uppermost surface

bedrock solid layers of rock beneath younger materials

biomass the total mass of living organisms within a specified habitat

biosphere the living portion of the Earth that interacts with all other geologic and biologic processes

black smoker superheated hydrothermal water rising to the surface at a midocean ridge; the water is supersaturated with metals, and when exiting through the seafloor, the water quickly cools and the dissolved metals precipitate, resulting in black, smokelike effluent

blueschist (BLUE-shist) metamorphosed rocks of subducted ocean crust exposed on land

breccia (BRE-cha) a rock composed of angular fragments in a fine-grained matrix

calcite a mineral composed of calcium carbonate

caldera (kal-DER-eh) a large, pitlike depression at the summits of some volcanoes and formed by great explosive activity and collapse

carbonaceous (KAR-beh-NAY-shes) a substance containing carbon, namely sedimentary rocks such as limestone and certain types of meteorites

carbonate a mineral containing calcium carbonate such as limestone

carbon cycle the flow of carbon into the atmosphere and ocean, the conversion to carbonate rock, and the return to the atmosphere by volcanoes

catastrophism a theory that ascribes to the belief that recurrent, violent, worldwide events are the reason for the sudden disappearance of some species and the abrupt rise of new ones

Cenozoic (SIN-eh-zoe-ik) an era of geologic time comprising the last 65 million years

chalk a soft form of limestone composed chiefly of calcite shells of microorganisms

chert an extremely hard, fine-grained quartz mineral

chondrite (KON-drite) the most common type of meteorite, composed mostly of rocky material with small spherical grains

circum-Pacific active seismic regions around the rim of the belt Pacific plate coinciding with the Ring of Fire

circumpolar traveling around the world from pole to pole

conductivity the property of transmitting a quality

conglomerate (kon-GLOM-er-ate) a sedimentary rock composed of welded fine-grained and coarse-grained rock fragments

continent a landmass composed of light, granitic rock that rides on the denser rocks of the upper mantle

continental an ice sheet covering a portion of a glacier continent

continental drift the concept that the continents have been drifting across the surface of the Earth throughout geologic time

continental margin the area between the shoreline and the abyss that represents the true edge of a continent

continental shelf the offshore area of a continent in a shallow sea

continental shield ancient crustal rocks upon which the continents grew

continental slope the transition from the continental shelf to the deep-sea basin

convection a circular, vertical flow of a fluid medium by heating from below; as materials are heated, they become less dense, rise, cool, become more dense, and sink

convergent plate boundary the boundary between crustal plates where the plates come together; generally, it corresponds to the deep-sea trenches where old crust is destroyed in subduction zones

coral any of a large group of shallow-water, bottom-dwelling marine invertebrates that are reef-building common colonies in warm waters

Cordillera (kor-dil-ER-ah) a range of mountains that includes the Rockies, Cascades, and Sierra Nevada in North America and the Andes in South America

core the central part of the Earth, consisting of a heavy, iron-nickel alloy; also, a cylindrical rock sample drilled through the crust

correlation (KOR-eh-LAY-shen) the tracing of equivalent rock exposures over distance usually with the aid of fossils

craton (CRAY-ton) the stable interior of a continent, usually composed of the oldest rocks

Cretaceous-Tertiary boundary see K-T boundary

crosscutting a body of rocks cutting across older rock units

crust the outer layers of a planet's or a moon's rocks

crustal plate a segment of the lithosphere involved in the interaction of other plates in tectonic activity

desiccated basin (deh-si-KAY-ted) a basin formed when an ancient sea evaporated

diapir (DIE-ah-per) the buoyant rise of a molten rock through heavier rock

diatom (DIE-ah-tom) any of numerous microscopic unicellular marine or freshwater algae having siliceous cell walls

diatomite an ultrafine-grained siliceous earth composed mainly of diatom cell walls

dike a tabular intrusive body that cuts across older strata

divergent plate boundary the boundary between crustal plates where the plates move apart; it generally corresponds to the midocean ridges where new crust is formed by the solidification of liquid rock rising from below

dolomite (DOE-leh-mite) a sedimentary rock formed by the replacement of calcium with magnesium in limestone

dolomitization the process by which limestone becomes dolomite by the substitution of magnesium for the original calcite; common in organisms whose original hard parts were composed of calcite or aragonite, such as corals, brachiopods, and echinoderms

down-drop the lowering of a fault block

downfault the down-dropping of the crust along a fault plane

dynamo effect the creation of the Earth's magnetic field by rotational, thermal, chemical, and electrical differences between the solid inner core and the liquid outer core

earthquake the sudden rupture of rocks along active faults in response to geologic forces within the Earth

East Pacific Rise a midocean spreading center that runs north-south along the eastern side of the Pacific and the predominant location upon which the hot springs and black smokers have been discovered

eon on the geologic time scale, the longest unit of time, comprised of several eras

erosion the wearing away of surface materials by natural agents such as wind and water

escarpment (es-KARP-ment) a mountain wall produced by the elevation of a block of land

evaporite (ee-VA-per-ite) the deposition of salt, anhydrite, and gypsum from evaporation in an enclosed basin of stranded seawater

exfoliation (eks-FOE-lee-A-shen) the weathering of rock causing the outer layers to flake off

extrusive (ik-STRU-siv) an igneous volcanic rock ejected onto the surface of a planet or moon

fault a break in crustal rocks caused by Earth movements

fault block mountain a mountain formed by a block faulting, that is, a mountain isolated by faulting and categorized as structural or tectonic

fissure a large crack in the crust through which magma might escape from a volcano

fluvial (FLUE-vee-al) pertaining to being deposited by a river

formation a combination of rock units that can be traced over distance

fossil any remains, impressions, or traces in rock of a plant or animal of a previous geologic age

fumarole (FUME-ah-role) a vent through which steam or other hot gases escape from underground such as a geyser

gabbro (GAH-broe) a dark, coarse-grained, intrusive igneous rock

Genesis Rock ancient moon rocks forming the original lunar crust

geochemical pertaining to the distribution and circulation of chemical elements in the Earth's soil, water, and atmosphere

geodetic pertaining to the study of the external shape of the Earth as a whole

geologic column a diagram showing in columns the total succession of stratigraphic units of a region

geomorphology (JEE-oh-more-FAH-leh-jee) the study of surface features of the Earth

geosphere the inorganic world, including the lithosphere (solid portions of the Earth), the hydrosphere (bodies of water), and the atmosphere (air)

geosyncline (JEE-oh-SIN-kline) a basinlike or elongated subsidence of the Earth's crust; its length might extend for several thousand miles and might contain sediments thousands of feet thick, representing millions of years of deposits; a geosyncline generally forms along continental edges and is destroyed during periods of crustal deformation

geothermal the generation of hot water or steam by hot rocks in the Earth's interior

geyser (GUY-sir) a spring that ejects intermittent jets of steam and hot water

glacier a thick mass of moving ice occurring where winter snowfall exceeds summer melting

Glossopteris (GLOS-op-ter-is) late Paleozoic genus of fern living on Gondwana

gneiss (nise) a banded, coarse-grained metamorphic rock with alternating layers of different minerals, consisting of essentially the same components as granite

Gondwana (gone-DWAN-ah) a southern supercontinent of Paleozoic time, comprising Africa, South America, India, Australia, and Antarctica, broken up into the present continents during the Mesozoic era

graben (GRA-bin) a valley formed by a down-dropped fault block

granite a coarse-grained, silica-rich rock consisting primarily of quartz and feldspars; the principal constituent; of the continents, believed to be derived from a molten state beneath the Earth's surface

granitic intrusion the injection of magma into the crust that forms granitic rocks

granulite (GRAN-yeh-lite) a metamorphic rock comprising continental interiors

greenhouse effect the trapping of heat in the atmosphere principally by water vapor, carbon dioxide, and methane

greenstone a green metamorphosed igneous rock of Archean age

groundwater water derived from the atmosphere that percolates and circulates below the surface

guyot a seamount of circular form and having a flat top; thought to be volcanic cones whose tops have been flattened by surface wave erosion

Hadean eon a term applied to the first half-billion years of Earth history

hiatus (hie-AY-tes) a break in geologic time due to a period of erosion or nondeposition of sedimentary rock

high-angle fault a fault with a dip exceeding 45 degrees

horst an elongated, uplifted block of crust bounded by faults

hot spot a volcanic center with no relation to a plate boundary; an anomalous magma generation site in the mantle

hydrocarbon a molecule consisting of carbon chains with attached hydrogen atoms

hydrologic cycle the flow of water from the ocean to the land and back to the sea

hydrology the study of water flow over the Earth

hydrosphere the water layer at the surface of the Earth

hydrothermal relating to the movement of hot water through the crust; also a mineral ore deposit emplaced by hot groundwater

hypocenter the point of origin of earthquakes, also called the focus

Iapetus Sea (EYE-ap-it-us) a former sea that occupied a similar area as the present Atlantic Ocean prior to Pangaea

ice age a period of time when large areas of the Earth were covered by massive glaciers

iceberg a portion of a glacier calved off upon entering the sea

ice cap a polar cover of snow and ice

igneous rocks all rocks solidified from a molten state

impact the point on the surface upon which a celestial object has landed, creating a crater

intrusive any igneous body that has solidified in place below the Earth's surface

island arc volcanoes landward of a subduction zone, parallel to a trench, and above the melting zone of a subducting plate

isostasy (eye-SOS-tah-see) a geologic principle that states that the Earth's crust is buoyant and rises and sinks depending on its density

isotope (I-seh-tope) a particular atom of an element that has the same number of electrons and protons as the other atoms of the element but a different number of neutrons; that is, the atomic numbers are the same, but the atomic weights differ

jointing the production of parallel fractures in rock formations

kimberlite (KIM-ber-lite) a volcanic rock composed mostly of peridotite, originating deep within the mantle and that brings diamonds to the surface

K-T boundary the boundary formation marking the end of the dinosaur era

laccolith (LA-KEH-lith) a dome-shaped intrusive magma body that arches the overlying sediments and sometimes forms mountains

lahar (LAH-har) a mudflow of volcanic material on the flanks of a volcano

landform a surface feature of the Earth

landslide a rapid downhill movement of earth materials triggered by earthquakes and severe weather

lapilli (leh-PI-lie) small, solid pyroclastic fragments

Laurasia (lure-AY-zha) a northern supercontinent of Paleozoic time consisting of North America, Europe, and Asia

Laurentia (lure-IN-tia) an ancient North American continent

lava molten magma that flows out onto the surface

leach (out) the dissolution of soluble substances in rocks by the percolation of meteoric water

limestone a sedimentary rock consisting mostly of calcite from shells of marine invertebrates

liquefaction (li-kwe-FAK-shen) the loss of support of sediments that liquefy during an earthquake

lithosphere (LI-tha-sfir) the rocky outer layer of the mantle that includes the terrestrial and oceanic crusts; the lithosphere circulates between the Earth's surface and mantle by convection currents

lithospheric a segment of the lithosphere involved in the plate interaction of other plates in tectonic activity

magma a molten rock material generated within the Earth and is the constituent of igneous rocks

magnetic field a reversal of the north-south polarity of the magnetic poles

mantle the part of a planet below the crust and above the core, composed of dense rocks that might be in convective flow

maria dark plains on the lunar surface caused by massive basalt flows

massive sulfides sulfide metals deposited from hydrothermal solutions

Mesozoic (meh-zeh-ZOE-ik) literally the period of middle life, referring to a period between 250 and 65 million years ago

metamorphism (me-teh-MORE-fi-zem) recrystallization of previous igneous, metamorphic, or sedimentary rocks created under conditions of intense temperatures and pressures without melting

meteorite a metallic or stony celestial body that enters the Earth's atmosphere and impacts on the surface

Mid-Atlantic Ridge the seafloor-spreading ridge that marks the extensional edge of the North and South American plates to the west and the Eurasian and African plates to the east

midocean ridge a submarine ridge along a divergent plate boundary where a new ocean floor is created by the upwelling of mantle material

Mohorovicic discontinuity/Moho (mo-hoe-row-VE-check) the boundary between the crust and discontinuity/mantle discovered by Andrija Mohorovicic

mountain roots the deeper crustal layers under mountains

natural selection the process by which evolution selects species for survival or extinction depending on the environment

Neopangaea a hypothetical future supercontinent formed when the present continents collide with each other

neptunism (NEP-tune-isem) the belief that rocks precipitated from seawater

net slip the distance between two previously adjacent points on either side of a fault, it defines both directions and size of displacement

nontransform offset small offsets with overlapping ridge tips that offset the Mid-Atlantic Ridge

ophiolite (OH-fi-ah-lite) masses of oceanic crust thrust onto the continents by plate collisions

original horizontality the theory that rocks were originally deposited into the ocean horizontally and later folding and faulting tilted the beds

orogen (ORE-ah-gin) an eroded root of ancient mountain range

orogeny (oh-RAH-ja-nee) a process of mountain building by tectonic activity

outgassing the loss of gas from within a planet as opposed to degassing, the loss of gas from meteorites

overthrust a thrust fault in which one segment of crust overrides another segment for a great distance

pahoehoe lava (pah-HOE-ay-hoe-ay) a lava that forms ropelike structures when cooled

paleomagnetism the study of the Earth's magnetic field, including the position and polarity of the poles in the past

paleontology (pay-lee-ON-TAL-o-gee) the study of ancient life forms based on the fossil record of plants and animals

Paleozoic (PAY-lee-eh-ZOE-ik) the period of ancient time between 570 and 250 million years ago

Pangaea (pan-GEE-a) an ancient supercontinent that included all the lands of the Earth

Panthalassa (PAN-the-lass-a) a great global ocean that surrounded Pangaea

period a division of geologic time longer than an epoch and included in an era

pillow lava lava extruded on the ocean floor giving rise to tabular shapes

placer (PLAY-ser) a deposit of rocks left behind by a melting glacier; any ore deposit that is enriched by stream action

plate tectonics the theory that accounts for the major features of the Earth's surface in terms of the interaction of lithospheric plates

pluton (PLUE-ton) an underground body of igneous rock younger than the rocks that surround it and formed where molten rock oozes into a space between older rocks

plutonism the belief that rocks formed from molten material from the Earth's interior

polarity a condition in which a substance exhibits opposite properties such as electrical charges or magnetics

precipitation products of condensation that fall from clouds as rain, snow, hail, or drizzle; also the deposition of rocks from seawater

pumice volcanic ejecta that is full of gas cavities and extremely light in weight

pyroclastic (PIE-row-KLAS-tik) the fragmental ejecta released explosively from a volcanic vent

radioactive decay the process by which radioactive isotopes decay into stable elements

radioactivity an atomic reaction releasing detectable radioactive particles

radiogenic pertaining to something produced by radioactive decay, such as heat

radiometric dating the age determination of an object by chemically analyzing stable versus unstable radioactive elements

redbed red-colored sedimentary rocks indicative of a terrestrial deposition

reef the biological community that lives at the edge of an island or continent; the shells from dead organisms form a limestone deposit

regolith (REE-geh-lith) unconsolidated rock material resting on bedrock found at and near a planet's or moon's surface

regression a fall in sea level, exposing continental shelves to erosion

residence time the time required for a factor to remain in a certain environment, for example, carbon dioxide in the ocean and atmosphere

resurgent caldera a large caldera that experiences renewed volcanic activity that domes up the caldera floor

rhyolite (RYE-oh-lite) a volcanic rock that is highly viscous in the molten state and usually ejected explosively as pyroclastics

rift valley the center of an extensional spreading, where continental or oceanic plate separation occurs

rock bed a rock unit distinguishable from beds beneath or above it, a rock correlation tracing equivalent rock exposures over distance

schist (shist) a finely layered metamorphic rock that tends to split readily into thin flakes

seafloor spreading a theory that the ocean floor is created by the separation of lithospheric plates along midocean ridges, with new oceanic crust formed from mantle material that rises from the mantle to fill the rift

seamount a submarine volcano

sediment the organic or inorganic debris transported and deposited by wind, water, or ice; it might form loose sediment-like sand or mud or become consolidated to form sedimentary rock

sedimentary rock a rock composed of fragments cemented together

seismic (SIZE-mik) pertaining to earthquake energy or other violent ground vibrations

seismic sea wave an ocean wave generated by an undersea earthquake or volcano; also called tsunami

shield areas of exposed Precambrian nucleus of a continent

shield volcano a broad, low-lying volcanic cone built up by lava flows of low viscosity

sial a lightweight layer of rock that lies below the continents

sill an intrusive magma body parallel to planes of weakness in the overlying rock

sima a dense rock that comprises the ocean floor and on which the sial floats

spherules small, spherical, glassy grains found on certain types of meteorites, lunar soils, and large meteorite impact sites

stock an intrusive body of deep-seated igneous rock, usually lacking conformity and resembling a batholith, except for its smaller size

strata layered rock formations; also called beds

stratification a pattern of layering in sedimentary rock, lava flows, or water; materials of different composition or density

striae (STRY-aye) scratches on bedrock made by rocks embedded in a moving glacier

stromatolite (STRO-mat-eh-lite) sedimentary formation in the shape of cushions or columns produced by lime-secreting blue-green algae (cyanobacteria)

subduction zone a region where an oceanic plate dives below a continental plate into the mantle; ocean trenches are the surface expression of a subduction zone

subsidence the compaction of sediments due to the removal of fluids

succession a sequence of rock formations

superposition in any sequence of sedimentary strata that is not strongly folded or tilted; the youngest strata is at the top and the oldest at the bottom

syncline (SIN-Kline) a fold in which the beds slope inward toward a common axis

tectonic activity the formation of the Earth's crust by large-scale movements throughout geologic time

tectonics (tek-TAH-niks) the history of the Earth's larger features (rock formations and plates) and the forces and movements that produce them; see PLATE TECTONICS

tephra (TEH-fra) all clastic material, from dust particles to large chunks, expelled from volcanoes during eruptions

terrain a region of the Earth's surface that is treated as a physical feature or as a type of environment

terrane (teh-RAIN) a unique crustal segment attached to a landmass

terrestrial planet the rock planets Mercury, Venus, Earth, and Mars

Tethys Sea (TEH-this) the hypothetical midlatitude region of the oceans separating the northern and southern continents of Laurasia and Gondwana

thermal conductivity the amount of heat conducted per unit of time through any cross section of a substance, dependent on the temperature gradient at that section and the area of the section

till sedimentary material deposited by a glacier

tillite a sedimentary deposit composed of glacial till

transform fault a fracture in the Earth's crust along which lateral movement occurs; they are common features of the midocean ridges created in the line of seafloor spreading

transgression a rise in sea level that causes flooding of the shallow edges of continental margins

trapps (traps) a series of massive lava flows that resembles a staircase

trench a depression on the ocean floor caused by plate subduction

troposphere the lowest 6 to 12 miles of the Earth's atmosphere, characterized by decreasing temperatures with increasing height

tsunami (sue-NAH-me) a seismic sea wave produced by an undersea or near-shore earthquake or volcanic eruption

tuff a rock formed of pyroclastic fragments

tundra permanently frozen ground at high latitudes and elevations

ultramafic an igneous rock rich in iron and magnesium and poor in silica

ultraviolet radiation the invisible light with a shorter wavelength than visible light and longer than X-rays

unconformity an erosional surface separating younger rock strata from older rocks

uniformitarianism the belief that the slow processes that shape the Earth's surface have acted essentially unchanged throughout geologic time

upfault a block of crust pushed upward along fault lines

upwelling the process of rising as in magma or an ocean current

volatile a substance in magma such as water and carbon dioxide that controls the type of volcanic eruption

volcano a fissure or vent in the crust through which molten rock rises to the surface to form a mountain

BIBLIOGRAPHY

CONTINENTAL DRIFT

Anderson, Don L. "The Earth as a Planet: Paradigms and Paradoxes." *Science* 223 (January 1984): 347–354.

Ballard, Robert D. *Exploring Our Living Planet.* Washington, D.C.: National Geologic Society, 1983.

Boucot, A. J., and Jane Gray. "A Paleozoic Pangaea." *Science* 222 (November 11, 1983): 571–580.

Courtillot, Vincent, and Gregory E. Vink. "How Continents Break Up." *Scientific American* 249 (July 1983): 43–49.

Harrington, John W. *Dance of the Continents.* New York: J. P. Tarcher, 1983.

Hoffman, Paul F. "Oldest Terrestrial Landscape." *Nature* 375 (June 15, 1995): 537–538.

Kerr, Richard A. "Puzzling Out the Tectonic Plates." *Science* 247 (February 16, 1990): 808.

Macdonald, Kenneth C., and Paul J. Fox. "The Mid-Ocean Ridge." *Scientific American* 262 (June 1990): 72–79.

Miller, Russell. *Continents in Collision.* Alexander, Va.: Time-Life Books, 1983.

Parker, Ronald B. *Inscrutable Earth; Explorations into the Science of Earth.* New York: Scribners, 1984.

Sclater, John G., and Christopher Tapscott. "The History of the Atlantic." *Scientific American* 240 (June 1979): 156–174.

Simon, Cheryl. "The Great Earth Debate." *Science News* 121 (March 13, 1982): 178–179.
Weiner, Jonathan. *Planet Earth.* New York: Bantam, 1986.

HISTORICAL TECTONICS

Allegre, Claud J., and Stephen H. Snider. "The Evolution of the Earth." *Scientific American* 271 (October 1994): 66–75.
Dalziel, Ian W. D. "Earth before Pangaea." *Scientific American* 272 (January 1995): 58–63.
Dickinson, William R. "Making Composite Continents." *Nature* 364 (July 22, 1993): 284–285.
Galer, Stephen J. G. "Oldest Rocks in Europe." *Nature* 370 (August 18, 1994): 505–506.
Jones, Richard C., and Anthony N. Stranges. "Unraveling Origins, the Archean." *Earth Science* 42 (Winter 1989): 20–22.
Knoll, Andrew H. "End of the Proterozoic Eon." *Scientific American* 265 (October 1991): 64–73.
Krogstad, E. J., et al. "Plate Tectonics 2.5 Billion Years Ago: Evidence at Kolar, South India." *Science* 243 (March 10, 1989): 1337–1339.
Kunzig, Robert. "Birth of a Nation." *Discover* 11 (February 1990): 26–27.
Moores, Eldridge. "The Story of Earth." *Earth* 6 (December 1996): 30–33.
Shurkin, Joel, and Tom Yulsman. "Assembling Asia." *Earth* 4 (June 1995): 52–59.
Vermeij, Geerat J. "The Biological History of a Seaway." *Science* 250 (November 23, 1990): 1078–1080.
Waters, Tom. "Greetings from Pangaea." *Discover* 13 (February 1992): 38–43.
Weiss, Peter. "Land Before Time." *Earth* 8 (February 1998): 29–33.
York, Derek. "The Earliest History of the Earth." *Scientific American* 268 (January 1993): 90–96.
Zimmer, Carl. "In Times of Ur." *Discover* 18 (January 1997): 18–19.

CONVECTION CURRENTS

Anderson, Don L., and Adam M. Dziewonkski. "Seismic Tomography." *Scientific American* 251 (October 1984): 60–68.
Bercovici, Dave, Gerald Schubert, and Gary A. Glatzmaier. "Three-Dimensional Spherical Models of Convection in the Earth's Mantle." *Science* 244 (May 26, 1989): 950–954.

Bonatti, Enrico. "The Rifting of Continents." *Scientific American* 256 (March 1987): 97–103.

Courtillot, Vincent, and Gregory E. Vink. "How Continents Break Up." *Scientific American* 249 (July 1983): 43–49.

Heppenheimer, T. A. "Journey to the Center of the Earth." *Discover* 8 (November 1987): 86–93.

Jeanloz, Raymond, and Thorn Lay. "The Core-Mantle Boundary." *Scientific American* 268 (May 1993): 48–55.

Kerr, Richard A. "Putting Stiffness in Earth's Mantle." *Science* 271 (February 23, 1996): 1053–1054.

———. "Having It Both Ways in the Mantle." *Science* 258 (December 1992): 1576–1578.

Monastersky, Richard. "Drilling Shortcut Penetrates Earth's Mantle." *Science News* 143 (February 20, 1993): 117.

Newsom, Horton E., and Kenneth W. W. Sims. "Core Formation During Early Accretion of the Earth." *Science* 252 (May 17, 1991): 926–933.

Powell, Corey S. "Perring Inward." *Scientific American* 264 (June 1991): 100–111.

Stager, Curt. "Africa's Great Rift." *National Geographic* 177 (May 1990): 10–41.

White, Robert S., and Dan P. McKenzie. "Volcanism at Rifts." *Scientific American* 261 (July 1989): 62–71.

Wickelgren, Ingrid. "Simmering Planet." *Discover* 11 (July 1990): 73–75.

Wysession, Michael E. "Journey to the Center of the Earth." *Earth* 5 (December 1996): 46–49.

CRUSTAL PLATES

Burchfiel, B. Clark. "The Continental Crust." *Scientific American* 249 (September 1983): 130–142.

Carter, William E., and Douglas S. Robertson. "Studying the Earth by Very-Long-Baseline-Interferometry." *Scientific American* 255 (November 1986): 46–54.

Gonzalez, Frank I. "Tsunami!" *Scientific American* 280 (May 1999): 56–65.

Green, Harry W., II. "Solving the Paradox of Deep Earthquakes." *Scientific American* 271 (September 1994): 64–71.

Hopkins, Ralph L. "Land Torn Apart." *Earth* 7 (February 1997): 37–41.

Howell, David G. "Terranes." *Scientific American* 253 (November 1985): 116–125.

Johnston, Arch C., and Lisa R. Kanter. "Earthquakes in Stable Continental Crust." *Scientific American* 262 (March 1990): 68–75.

Jordan, Thomas H., and J. Bernard Minster. "Measuring Crustal Deformation in the American West." *Scientific American* 259 (August 1988): 48–58.
Kerr, Richard A. "Iceland's Fires Trap the Heart of the Planet." *Science* 284 (May 14, 1999): 1095–1096.
Larson, Roger L. "The Mid-Cretaceous Superplume Episode." *Scientific American* 272 (February 1995): 82–86.
Monastersky, Richard. "A Stirring Tale from Inside the Earth." *Science News* 155 (March 20, 1999): 250–252.
Mutter, John C. "Seismic Images of Plate Boundaries." *Scientific American* 254 (February 1986): 66–75.
Stein, Ross S., and Robert S. Yeats. "Hidden Earthquakes." *Scientific American* 260 (June 1989): 48–57.
Taylor, S. Ross, and Scott M. McLennan. "The Evolution of Continental Crust." *Scientific American* 274 (January 1996): 76–81.
Vink, Gregory E., W. Jason Morgon, and Peter R. Vogt. "The Earth's Hot Spots." *Scientific American* 252 (April 1985): 50–57.

SEAFLOOR SPREADING

Bonatti, Enrico, and Kathleen Crane. "Ocean Fracture Zones." *Scientific American* 250 (May 1984): 40–51.
Broecker, Wallace S. "The Ocean." *Scientific American* 249 (September 1983): 146–160.
Cann, Joe, and Cherry Walker. "Breaking New Ground on the Ocean Floor." *New Scientist* 139 (October 30, 1993): 24–29.
Edmond, John M., and Karen Von Damm. "Hot Springs on the Ocean Floor." *Scientific American* 248 (April 1983): 78–93.
Gurnis, Michael. "Ridge Spreading, Subduction, and Sea Level Fluctuations." *Science* 150 (November 16, 1990): 970–972.
Hekinian, Roger. "Undersea Volcanoes." *Scientific American* 251 (July 1984): 46–55.
Hoffman, Kenneth. "Ancient Magnetic Reversals: Clues to the Geodynamo." *Scientific American* 258 (May 1988): 76–83.
Kerr, Richard A. "Sea-Floor Spreading Is Not So Variable." *Science* 223 (February 3, 1984): 472–473.
Macdonald, Ken C., and Bruce P. Luyendyk. "The Crest of the East Pacific Rise." *Scientific American* 224 (May 1981): 100–116.
Monastersky, Richard. "A New View of Earth." *Science News* 148 (December 16, 1995): 410–411.
———. "Mid-Atlantic Ridge Survey Hits Bull's-Eye." *Science News* 135 (May 13, 1989): 295.

Rona, Peter A. "Mineral Deposits from Sea-Floor Hot Springs." *Scientific American* 254 (January 1986): 84–92.

Yulsman, Tom. "Plate Tectonics Revised." *Science Digest* 93 (November 1985): 35

SUBDUCTION ZONES

Appenzeller, Tim. "How Vanished Oceans Drop an Anchor." *Science* 270 (November 17, 1995): 1122.

Frohlich, Cliff. "Deep Earthquakes." *Scientific American* 260 (January 1989): 48–55.

Fryer, Patricia. "Mud Volcanoes of the Marianas." *Scientific American* 266 (February 1992): 46–52.

Gordon, Richard G., and Seth Stein. "Global Tectonics and Space Geodesy." *Science* 256 (April 17, 1992): 333–341.

Gurnis, Michael. "Ridge Spreading, Subduction, and Sea Level Fluctuations." *Science* 250 (November 16, 1990): 970–972.

Heaton, Thomas H., and Stephen H. Hartzell. "Earthquake Hazards on the Cascadia Subduction Zone." *Science* 236 (April 10, 1987): 162–168.

Kerr, Richard A. "Deep-Sinking Slabs Stir the Mantle." *Science* 275 (January 31, 1997): 613–615.

———. "Earth's Surface May Move Itself." *Science* 269 (September 1, 1995): 1214–1216.

Monastersky, Richard. "Birth of a Subduction Zone." *Science News* 136 (December 16, 1989): 396.

———. "Catching Subduction in the Act." *Science News* 133 (January 2, 1988): 8

Pecock, Simon M. "Fluid Processes in Subduction Zones." *Science* 248 (April 20, 1990): 329–336.

Pratson, Lincoln F., and William F. Haxby. "Panoramas of the Seafloor." *Scientific American* 276 (June 1997): 82–87.

Sullivan, Walter. "Earth's Crust Sinks Deep, Only to Rise in Plumes of Lava." *The New York Times* (June 15, 1993): C1 & C8.

Weisburd, Stefi. "Sea-Surface Shape by Satellite." *Science* 129 (January 18, 1986): 37.

Zimmer, Carl. "The Ocean Within." *Discover* 15 (October 1994): 20–21.

MOUNTAIN BUILDING

Barnes, H. L., and A. W. Rose. "Origins of Hydrothermal Ores." *Science* 279 (March 27, 1998): 2064–2065.

Bird, Peter. "Formation of the Rocky Mountains, Western United States: A Continuum Computer Model." *Science* 239 (March 25, 1988): 1501–1507.

Brimhall, George. "The Genesis of Ores." *Scientific American* 264 (May 1991): 84–91.

Gass, Ian G. "Ophiolites." *Scientific American* 247 (August 1982): 122–131.

Harrison, T. Mark, et al. "Raising Tibet." *Science* 255 (March 27, 1992): 1663–1670.

Kerr, Richard A. "Urals Yield Secret of a Lasting Bond." *Science* 274 (October 11, 1996): 181.

———. "Did Deeper Forces Act to Uplift the Andes?" *Science* 269 (September 1, 1995): 1215–1216.

Molnar Peter. "The Structure of Mountain Ranges." *Scientific American* 255 (July 1986): 70–79.

Monastersky, Richard. "Mountains Frozen in Time." *Science News* 148 (December 23 & 30, 1995): 431.

———. "What's Holding Up the High Sierras?" *Science News* 138 (December 15, 1990): 380.

Murphy, J. Brendan, and R. Damian Nance. "Mountain Belts and the Supercontinent Cycle." *Scientific American* 266 (April 1992): 84–91.

Pinter, Nicholas, and Mark T. Brandon. "How Erosion Builds Mountains." *Scientific American* 276 (April 1997): 74–79.

Ruddiman, William F., and John E. Kutzbach. "Plateau Uplift and Climate Change." *Scientific American* 264 (March 1991): 66–74.

Svitil, Kathy A. "The Coming Himalayan Catastrophe." *Discover* 16 (July 1995): 81–85.

THE ROCK CYCLE

Berner, Robert A., and Antonio C. Lasaga. "Modeling the Geochemical Carbon Cycle." *Scientific American* 260 (March 1989): 74–81.

Broecker, Wallace S. "Carbon Dioxide Circulation through Ocean and Atmosphere." *Nature* 308 (April 12, 1984): 602.

———. "What Drives Glacial Cycles." *Scientific American* 262 (January 1990): 49–56.

Cathles, Lawrence M. III. "Scales and Effects of Fluid Flow in the Upper Crust." *Science* 248 (April 20, 1990): 323–328.

Fisher, Arthur. "Too Much Nitrogen." *Popular Science* 250 (July 1997): 33.

Gore, Rick. "Living On Fire." *National Geographic* 173 (May 1998): 4–37.

Kasting, James F., Owen B. Toon, and James B. Pollack. "How Climates Evolved on the Terrestrial Planets." *Scientific American* 258 (February 1988): 90–97.

Kerr, Richard A. "The Carbon Cycle and Climate Warming." *Science* 222 (December 9, 1983): 1107–1108.

Kunzig, Robert. "Ice Cycles." *Discover* 10 (May 1989): 74–79.

Monastersky, Richard. "The Case of the Missing Carbon Dioxide." *Science News* 155 (June 12, 1999): 383.

———. "Spinning the Supercontinent Cycle." *Science News* 135 (June 3, 1989): 344–346.

Nance, R. Damian, Thomas R. Worsley, and Judith B. Moody. "The Supercontinent Cycle." *Scientific American* 259 (July 1988): 72–79.

Schnider, Stephen H. "Climate Modeling." *Scientific American* 256 (May 1987): 72–80.

Williams, George E. "The Solar Cycle in Precambrian Time." *Scientific American* 255 (August 1986): 88–96.

TECTONICS AND LIFE

Broad, William J. "Life Springs Up in Ocean's Volcanic Vents, Deep Divers Find." *The New York Times* (October 19, 1993): C4.

Coffin, Millard F., and Olav Eldholm. "Large Igneous Provinces." *Scientific American* 269 (October 1993): 42–49.

Eldredge, Niles. "What Drives Evolution?" *Earth* 5 (December 1996): 34–37.

Erwin, Douglas H. "The Mother of Mass Extinctions." *Scientific American* 275 (July 1996): 72–78.

Fischman, Josh. "In Search of the Elusive Megaplume." *Discover* 20 (March 1999): 108–115.

Flannery, Tim. "Debating Extinction." *Science* 283 (January 8, 1999): 182–183.

Gould, Stephen Jay. "The Evolution of Life on the Earth." *Scientific American* 271 (October 1994): 85–91.

Hoffman, Paul F., and David P. Schrag. "Snowball Earth." *Scientific American* 282 (January 2000): 68–75.

Krajick, Kevin. "To Hell and Back." *Discover* 20 (July 1999): 76–82.

Levinton, Jeffrey S. "The Big Bang of Animal Evolution." *Scientific American* 267 (November 1992): 84–91.

Monastersky, Richard. "The Rise of Life on Earth." *National Geographic* 194 (March 1998): 54–81.

Morris, S. Conway. "Burgess Shale Faunas and the Cambrian Explosion." *Science* 246 (October 20, 1989): 339–345.

Orgel, Leslie E. "The Origin of Life on the Earth." *Scientific American* 271 (October 1994): 77–83.

Parfit, Michael. "Timeless Valleys of the Antarctic Desert." *National Geographic* 194 (October 1998): 120–135.

Rampino, Michael R., and Richard B. Stothers. "Flood Basalt Volcanism During the Past 250 Million Years." *Science* 241 (August 5, 1988): 663–667.

Stover, Dawn. "Creatures of the Thermal Vents." *Popular Science* 246 (May 1995): 55–57.

TECTONICS IN SPACE

Bullock, Mark A., and David H. Grinspoon. "Global Climate Change on Venus." *Scientific American* 280 (March 1999): 50–57.

Carroll, Michael. "Assault on the Red Planet." *Popular Science* 250 (January 1997): 44–49.

Chaikin, Andrew. "Life on Mars: The Great Debate." *Popular Science* 251 (July 1997): 60–65.

Coswell, Ken. "Io, Jupiter's Fiery Satellite." *Space World* 295 (July 1988): 12–15.

Cowen, Ron. "Plate Tectonics . . . on Mars." *Science News* 155 (May 1, 1999): 284–286.

———. "Galileo Explores the Galilean Moons." *Science News* 152 (August 9, 1997): 90–91.

Golombek, M. P., et al. "Overview of the Pathfinder Mission and Assessment of Landing Site Predictions." *Science* 278 (December 5, 1997): 1743–1748.

Johnson, Torrence V., Robert Hamilton Brown, and Laurence A. Soderblom. "The Moons of Uranus." *Scientific American* 256 (April 1987): 48–60.

Kaula, William M. "Venus: A Contrast in Evolution to Earth." *Science* 247 (March 9, 1990): 1191–1196.

Kerr, Richard A. "The Solar System's New Diversity." *Science* 265 (September 2, 1994): 1360–1362.

———. "Triton Steals Voyager's Last Show." *Science* 245 (September 1, 1989): 928–930.

Nelson, Robert M. "Mercury: The Forgotten Planet." *Scientific American* 277 (November 1997): 56–67.

Papparlardo, Robert T., James W. Head, and Ronald Greeley. "The Hidden Ocean of Europa." *Scientific American* 281 (October 1999): 54–63.

Prinn, Ronald G. "The Volcanoes and Clouds of Venus." *Scientific American* 252 (March 1985): 46–53.

Taylor, G. Jeffrey. "The Scientific Legacy of Apollo." *Scientific American* 271 (July 1994): 40–47.

INDEX

Boldface page numbers indicate extensive treatment of a topic. *Italic* page numbers indicate illustrations or captions. Page numbers followed by *m* indicate maps; *t* indicate tables; *g* indicate glossary.